D1519806

STRUCTURAL STABILITY OF STEEL: CONCEPTS AND APPLICATIONS FOR STRUCTURAL ENGINEERS

STRUCTURAL STABILITY OF STEEL: CONCEPTS AND APPLICATIONS FOR STRUCTURAL ENGINEERS

THEODORE V. GALAMBOS
ANDREA E. SUROVEK

JOHN WILEY & SONS, INC.

This book is printed on acid-free paper. ♾

Published by John Wiley & Sons, Inc., Hoboken, New Jersey
Published simultaneously in Canada

Library of Congress Cataloging-in-Publication Data:

Galambos, T. V. (Theodore V.)

Structural stability of steel : concepts and applications for structural engineers / Theodore Galambos, Andrea Surovek.

p. cm.

Includes bibliographical references and index.

ISBN 978-0-470-03778-2 (cloth)

1. Building, Iron and steel–Congresses. 2. Structural stability–Congresses. I. Surovek, Andrea. II. Title.

TA684.G26 2005

624.1′821–dc22

2007035514

ISBN: 978-0-470-03778-2

CONTENTS

PREFACE

In order to truly understand the behavior and design of metal structures, an engineer needs to have a fundamental understanding of structural stability. More so than structures designed using other construction materials, steel structures are governed to a great extent on stability limit states. All major international design specifications include provisions based on stability theory. The purpose of this book is to provide students and practicing engineers with both the theory governing stability of steel structures and a practical look at how that theory translates into design methodologies currently implemented in steel design specifications.

The topics presented in the text pertain to various aspects of elastic buckling and inelastic instability. An understanding of stability limits is very important in the design of structures: Catastrophic failures can, and tragically have, resulted from violating fundamental principles of stability in design. Maintaining stability is particularly important during the erection phase of construction, when the structural skeleton is exposed prior to the installation of the final stabilizing features, such as slabs, walls and/or cladding.

The book contains a detailed treatment of the elastic and inelastic stability analysis of columns, beams, beam-columns, and frames. In addition, it provides numerous worked examples. Practice problems are included at the end of each chapter. The first six chapters of this book are based on lecture notes of the first author, used in his teaching of structural engineering graduate courses since 1960, first at Lehigh University in Bethlehem, Pennsylvania, (1960–1965), then at Washington University in St. Louis, Missouri, (1966–1981), and finally at the University of Minnesota in Minneapolis, Minnesota.

The genesis of the course material was in lectures at Lehigh University given by Professors Bruce Johnston, Russell Johnson, and Bruno Thurlimann in the 1950s. The material in the last two chapters is concerned with the application of stability theory in the practical design of steel structures, with special emphasis on examples based on the 2005 Specification for Structural Steel Buildings of the American Institute of Steel Construction (AISC). Chapter 7 is based heavily on the work performed by Professors Joe Yura and Todd Helwig of the University of Texas in developing Appendix 6 of the 2005 AISC Specification. A portion of the material in Chapter 8 is based on the work of the second author and Professor Don White of Georgia Tech, as well as verification studies and design examples developed by members of AISC TC 10, chaired by Dr. Shankar Nair.

The material in the book is suitable for structural engineering students at the graduate level. It is also useful for design engineers who wish to understand the background of the stability design criteria in structural specifications, or for those who may have a need to investigate special stability problems. Since the fundamental mechanics governing the behavior of beams, columns, beam-columns, and frames is discussed in the book, it is also useful for an international structural engineering constituency. A background in both structural analysis approaches and differential equations is essential in understanding the derivations included in the first six chapters.

Chapter 1 is an introduction to the principles of stability theory. The various aspects of behavior at the limits of instability are defined on hand of simple spring-bar examples. Chapter 2 deals with the stability of axially loaded planar elastic systems. Individual columns, simple frames, and subassemblies of members are analyzed. The background for the effective length concept of designing metal structures is also presented. Chapter 3 expands the analysis to the nonlinear material behavior. Tangent modulus, reduced modulus, and maximum strength theories are introduced. Derivations are presented that lead to an understanding of modern column design formulas in structural codes. The subject of Chapter 4 is the elastic and inelastic stability limit of planar beam-columns. Various aspects of the interaction between axial force and bending moment are presented, and the interaction formulas in design specifications are evaluated. Chapter 5 illustrates many features of elastic and inelastic instability of planar frames using as example a one-story two-bay structure.

In Chapter 6 the out-of-plane lateral-torsional buckling of beams, columns, and beam-columns is presented. Since stability of the structure is vitally dependent on the strength and stiffness of the bracing systems that are provided during erection and in the final stage of construction, Chapter 7 is devoted entirely to this subject. Modern design standards for structural steel design require an analysis procedure that provides stability through the direct inclusion of the destabilizing effects of structural imperfections, such as residual stresses and unavoidable out-of-plumb geometry. The topic of Chapter 8 is the analysis and design of steel frames according to the 2005 Specification of the AISC.

CHAPTER ONE

FUNDAMENTALS OF STABILITY THEORY

1.1 INTRODUCTION

It is not necessary to be a structural engineer to have a sense of what it means for a structure to be stable. Most of us have an inherent understanding of the definition of instability—that a small change in load will cause a large change in displacement. If this change in displacement is large enough, or is in a critical member of a structure, a local or member instability may cause collapse of the entire structure. An understanding of stability theory, or the mechanics of why structures or structural members become unstable, is a particular subset of engineering mechanics of importance to engineers whose job is to design safe structures.

The focus of this text is not to provide in-depth coverage of all stability theory, but rather to demonstrate how knowledge of structural stability theory assists the engineer in the design of safe steel structures. Structural engineers are tasked by society to design and construct buildings, bridges, and a multitude of other structures. These structures provide a load-bearing skeleton that will sustain the ability of the constructed artifact to perform its intended functions, such as providing shelter or allowing vehicles to travel over obstacles. The structure of the facility is needed to maintain its shape and to keep the facility from falling down under the forces of nature or those made by humans. These important characteristics of the structure are known as *stiffness* and *strength.*

This book is concerned with one aspect of the strength of structures, namely their stability. More precisely, it will examine how and under what loading condition the structure will pass from a stable state to an unstable one. The reason for this interest is that the structural engineer, knowing the circumstances of the limit of stability, can then proportion a structural scheme that will stay well clear of the zone of danger and will have an adequate margin of safety against collapse due to instability. In a well-designed structure, the user or occupant will never have to even think of the structure's existence. Safety should always be a given to the public.

Absolute safety, of course, is not an achievable goal, as is well known to structural engineers. The recent tragedy of the World Trade Center collapse provides understanding of how a design may be safe under any expected circumstances, but may become unstable under extreme and unforeseeable circumstances. There is always a small chance of failure of the structure.

The term *failure* has many shades of meaning. Failure can be as obvious and catastrophic as a total collapse, or more subtle, such as a beam that suffers excessive deflection, causing floors to crack and doors to not open or close. In the context of this book, *failure* is defined as the behavior of the structure when it crosses a *limit state*—that is, when it is at the limit of its structural usefulness. There are many such limit states the structural design engineer has to consider, such as excessive deflection, large rotations at joints, cracking of metal or concrete, corrosion, or excessive vibration under dynamic loads, to name a few. The one limit state that we will consider here is the limit state where the structure passes from a stable to an unstable condition.

Instability failures are often catastrophic and occur most often during erection. For example, during the late 1960s and early 1970s, a number of major steel box-girder bridges collapsed, causing many deaths among erection personnel. The two photographs in Figure 1.1 were taken by author Galambos in August 1970 on the site two months before the collapse of a portion of the Yarra River Crossing in Melbourne, Australia. The left picture in Figure 1.1 shows two halves of the multi-cell box girder before they were jacked into place on top of the piers (see right photo), where they were connected with high-strength bolts. One of the 367.5 ft. spans collapsed while the ironworkers attempted to smooth the local buckles that had formed on the top surface of the box. Thirty-five workers and engineers perished in the disaster.

There were a number of causes for the collapse, including inexperience and carelessness, but the Royal Commission (1971), in its report pinpointed the main problem: "We find that [the design organization] made assumptions about the behavior of box girders which extended beyond the range of engineering knowledge." The Royal Commission concluded " . . . that the design firm "failed altogether to give proper and careful regard to the

Fig. 1.1 Stability-related failures.

process of structural design." Subsequent extensive research in Belgium, England, the United States, and Australia proved that the conclusions of the Royal Commission were correct. New theories were discovered, and improved methods of design were implemented. (See Chapter 7 in the Stability Design Criteria for Metal Structures (Galambos 1998)).

Structural instability is generally associated with the presence of compressive axial force or axial strain in a plate element that is part of a cross-section of a beam or a column. *Local instability* occurs in a single portion of a member, such as local web buckling of a steel beam. *Member instability* occurs when an isolated member becomes unstable, such as the buckling of a diagonal brace. However, member instability may precipitate a *system instability*. System instabilities are often catastrophic.

This text examines the stability of some of these systems. The topics include the behavior of columns, beams, and beam-columns, as well as the stability of frames and trusses. Plate and shell stability are beyond the scope of the book. The presentation of the material concentrates on steel structures, and for each type of structural member or system, the recommended design rules will be derived and discussed. The first chapter focuses on basic stability theory and solution methods.

1.2 BASICS OF STABILITY BEHAVIOR: THE SPRING-BAR SYSTEM

A *stable* elastic structure will have displacements that are proportional to the loads placed on it. A *small* increase in the load will result in a *small* increase of displacement.

As previously mentioned, it is intuitive that the basic idea of instability is that a *small* increase in load will result in a *large* change in the displacement. It is also useful to note that, in the case of axially loaded members,

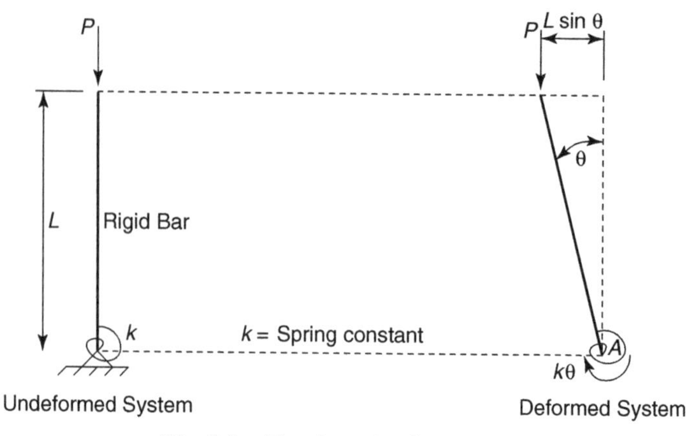

Fig. 1.2 Simple spring-bar system.

the large displacement related to the instability is not in the same direction as the load causing the instability.

In order to examine the most basic concepts of stability, we will consider the behavior of a spring-bar system, shown in Figure 1.2. The left side in Figure 1.2 shows a straight vertical rigid bar of length L that is restrained at its bottom by an elastic spring with a spring constant k. At the top of the bar there is applied a force P acting along its longitudinal axis. The right side shows the system in a deformed configuration. The moment caused by the axial load acting through the displacement is resisted by the spring reaction $k\theta$. The symbol θ represents the angular rotation of the bar in radians.

We will begin with the most basic solution of this problem. That is, we will find the *critical load* of the structure. The critical load is the load that, when placed on the structure, causes it to pass from a stable state to an unstable state. In order to solve for the critical load, we must consider a deformed shape, shown on the right in Figure 1.2. Note that the system is slightly perturbed with a rotation θ. We will impose equilibrium on the deformed state. Summing moments about point A we obtain

$$\sum M_A = 0 = PL\sin\theta - k\theta \tag{1.1}$$

Solving for P at equilibrium, we obtain

$$P_{\text{cr}} = \frac{k\theta}{L\sin\theta} \tag{1.2}$$

If we consider that the deformations are very small, we can utilize *small displacement theory* (this is also referred to in mechanics texts as small strain theory). Small displacement theory allows us to simplify the math by recognizing that for very small values of the angle, θ, we can use the simplifications that

$$\sin\theta = \theta$$
$$\tan\theta = \theta$$
$$\cos\theta = 1$$

Substituting $\sin\theta = \theta$, we determine the critical load P_{cr} of the spring-bar model to be:

$$P_{cr} = \frac{k\theta}{L\theta} = \frac{k}{L} \tag{1.3}$$

The equilibrium is in a neutral position: it can exist both in the undeformed and the deformed position of the bar. The small displacement response of the system is shown in Figure 1.3. The load ratio $PL/k = 1$ is variously referred in the literature as the *critical load*, the *buckling load*, or the *load at the bifurcation of the equilibrium*. The bifurcation point is a branch point; there are two equilibrium paths after P_{cr} is reached, both of which are unstable. The upper path has an increase in P with no displacement. This equilibrium path can only exist on a perfect system with no perturbation and is therefore not a practical solution, only a theoretical one.

Another means of solving for the critical load is through use of the *principle of virtual work*. Energy methods can be very powerful in describing structural behavior, and have been described in many structural analysis

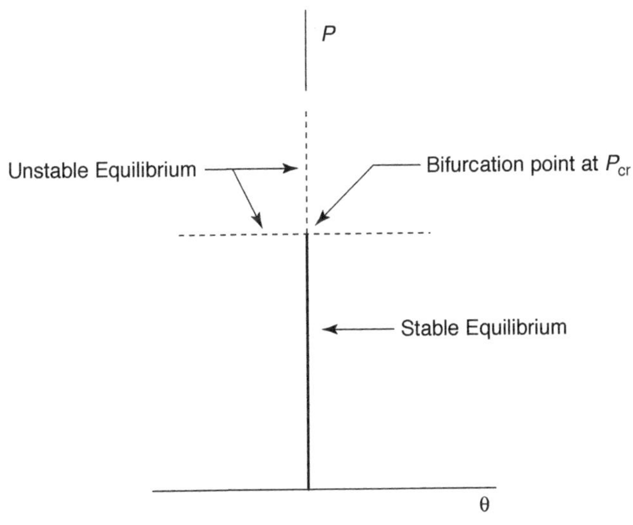

Fig. 1.3 Small displacement behavior of spring-bar system.

and structural mechanics texts. Only a brief explanation of the method will be given here. The total potential Π of an elastic system is defined by equation 1.4 as

$$\Pi = U + V_p \tag{1.4}$$

1. *U is the elastic strain energy of a conservative system.* In a conservative system the work performed by both the internal and the external forces is independent of the path traveled by these forces, and it depends only on the initial and the final positions. U is the *internal work* performed by the internal forces; $U = W_i$
2. *V_p is the potential of the external forces*, using the original deflected position as a reference. V_p is *the external work*; $V_p = -W_e$.

Figure 1.4 shows the same spring-bar system we have considered, including the distance through which the load P will move when the bar displaces.
The strain energy is the work done by the spring,

$$U = W_i = \frac{1}{2}k\theta^2. \tag{1.5}$$

The potential of the external forces is equal to

$$V_p = -W_e = -PL(1 - \cos\theta) \tag{1.6}$$

The total potential in the system is then given by:

$$\Pi = U + V_p = \frac{1}{2}k\theta^2 - PL(1 - \cos\theta) \tag{1.7}$$

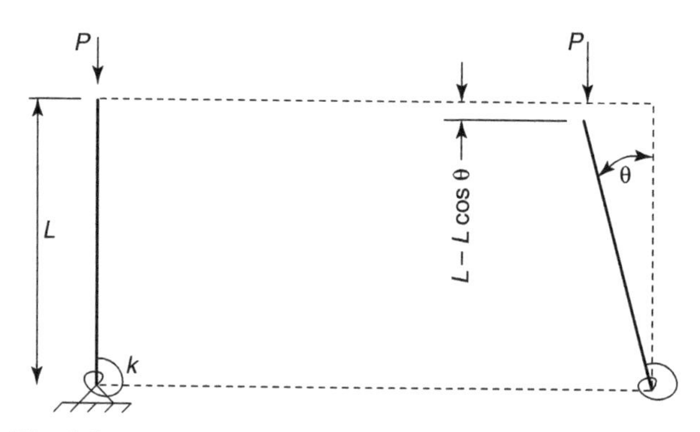

Fig. 1.4 Simple spring-bar system used in energy approach.

According to the principle of virtual work the maxima and minima are equilibrium positions, because if there is a small change in θ, there is no change in the total potential. In the terminology of structural mechanics, the total potential is *stationary*. It is defined by the derivative

$$\frac{d\Pi}{d\theta} = 0 \tag{1.8}$$

For the spring bar system, equilibrium is obtained when

$$\frac{d\Pi}{d\theta} = 0 = k\theta - PL\sin\theta \tag{1.9}$$

To find P_{cr}, we once again apply small displacement theory $(\sin\theta = \theta)$ and obtain

$$P_{\mathrm{cr}} = k/L$$

as before.

Summary of Important Points

- Instability occurs when a small change in load causes a large change in displacement. This can occur on a local, member or system level.
- The critical load, or buckling load, is the load at which the system passes from a stable to an unstable state.
- The critical load is obtained by considering *equilibrium* or *potential energy* of the system in a deformed configuration.
- Small displacement theory may be used to simplify the calculations if only the critical load is of interest.

1.3 FUNDAMENTALS OF POST-BUCKLING BEHAVIOR

In section 1.2, we used a simple example to answer a fundamental question in the study of structural stability: At what load does the system become unstable, and how do we determine that load? In this section, we will consider some basic principles of stable and unstable behavior. We begin by reconsidering the simple spring-bar model in Figure 1.2, but we introduce a disturbing moment, M_o at the base of the structure. The new system is shown in Figure 1.5.

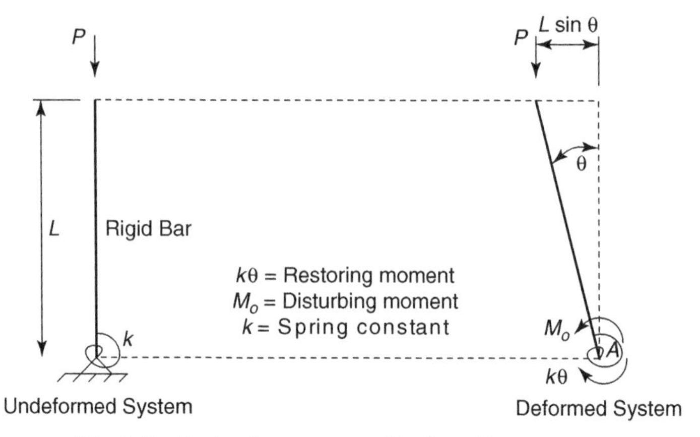

Fig. 1.5 Spring-bar system with disturbing moment.

Similar to Figure 1.1, the left side of Figure 1.5 shows a straight, vertical rigid bar of length L that is restrained at its bottom by an infinitely elastic spring with a spring constant k. At the top of the bar there is applied a force P acting along its longitudinal axis. The right sketch shows the deformation of the bar if a disturbing moment M_o is acting at its base. This moment is resisted by the spring reaction $k\theta$, and it is augmented by the moment caused by the product of the axial force times the horizontal displacement of the top of the bar. The symbol θ represents the angular rotation of the bar (in radians).

1.3.1 Equilibrium Solution

Taking moments about the base of the bar (point A) we obtain the following equilibrium equation for the displaced system:

$$\sum M_A = 0 = PL \sin\theta + M_o - k\theta$$

Letting $\theta_o = M_o/k$ and rearranging, we can write the following equation:

$$\frac{PL}{k} = \frac{\theta - \theta_o}{\sin\theta} \tag{1.10}$$

This expression is displayed graphically in various contexts in Figure 1.6.

The coordinates in the graph are the load ratio PL/k as the abscissa and the angular rotation θ (radians) as the ordinate. Graphs are shown for three values of the disturbing action

$$\theta_o = 0,\ \theta_o = 0.01,\ \text{and}\ \theta_o = 0.05.$$

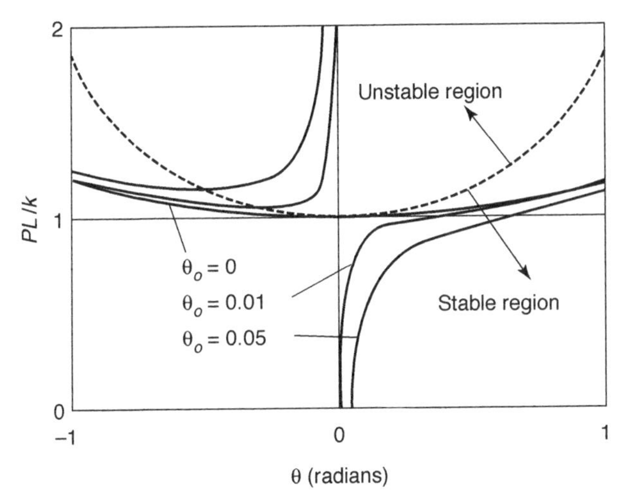

Fig. 1.6 Load-deflection relations for spring-bar system with disturbing moment.

When $\theta_o = 0$, that is $\frac{PL}{k} = \frac{\theta}{\sin\theta}$, there is no possible value of PL/k less than unity since θ is always larger than $\sin\theta$. Thus no deflection is possible if $PL/k < 1.0$. At $PL/k > 1.0$ deflection is possible in either the positive or the negative direction of the bar rotation. As θ increases or decreases the force ratio required to maintain equilibrium becomes larger than unity. However, at relatively small values of θ, say, below 0.1 radians, or about 5°, the load-deformation curve is flat for all practical purposes. Approximately, it can be said that equilibrium is possible at $\theta = 0$ and at a small adjacent deformed location, say $\theta < 0.1$ or so. The load $PL/k = 1.0$ is thus a special type of load, when the system can experience two adjacent equilibrium positions: one straight and one deformed. The equilibrium is thus in a neutral position: It can exist both in the undeformed and the deformed position of the bar. The load ratio $PL/k = 1$ is variously referred in the literature as the *critical load*, the *buckling load*, or the *load at the bifurcation of the equilibrium*. We will come back to discuss the significance of this load after additional features of behavior are presented next.

The other two sets of solid curves in Figure 1.6 are for specific small values of the disturbing action θ_o of 0. 01 and 0.05 radians. These curves each have two regions: When θ is positive, that is, in the right half of the domain, the curves start at $\theta = \theta_o$ when $PL/k = 0$ and then gradually exhibit an increasing rotation that becomes larger and larger as $PL/k = 1.0$ is approached, finally becoming affine to the curve for $\theta_o = 0$ as θ becomes very large. While this in not shown in Figure 1.6, the curve for smaller and smaller values of θ_o will approach the curve of the bifurcated equilibrium. The other branches of the two curves are for negative values of θ. They are

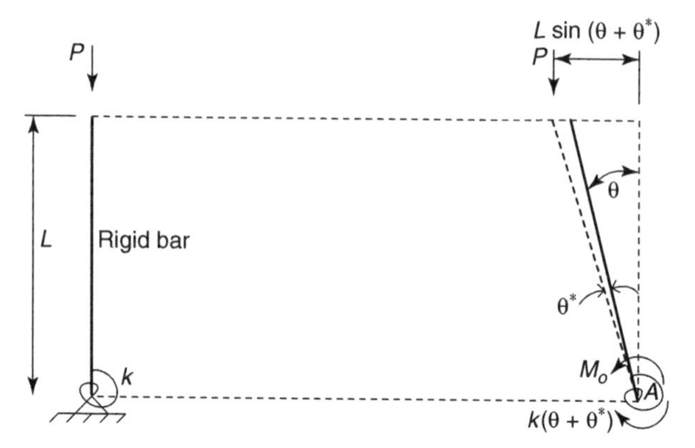

Fig. 1.7 Disturbed equilibrium configuration.

in the left half of the deformation domain and they lie above the curve for $\theta_o = 0$. They are in the unstable region for smaller values of $-\theta$, that is, they are above the dashed line defining the region between stable and unstable behavior, and they are in the stable region for larger values of $-\theta$. (*Note*: The stability limit will be derived later.) The curves for $-\theta$ are of little practical consequence for our further discussion.

The nature of the equilibrium, that is, its stability, is examined by disturbing the already deformed system by an additional small rotation θ^*, as shown in Figure 1.7.

The equilibrium equation of the disturbed geometry is

$$\sum M_A = 0 = PL\sin(\theta + \theta^*) + M_o - k(\theta + \theta^*)$$

After rearranging we get, noting that $\theta_o = \frac{M_o}{k}$

$$\frac{PL}{k} = \frac{\theta + \theta^* - \theta_o}{\sin(\theta + \theta^*)} \tag{1.11}$$

From trigonometry we know that $\sin(\theta + \theta^*) = \sin\theta\cos\theta^* + \cos\theta\sin\theta^*$. For small values of θ^* we can use $\cos\theta^* \approx 1.0$; $\sin\theta^* \approx \theta^*$, and therefore

$$\frac{PL}{k} = \frac{\theta + \theta^* - \theta_o}{\sin\theta + \theta^*\cos\theta} \tag{1.12}$$

This equation can be rearranged to the following form: $\frac{PL}{k}\sin\theta - \theta + \theta_o + \theta^*(\frac{PL}{k}\cos\theta - 1) = 0$. However, $\frac{PL}{k}\sin\theta - \theta + \theta_o = 0$ as per equation 1.10, $\theta^* \neq 0$, and thus

$$\frac{PL}{k}\cos\theta - 1 = 0 \tag{1.13}$$

Equation 1.13 is the locus of points for which $\theta^* \neq 0$ while equilibrium is just maintained, that is the equilibrium is *neutral*. The same result could have been obtained by setting the derivative of $F = \frac{PL}{k} \sin\theta - \theta + \theta_o$ with respect to θ equal to zero:

$$\frac{dF}{d\theta} = \frac{PL}{k}\cos\theta - 1.$$

The meaning of the previous derivation is that when

1. $\cos\theta < \frac{1}{PL/k}$, the equilibrium is *stable*—that is, the bar returns to its original position when q^* is removed; energy must be added.
2. $\cos\theta = \frac{1}{PL/k}$, the equilibrium is *neutral*—that is, no force is required to move the bar a small rotation θ^*.
3. $\cos\theta > \frac{1}{PL/k}$, the equilibrium is *unstable*—that is, the configuration will snap from an unstable to a stable shape; energy is released.

These derivations are very simple, yet they give us a lot of information:

1. The load-deflection path of the system that sustains an applied action θ_o from the start of loading. This will be henceforth designated as an *imperfect* system, because it has some form of deviation in either loading or geometry from the ideally *perfect* structure that is straight or unloaded before the axial force is applied.
2. It provides the critical, or buckling, load at which the equilibrium become neutral.
3. It identifies the character of the equilibrium path, whether it is neutral, stable, or unstable.

It is good to have all this information, but for more complex actual structures it is often either difficult or very time-consuming to get it. We may not even need all the data in order to design the system. Most of the time it is sufficient to know the buckling load. For most practical structures, the determination of this critical value requires only a reasonably modest effort, as shown in section 1.2.

In the discussion so far we have derived three hierarchies of results, each requiring more effort than the previous one:

1. Buckling load of a perfect system (Figure 1.2)
2. The post-buckling history of the perfect system (Figure 1.5)
3. The deformation history of the "imperfect" system (Figure 1.7)

In the previous derivations the equilibrium condition was established by utilizing the statical approach. Equilibrium can, however, be determined by using the theorem of virtual work. It is sometimes more convenient to use this method, and the following derivation will feature the development of this approach for the spring-bar problem.

1.3.2 Virtual Work Solution

We also examine the large displacement behavior of the system using the energy approach described in section 1.2. The geometry of the system is shown in Figure 1.8

For the spring-bar system the strain energy is the work done by the spring, $U = W_i = \frac{1}{2}k\theta^2$. The potential of the external forces is equal to $V_p = W_e = -PL(1 - \cos\theta) - M_o\theta$. With $\theta_o = \frac{M_o}{k}$ the total potential becomes

$$\frac{\Pi}{k} = \frac{\theta^2}{2} - \frac{PL}{k}(1 - \cos\theta) - \theta_o\theta \tag{1.14}$$

The total potential is plotted against the bar rotation in Figure 1.9 for the case of $\theta_o = 0.01$ and $PL/k = 1.10$. In the range $-1.5 \leq \theta \leq 1.5$ the total potential has two minima (at approximately $\theta = 0.8$ and -0.7) and one maximum (at approximately $\theta = -0.1$). According to the *Principle of Virtual Work*, the maxima and minima are equilibrium positions, because if there is a small change in θ, there is no change in the total potential. In the terminology of structural mechanics, the total potential is *stationary*. It is defined by the derivative $\frac{d\Pi}{d\theta} = 0$. From equation 1.6, $\frac{d\Pi}{d\theta} = 0 = \frac{2\theta}{2} - \frac{PL}{k}\sin\theta - \theta_o$, or

$$\frac{PL}{k} = \frac{\theta - \theta_o}{\sin\theta} \tag{1.15}$$

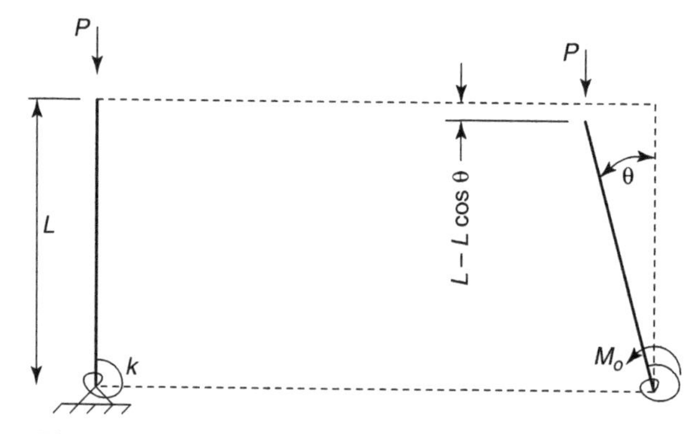

Fig. 1.8 Geometry for the total potential determination.

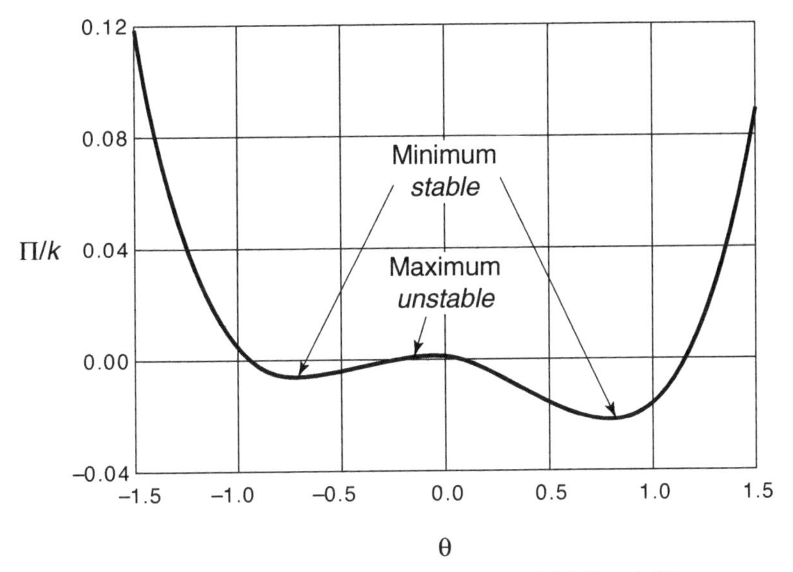

Fig. 1.9 Total potential for $\theta_o = 0.01$ and $PL/k = 1.10$.

This equation is identical to equation 1.10. The status of stability is illustrated in Figure 1.10 using the analogy of the ball in the cup (stable equilibrium), the ball on the top of the upside-down cup (unstable equilibrium), and the ball on the flat surface.

The following summarizes the problem of the spring-bar model's energy characteristics:

$$\frac{\Pi}{k} = \frac{\theta^2}{2} - \frac{PL}{k}(1 - \cos\theta) - \theta_o\theta \rightarrow \textit{Total potential}$$

$$\frac{d(\Pi/k)}{d\theta} = \theta - \theta_o - \frac{PL}{k}\sin\theta = 0 \rightarrow \frac{PL}{k} = \frac{\theta - \theta_o}{\sin\theta} \rightarrow \textit{Equilibrium} \qquad (1.16)$$

$$\frac{d^2(\Pi/k)}{d\theta^2} = 1 - \frac{PL}{k}\cos\theta = 0 \rightarrow \frac{PL}{k} = \frac{1}{\cos\theta} \rightarrow \textit{Stability}$$

These equations represent the energy approach to the large deflection solution of this problem.

For the small deflection problem we set $\theta_o = 0$ and note that $1 - \cos\theta \approx \frac{\theta^2}{2}$. The total potential is then equal to $\Pi = \frac{k\theta^2}{2} - \frac{PL\theta^2}{2}$. The derivative with respect to θ gives the critical load:

$$\frac{d\Pi}{d\theta} = 0 = \theta(k - PL) \rightarrow P_{\text{cr}} = k/L \qquad (1.17)$$

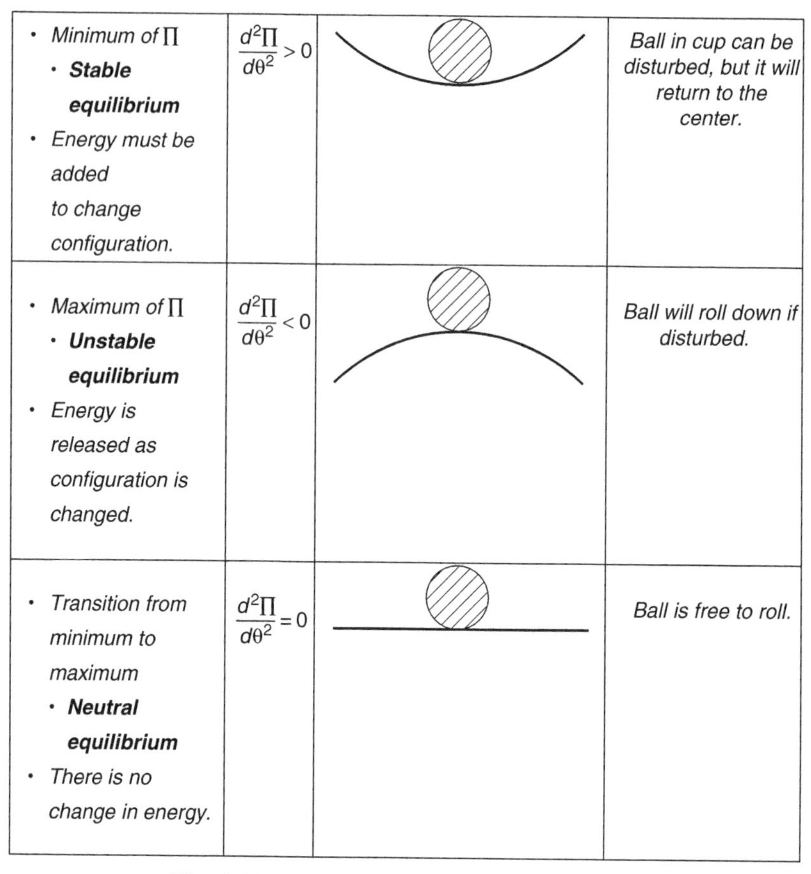

• *Minimum of* Π • ***Stable equilibrium*** • *Energy must be added to change configuration.*	$\frac{d^2\Pi}{d\theta^2} > 0$		*Ball in cup can be disturbed, but it will return to the center.*
• *Maximum of* Π • ***Unstable equilibrium*** • *Energy is released as configuration is changed.*	$\frac{d^2\Pi}{d\theta^2} < 0$		*Ball will roll down if disturbed.*
• *Transition from minimum to maximum* • ***Neutral equilibrium*** • *There is no change in energy.*	$\frac{d^2\Pi}{d\theta^2} = 0$		*Ball is free to roll.*

Fig. 1.10 Table illustrating status of stability.

Thus far, we have considered three methods of stability evaluation:

1. The small deflection method, giving only the buckling load.
2. The large deflection method for the perfect structure, giving information about post-buckling behavior.
3. The large deflection method for the imperfect system, giving the complete deformation history, including the reduction of stiffness in the vicinity of the critical load.

Two methods of solution have been presented:

1. Static equilibrium method
2. Energy method

Such stability-checking procedures are applied to analytically exact and approximate methods for real structures in the subsequent portions of this book.

The spring-bar system of Figure 1.5 exhibited a particular post-buckling characteristic: The post-buckling deflections increased as the load was raised above the bifurcation point, as seen in Figure 1.6. Such *hardening* behavior is obviously desirable from the standpoint of safety. However, there are structural systems where the post-buckling exhibits a *softening* character. Such a spring-bar structure will be considered next for the system of Figure 1.11.

Equilibrium is obtained by taking moments about the pinned base of the rigid bar that is restrained by a horizontal spring a distance a above its base and is disturbed by a moment M_o:

$$(ka \sin\theta)\, a \cos\theta - M_o - PL \sin\theta = 0$$

Rearrangement and introduction of the imperfection parameter $\theta_o = \frac{M_o}{ka^2}$ gives the following equation:

$$\frac{PL}{ka^2} = \frac{\sin\theta \cos\theta - \theta_o}{\sin\theta} \tag{1.18}$$

The *small deflection ideal geometry* assumption $(\theta_o = 0;\ \sin\theta = \theta;\ \cos\theta = 1)$ leads to the buckling load

$$P_{\text{cr}} = \frac{ka^2}{L} \tag{1.19}$$

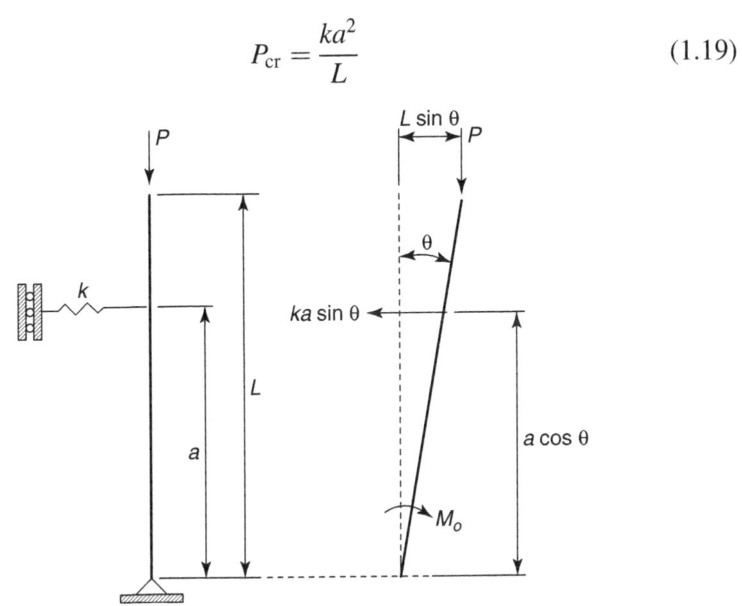

Fig. 1.11 Softening spring-bar structure.

From the *large deflection-ideal geometry* assumption ($\theta_o = 0$) we get the post-buckling strength:

$$P_{\mathrm{cr}} = \frac{ka^2}{L}\cos\theta \tag{1.20}$$

The load-rotation curves from equations 1.18 and 1.20 are shown in Figure 1.12 for the perfect ($\theta_o = 0$) and the imperfect ($\theta_o = 0.01$) system. The post-buckling behavior is *softening*—that is, the load is decreased as the rotation increases. The deflection of the imperfect system approaches that of the perfect system for large bar rotations. However, the strength of the imperfect member will never attain the value of the ideal critical load. Since in actual structures there will always be imperfections, the theoretical buckling load is upper bound.

The nature of stability is determined from applying a virtual rotation to the deformed system. The resulting equilibrium equation then becomes equal to

$$[ka\sin(\theta+\theta^*)]a\cos(\theta+\theta^*) - M_o - PL\sin(\theta+\theta^*) = 0$$

Noting that θ^* is small, and so $\sin\theta^* = \theta^*$; $\cos\theta^* = 1$. Also making use of the trigonometric relationships

$$\sin(\theta+\theta^*) = \sin\theta\cos\theta^* + \cos\theta\sin\theta^* = \sin\theta + \theta^*\cos\theta$$

$$\cos(\theta+\theta^*) = \cos\theta\cos\theta^* - \sin\theta\sin\theta^* = \cos\theta - \theta^*\sin\theta$$

we can arrive at the following equation:

$$\begin{aligned}&[ka^2\sin\theta\cos\theta - M_o - PL\sin\theta]\\&\quad + \theta^*[ka^2(\cos^2\theta - \sin^2\theta) - PL\cos\theta]\\&\quad - \theta^*[ka^2\cos\theta\sin\theta] = 0\end{aligned}$$

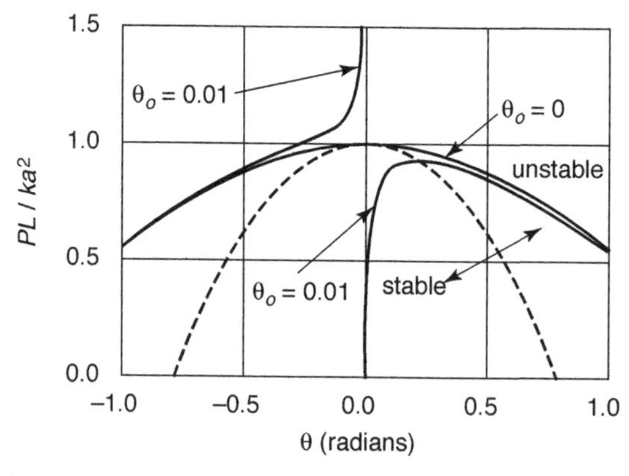

Fig. 1.12 Load-rotation curves for a softening system.

The first line is the equilibrium equation, and it equals zero, as demonstrated above. The bracket in the third line is multiplied by the square of a small quantity ($\theta^* \gg \theta^2$) and so it can be neglected. From the second line we obtain the stability condition that is shown in Figure 1.12 as a dashed line:

$$\frac{PL}{ka^2} = \frac{\cos^2\theta - \sin^2\theta}{\cos\theta} = \frac{2\cos^2\theta - 1}{\cos\theta} \tag{1.21}$$

This problem is solved also by the energy method, as follows:

$$\text{Total potential: } \Pi = \frac{k(a\sin\theta)^2}{2} - M_o\theta - PL(1 - \cos\theta)$$

$$\text{Equilibrium: } \frac{\partial\Pi}{\partial\theta} = ka^2 \sin\theta\cos\theta - M_o - PL\sin\theta = 0 \rightarrow \frac{PL}{ka^2} = \frac{\sin\theta\cos\theta}{\sin\theta}$$

$$\text{Stability: } \frac{\partial\Pi}{\partial\theta} = ka^2[\cos^2\theta - \sin^2\theta] - PL\cos\theta = 0 \rightarrow \frac{PL}{ka^2} = \frac{2\cos^2\theta - 1}{\cos\theta}$$

The two spring-bar problems just discussed illustrate three post-buckling situations that occur in real structures: hardening post-buckling behavior, softening post-buckling behavior, and the transitional case where the post-buckling curve is flat for all practical purposes. These cases are discussed in various contexts in subsequent chapters of this book. The drawings in Figure 1.13 summarize the different post-buckling relationships, and indicate the applicable real structural problems. Plates are insensitive to initial imperfections, exhibiting reliable additional strength beyond the buckling load. Shells and columns that buckle after some parts of their cross section have yielded are imperfection sensitive. Elastic buckling of columns, beams, and frames have little post-buckling strength, but they are not softening, nor are they hardening after buckling.

Before leaving the topic of spring-bar stability, we will consider two more topics: the snap-through buckling and the multidegree of freedom column.

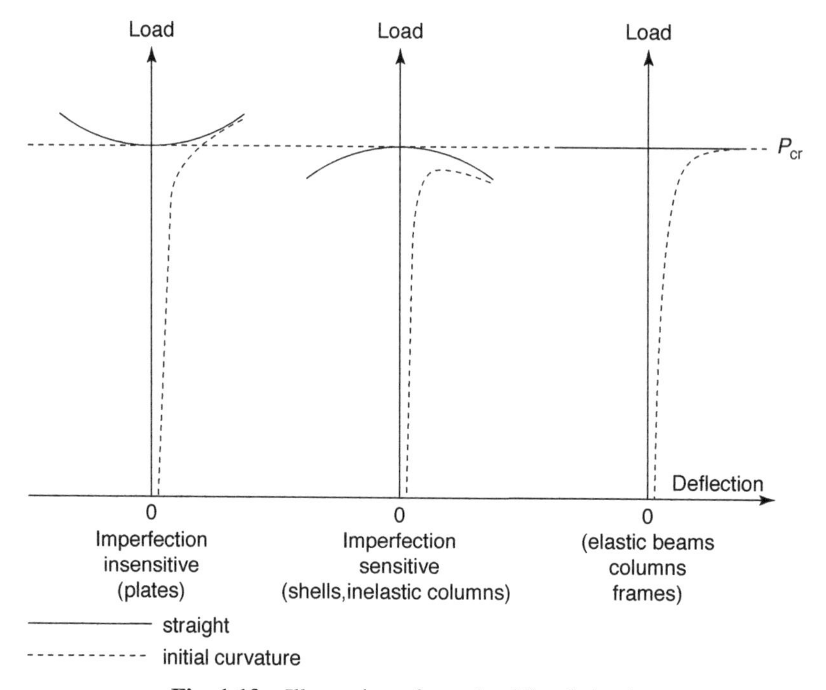

Fig. 1.13 Illustration of post-buckling behavior.

1.4 SNAP-THROUGH BUCKLING

Figure 1.14 shows a two-bar structure where the two rigid bars are at an angle to each other. One end of the right bar is on rollers that are restrained by an elastic spring. The top Figure 1.14 shows the loading and geometry, and the bottom features the deformed shape after the load is applied. Equilibrium is determined by taking moments of the right half of the deformed structure about point A.

$$\sum M_A = 0 = \frac{P}{2}[L\cos(\alpha - \theta)] - \Delta kL\sin(\alpha - \theta)$$

From the deformed geometry of Figure 1.14 it can be shown that

$$\Delta = 2L\cos(\alpha - \theta) - 2L\cos\theta$$

The equilibrium equation thus is determined to be

$$\frac{P}{kL} = 4[\sin(\alpha - \theta) - \tan(\alpha - \theta)\cos\alpha] \tag{1.22}$$

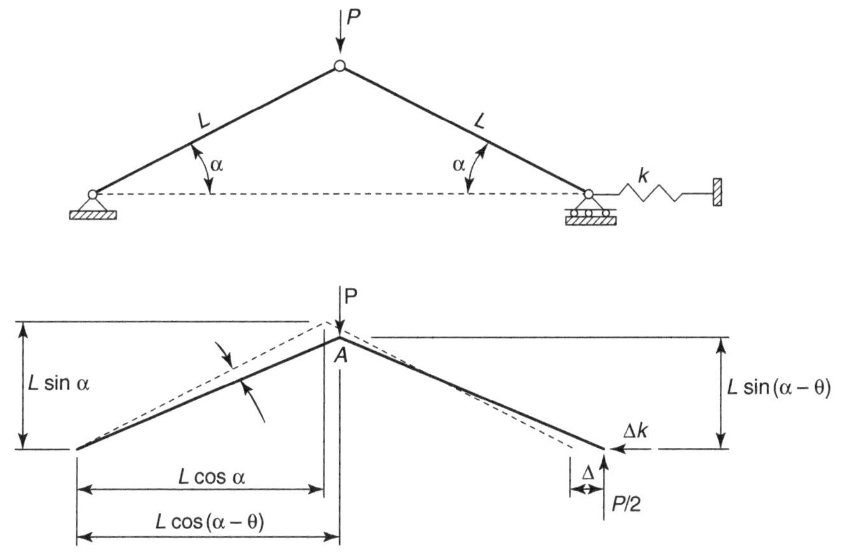

Fig. 1.14 The snap-through structure.

The state of the equilibrium is established by disturbing the deflected structure by an infinitesimally small virtual rotation θ^*. After performing trigonometric and algebraic manipulations it can be shown that the curve separating stable and unstable equilibrium is

$$\frac{P}{kL} = 4\left[\frac{1 - 2\cos^2(\alpha - \theta) + \cos(\alpha - \theta)\cos\alpha}{\sin(\alpha - \theta)}\right] \tag{1.23}$$

If we substitute PL/k from equation 1.22 into equation 1.23, we get, after some elementary operations, the following equation that defines the angle θ at the limit of stable equilibrium:

$$\cos^3(\alpha - \theta) - \cos\alpha = 0 \tag{1.24}$$

The curve shown in Figure 1.15 represents equilibrium for the case of $\alpha = 30°$. Bar rotation commences immediately as load is increased from zero. The load-rotation response is nonlinear from the start. The slope of the curve increases until a peak is reached at $P/kl = 0.1106$ and $\theta = 0.216$ radians. This is also the point of passing from stable to unstable equilibrium as defined by equations 1.23 and 1.24. The deformation path according to equation 1.22 continues first with a negative slope until minimum is reached, and then it moves up with a positive slope. However, the actual path of the deflection of the structure does not follow this unstable path, but the structure snaps through to $\theta = 1.12$ radians. Such behavior is typical of shell-type structures, as can be easily demonstrated by standing on the top of

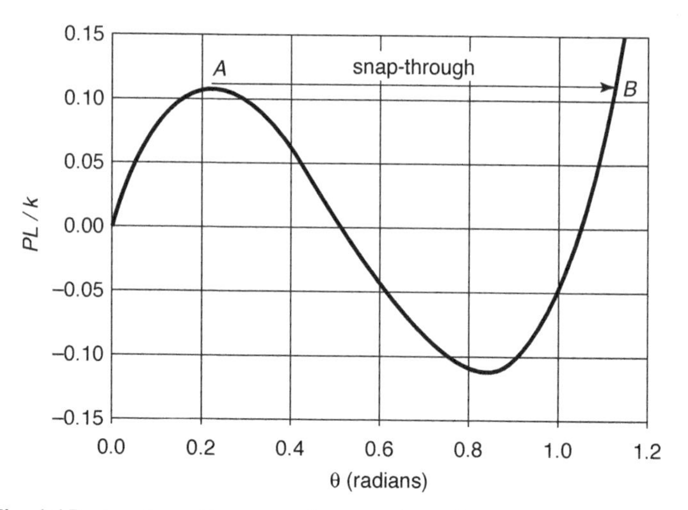

Fig. 1.15 Load-rotation curve for snap-through structure for $\alpha = 30°$.

an empty aluminum beverage can and having someone touch the side of the can. A similar event takes place any time a keyboard of a computer is pushed. Snap-through is sudden, and in a large shell structure it can have catastrophic consequences.

Similarly to the problems in the previous section, the energy approach can be also used to arrive at the equilibrium equation of equation 1.22 and the stability limit of equation 1.23 by taking, respectively, the first and second derivative of the total potential with respect to θ. The total potential of this system is

$$\Pi = \frac{1}{2}k\{2L[\cos(\alpha - \theta) - \cos\alpha]\}^2 - PL[\sin\alpha - \sin(\alpha - \theta)] \quad (1.25)$$

The reader can complete to differentiations to verify the results.

1.5 MULTI-DEGREE-OF-FREEDOM SYSTEMS

The last problem to be considered in this chapter is a structure made up of three rigid bars placed between a roller at one end and a pin at the other end. The center bar is connected to the two edge bars with pins. Each interior pinned joint is restrained laterally by an elastic spring with a spring constant k. The structure is shown in Figure 1.16a. The deflected shape at buckling is presented as Figure 1.16b. The following buckling analysis is performed by assuming small deflections and an initially perfect geometry. Thus, the only information to be gained is the critical load at which a straight and a buckled configuration are possible under the same force.

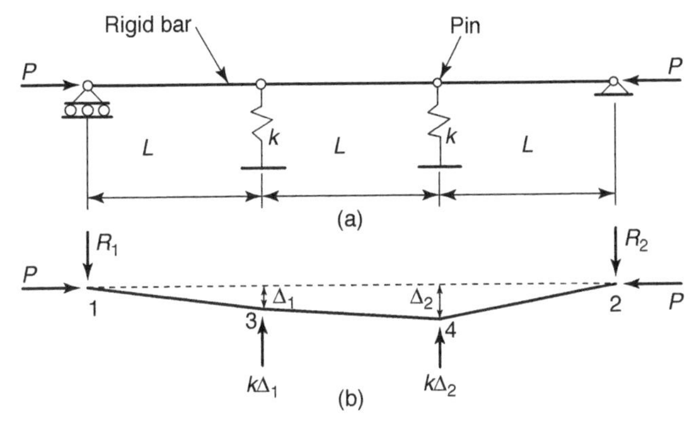

Fig. 1.16 Three-bar structure with intermediate spring supports.

Equilibrium equations for this system are obtained as follows:

Sum of moments about Point 1: $\sum M_1 = 0 = k\Delta_1 L + k\Delta_2(2L) - R_2(3L)$
Sum of vertical forces: $\sum F_y = 0 = R_1 + R_2 - k\Delta_1 - k\Delta_2$
Sum of moments about point 3, to the left: $\sum M_3 = 0 = P\Delta_1 - R_1 L$
Sum of moments about point 4, to the right: $\sum M_4 = 0 = P\Delta_2 - R_2 L$

Elimination of R_1 and R_2 from these four equations leads to the following two homogeneous simultaneous equations:

$$\begin{bmatrix} P - \frac{2kL}{3} & -\frac{kL}{3} \\ -\frac{kL}{3} & P - \frac{2kL}{3} \end{bmatrix} \begin{bmatrix} \Delta_1 \\ \Delta_2 \end{bmatrix} = 0 \tag{1.26}$$

The deflections Δ_1 and Δ_2 can have a nonzero value only if the determinant of their coefficients becomes zero:

$$\begin{vmatrix} P - \frac{2kL}{3} & -\frac{kL}{3} \\ -\frac{kL}{3} & P - \frac{2kL}{3} \end{vmatrix} = 0 \tag{1.27}$$

Decomposition of the determinant leads to the following quadratic equation:

$$3\left(\frac{P}{kL}\right)^2 - 4\frac{P}{kL} + 1 = 0 \tag{1.28}$$

This equation has two roots, giving the following two critical loads:

$$P_{\mathrm{cr1}} = kL$$
$$P_{\mathrm{cr2}} = \frac{kL}{3} \tag{1.29}$$

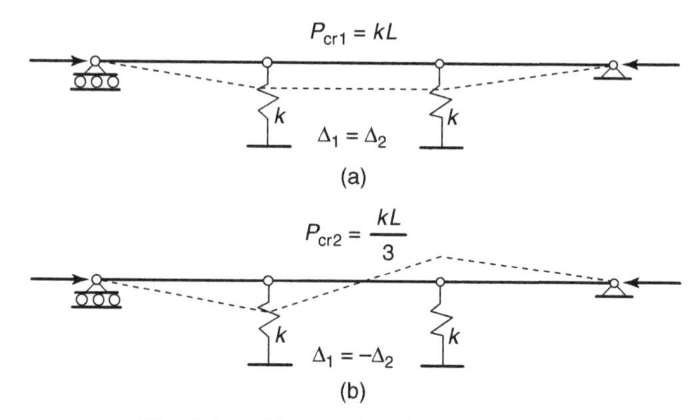

Fig. 1.17 Shapes of the buckled modes.

The smaller of the two critical loads is then the buckling load of interest to the structural engineer. Substitution each of the critical loads into equation 1.26 results in the mode shapes of the buckled configurations, as illustrated in Figure 1.17.

Finally then, $P_{cr} = \frac{kL}{3}$ is the governing buckling load, based on the small deflection approach.

The energy method can also be used for arriving at a solution to this problem. The necessary geometric relationships are illustrated in Figure 1.18, and the small-deflection angular and linear deformations are given as follows:

$$\Delta_1 = \psi L \quad \text{and} \quad \Delta_2 = \theta L$$

$$\frac{\Delta_1 - \Delta_2}{L} = \gamma = \psi - \theta$$

$$\varepsilon_3 = L - L\cos\theta \approx \frac{L\theta^2}{2}$$

$$\varepsilon_2 = \varepsilon_3 + L[1 - \cos(\psi - \theta)] = \frac{L}{2}(2\theta^2 + \psi^2 - 2\psi\theta)$$

$$\varepsilon_3 = \varepsilon_2 + \frac{L\psi^2}{2} = L(\theta^2 + \psi^2 - \psi\theta)$$

The strain energy equals $U_P = \frac{k}{2}(\Delta_1^2 + \Delta_2^2) = \frac{kL^2}{2}(\psi^2 + \theta^2)$.

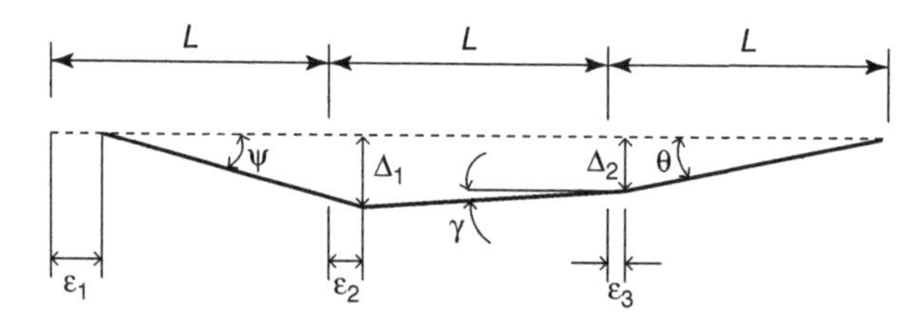

Fig. 1.18 Deflections for determining the energy solution.

The potential of the external forces equals $V_P = -P\varepsilon_1 = -PL(\theta^2 + \psi^2 - \psi\theta)$

The total potential is then

$$\Pi = U + V_P = \frac{kL^2}{2}(\psi^2 + \theta^2) - PL(\theta^2 + \psi^2 - \psi\theta) \quad (1.30)$$

For equilibrium, we take the derivatives with respect to the two angular rotations:

$$\frac{\partial \Pi}{\partial \psi} = 0 = \frac{kL^2}{2}(2\psi) - 2PL\psi + PL\theta$$

$$\frac{\partial \Pi}{\partial \theta} = 0 = \frac{kL^2}{2}(2\theta) - 2PL\theta + PL\psi$$

Rearranging, we get

$$\begin{bmatrix} (kL^2 - 2PL) & PL \\ PL & (kL^2 - 2PL) \end{bmatrix} \begin{bmatrix} \theta \\ \psi \end{bmatrix} = 0$$

Setting the determinant of the coefficients equal to zero results in the same two critical loads that were already obtained.

1.6 SUMMARY

This chapter presented an introduction to the subject of structural stability. Structural engineers are tasked with designing and building structures that are safe under the expected loads throughout their intended life. Stability is particularly important during the erection phase in the life of the structure, before it is fully braced by its final cladding. The engineer is especially interested in that critical load magnitude where the structure passes from a stable to an unstable configuration. The structure must be proportioned so that the expected loads are smaller than this critical value by a safe margin.

The following basic concepts of stability analysis are illustrated in this chapter by several simple spring-bar mechanisms:

- The critical, or buckling load, of geometrically perfect systems
- The behavior of structures with initial geometric or statical imperfections
- The amount of information obtained by small deflection and large deflection analyses
- The equivalence of the geometrical and energy approach to stability analysis

- The meaning of the results obtained by a *bifurcation* analysis, a computation of the *post-buckling* behavior, and by a *snap-through* investigation
- The hardening and the softening post-buckling deformations
- The stability analysis of multi-degree-of-freedom systems

We encounter each of these concepts in the subsequent parts of this text, as much more complex structures such as columns, beams, beam-columns, and frames are studied.

PROBLEMS

1.1. Derive an expression for the small deflection bifurcation load in terms of $\frac{EI}{L^2}$.

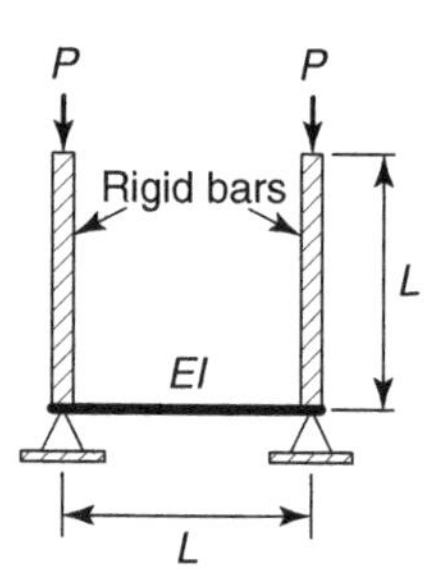

Fig. p1.1

1.2. Determine the critical load of this planar structural system if

$$a = L,\ L_1 = L \text{ and } L_2 = 3L.$$

Hint: The flexible beam provides a rotational and translational spring to the rigid bar compression member.

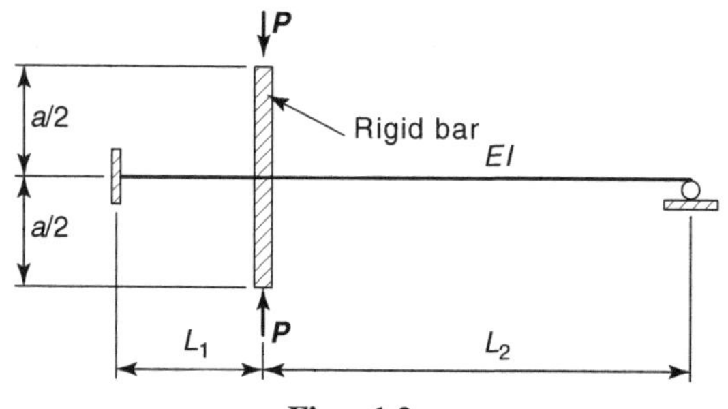

Fig. p1.2

1.3. Determine the critical load of this planar structural system.
Hint: The flexible beam provides a rotational and translational spring to the rigid bar compression member.

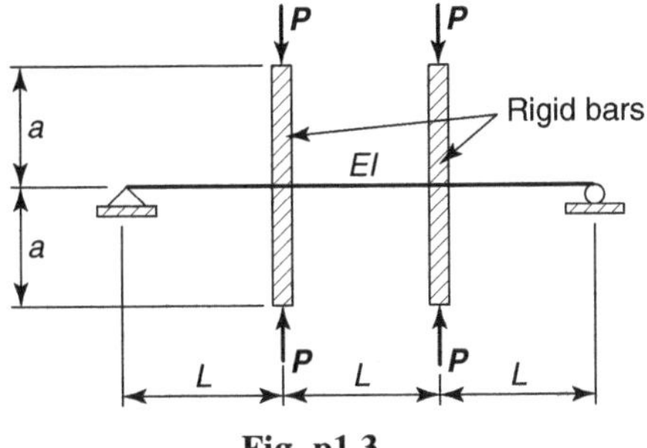

Fig. p1.3

1.4. In the mechanism a weightless infinitely stiff bar is pinned at the point shown. The load P remains vertical during deformation. The weight W does not change during buckling. The spring is unstretched when the bar is vertical. The system is disturbed by a moment M_o at the pin.

a. Determine the critical load P according to small deflection theory.

b. Calculate and plot the equilibrium path $p - \theta$ for $0 \leq \theta \leq \frac{\pi}{2}$ when $\theta_o = 0$ and

$$\theta_o = 0.01, \; p = \frac{PL - Wb}{ka^2} \quad \text{and} \quad \theta_o = \frac{M_o}{ka^2}, \; a = 0.75L \quad \text{and} \quad b = 1.5L.$$

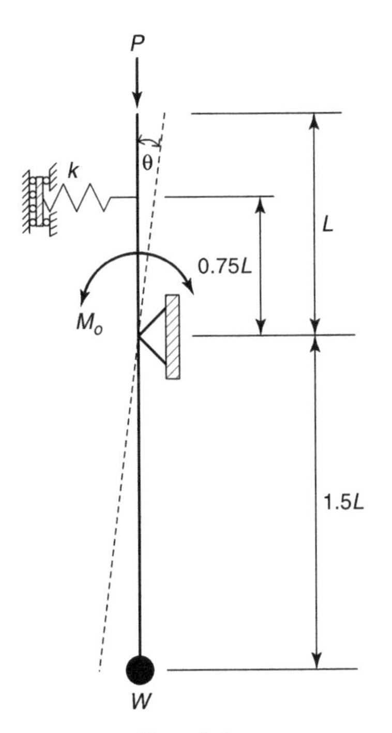

Fig. p1.4

c. Investigate the stability of the equilibrium path.

d. Discuss the problem.

Note: This problem was adapted from Chapter 2, Simitses "An introduction to the elastic stability of structures."

1.5. Develop an expression for the critical load using the small-deflection assumption. Employ both the equilibrium and the energy method. *Note:* that the units of K_1 are inch-kip/radian, and the units of K_2 are kip/inch

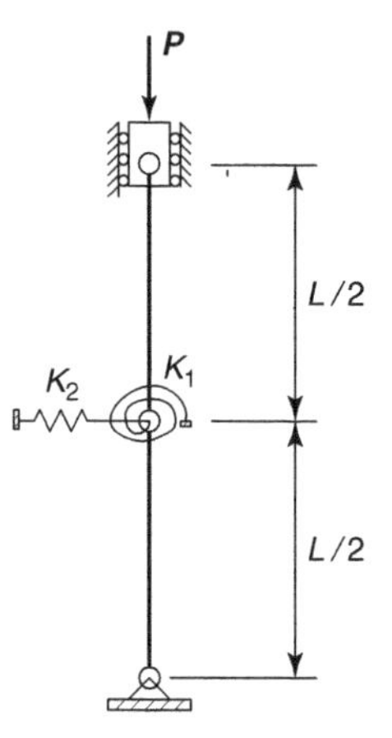

Fig. p1.5

1.6. Develop an expression for the critical load using the small-deflection assumption. The structure is made up of rigid bars and elastic springs. Employ both the equilibrium and the energy method.

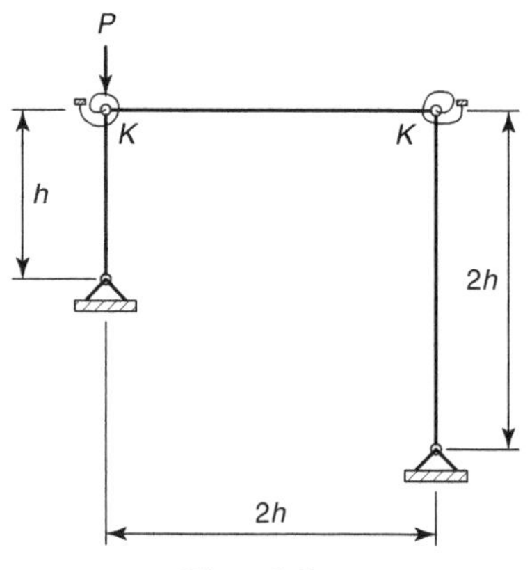

Fig. p1.6

1.7. The length of the bar is L, and it is in an initially rotated condition ϕ_i from the vertical. The spring is undistorted in this initial configuration. A vertical load P is applied to the system, causing it to deflect an angle ϕ from the vertical. The load P remains vertical at all times. Derive equations for equilibrium and stability, using the equilibrium and the energy methods. Plot P versus ϕ for $\phi_i = 0.05$ radians.

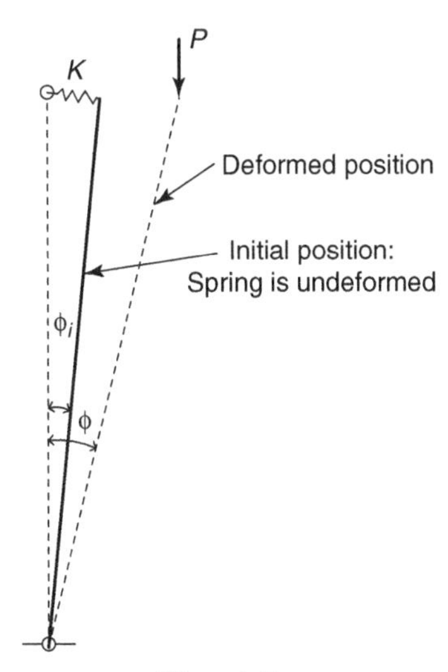

Fig. p1.7

CHAPTER TWO

ELASTIC BUCKLING OF PLANAR COLUMNS

2.1 INTRODUCTION

In Chapter 1, basic principles of stability were illustrated using simple, spring-bar models. The systems considered in the Chapter 1 were composed of discrete parts, and thus an algebraic solution to find P_{cr} was possible if small displacement theory was employed.

This chapter examines the continuous case of columns subjected to axial loads. These problems no longer are discrete, but instead consider stability of a continuous member. Therefore, the solutions for these problems are differential rather than algebraic in nature. Specifically, we will consider the elastic buckling of columns. Many classical textbooks on elasticity and structural stability discuss the topic of column buckling. The list of textbooks at the end of this chapter is a sampling of what has been published since the 1960s.

In discussing the buckling of columns, it is helpful to start with the basic case in which the Euler buckling equation, familiar to students and engineers, is derived. First, we consider the large displacement solution. From there, we can use small displacement assumptions to show the derivation of the Euler load. We then extend the problem to the generic case of planar flexure. This very general case can be used to examine the numerous parametric variations of the basic problem, such as effects of imperfections and boundary conditions.

2.2 LARGE-DEFLECTION SOLUTION OF AN ELASTIC COLUMN

We start the study of the buckling of elastic compression elements by considering a pinned-end prismatic column of length L and moment of inertia I, subject to a concentric axial force P, as shown in Figure 2.1. We assume that the column is:

- Perfectly straight
- Elastic
- Prismatic

Until the buckling load is reached, the column remains perfectly straight. At the point of buckling, there is a bifurcation of the deformation path. Equilibrium is considered in the deflected position of the member.

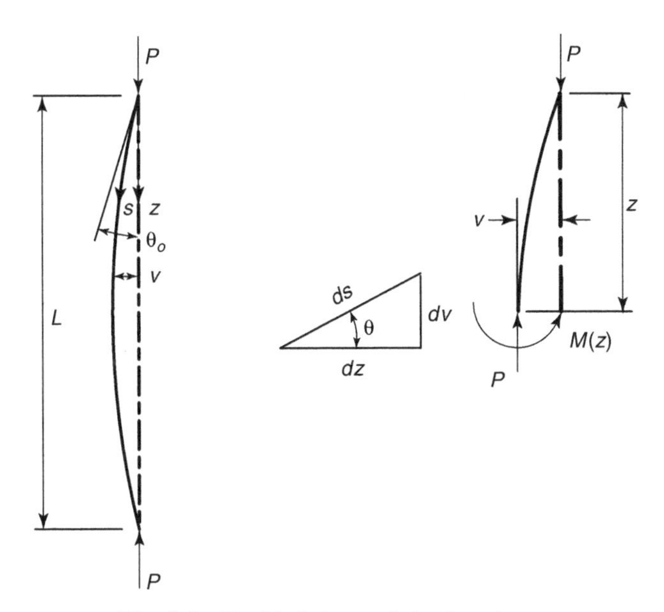

Fig. 2.1 Buckled shape of elastic column.

The deflection at any location z is v and the end slope of the deformed column at $z = 0$ is θ_o. The coordinate s is along the deformed deflection path. The bending moment at the coordinate z is equal to

$$M(z) = Pv = -EI\phi \tag{2.1}$$

In this expression Pv is the external moment and $-EI\phi = -EI\frac{d\theta}{ds}$ is the internal moment that is the product of the stiffness $-EI$ and the curvature $\phi = \frac{d\theta}{ds}$, and $\theta = \theta(z)$ is the slope along the deflected column. Rearranging equation 2.1, differentiating once, and noting that $\frac{dv}{ds} = \sin\theta$, the following differential equation of the deformation after buckling is obtained:

$$\frac{d^2\theta}{ds^2} + k^2 \sin\theta = 0 \tag{2.2}$$

where

$$k^2 = \frac{P}{EI} \tag{2.3}$$

The differential equation of equation 2.2 results in the following integral equation, as derived in Appendix 2.1:

$$\frac{kL}{2} = \int_0^{\frac{\pi}{2}} \frac{d\alpha}{\sqrt{1 - p^2 \sin^2\alpha}} \quad \text{where} \quad p = \sin\frac{\theta_o}{2} \tag{2.4}$$

The integral is a complete elliptic integral of the first kind, and it can be evaluated from tables of such integrals for assumed values of θ_o, or it can be calculated by any of a number of mathematical computer programs. The curve in Figure 2.2 shows the relationship between the axial load ratio $\frac{P}{P_{\mathrm{E}}}$ and the end-slope θ_o, where $P_{\mathrm{E}} = \frac{\pi^2 EI}{L^2}$, the *Euler* buckling load that will be introduced in section 2.3. The solution just described was first presented by the mathematician Leonard Euler, and the resulting relationship is also known as *Euler's Elastica*.

The post-buckling behavior of an elastic pinned-end column was introduced here to demonstrate that the relationship is *hardening*, that is, addition of load is required to increase deflection. Such a phenomenon was discussed in Chapter 1. However, for the elastic column the increase of load

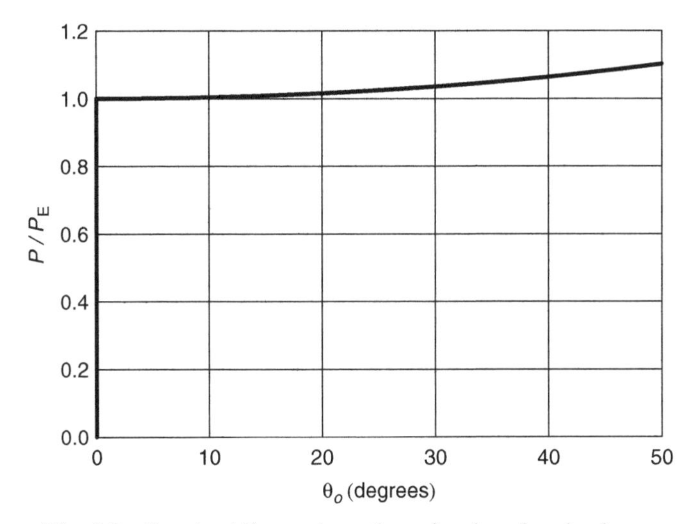

Fig. 2.2 Post-buckling end-rotation of a pinned-end column.

is sensibly noticeable only for very large end rotations, say 20°. For all practical purposes it can be assumed for the elastic buckling of columns that, at the instant of instability, there is an infinitesimally close deflected shape on which equilibrium can be formulated in order to calculate the buckling load. Small deflections may thus be assumed in the derivations. Having demonstrated that this simpler approach is defensible, we derive the governing differential equation that will be solved for a variety of useful column applications in section 2.3. First, however, we use small displacement assumptions to derive the familiar Euler buckling equation in its well-known form.

In order to derive the Euler buckling equation, given by

$$P_{\mathrm{cr}} = \frac{\pi^2 EI}{L^2}$$

we can use equation 2.2 noting that, for very small displacements, where $\sin\theta = \theta$

$$\frac{dv}{ds} = \theta \quad \text{and} \quad -EI\phi = -EI\frac{d^2v}{ds^2}$$

The governing differential equation becomes equal to

$$EI\frac{d^2v}{ds^2} + Pv = 0$$

Remembering that $k^2 = \frac{P}{EI}$ we obtain

$$v'' + k^2v = 0$$

The general solution for a homogenous differential equation of this form is given by

$$v = A\sin ks + B\cos ks$$

for small deflections we can set $s = z$ (see Figure 2.1). We know two boundary conditions for this column, based on the end restraint

$$v(0) = 0$$
$$v(L) = 0$$

From the first condition, we find that $B = 0$, leaving us with $v = A\sin kz$. Substituting in the second boundary condition gives this equation:

$$A\sin kL = 0$$

Since $A = 0$ is a trivial solution, we know that $\sin kL = 0$. For now, we will take the first possible solution for which $\sin kL = 0$, namely $kL = \pi$. Substituting in $k^2 = \frac{P}{EI}$, we obtain the classic Euler buckling equation:

$$P = \frac{\pi^2 EI}{L^2}$$

We discuss this solution in more depth in section 2.4.

2.3 DIFFERENTIAL EQUATION OF PLANAR FLEXURE

The solutions presented in section 2.2 are specific to a column that meets the rather stringent assumptions of that section. In order to consider several parameters and their effect on the column buckling strength, we derive the general equation of planar flexure for the beam-column shown in Figure 2.3a. There are two possible ways to derive the differential equation of planar buckling: (1) equilibrium formulation based on the deformed geometry of an element dx along the length of the member, and (2) using the calculus of variations approach. The former method is used in the ensuing parts of this section, since it illustrates the physical behavior of the beam-column more clearly (as all engineers are familiar with the concept of equilibrium).

The sketch in Figure 2.3a shows a column member that has a length an order of magnitude larger than its depth. Following are the 15 conditions and assumptions of the derivation:

1. The member may have a variation of its cross-sectional and material properties along its longitudinal axis z.
2. The axial force P acts along the centroidal axis of the column, and it does not change direction during buckling.
3. The deflection is due to a distributed load $q(z)$ that acts in the z-y plane.
4. The deflection v is restricted to the z-y plane.
5. The deflection v is resisted by an elastic foundation resulting in a distributed force αv, where α is the *foundation modulus.* (Such a foundation is known as a *Winkler Foundation.*)
6. Equilibrium is formulated on the deformed axis of the member.
7. The cross-section is singly or doubly symmetric. That is, the y-axis is an axis of symmetry. (The cross-section is defined by its conventional x-y plane.)
8. Deflections are small compared to the depth of the cross-section.
9. The stress-strain law of the material is known.
10. Plane sections before bending will remain plane after bending. (This is also known as *Navier's hypothesis.*)
11. Strains are proportional to the distance from the neutral axis of bending. That is, the member cannot be strongly curved in its plane of symmetry, such as, for example, a crane hook.
12. The cross-section does not change shape during deflection. That is, no local buckling occurs.
13. Shear deformations are neglected.
14. Lateral-torsional buckling is prevented by bracing.
15. Local effects are neglected. For example, *Saint Venant's principle* is a local effect.

The list of assumptions and restrictions is lengthy, but many of them will eventually be lifted. We start with the derivation by considering the

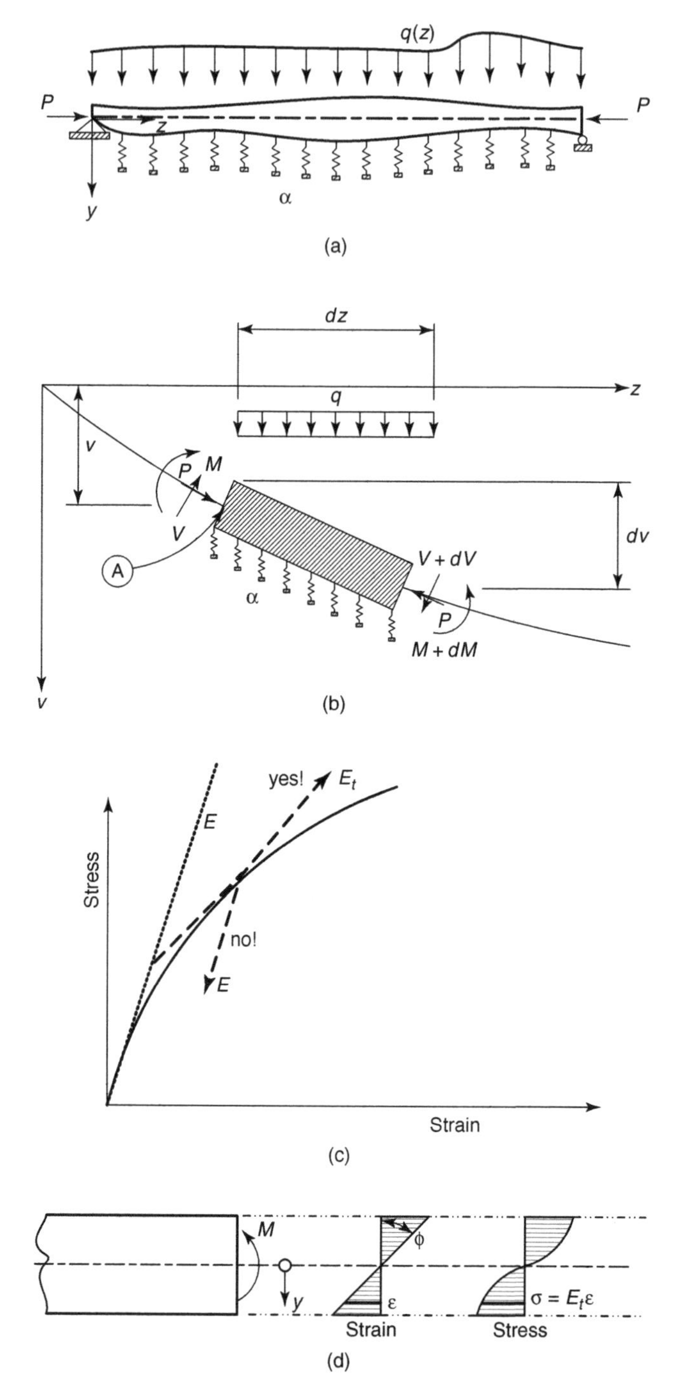

Fig. 2.3 Assumptions for deriving the differential equation for column buckling.

deformed differential length dz of the column (Figure 2.3b). Taking moments of forces about the point A:

$$\sum M_A = 0 = q \times dz \times \frac{dz}{2} - \alpha \times v \times dz \times \frac{dz}{2} + M - (M + dM) +$$
$$(V + dV)\cos\theta \times dz + (V + dV)\sin\theta \times dv + P \times \cos\theta \times dv - P \times \sin\theta \times dz$$

Because of the assumption of small deflections we can set

$$\cos\theta \approx 1.0;\ \sin\theta \approx \tan\theta \approx \theta = \frac{dv}{dz}$$
$$(dx)^2 \approx 0;\ dV \times dz \approx 0;\ \frac{dv}{dz} \times dv \approx 0;\ \frac{dv}{dz} \times dz \approx 0$$

Substitution results in the following equation:

$$-dM + Vdz + Pdv = 0 \rightarrow V + P\frac{dv}{dz} - \frac{dM}{dz} = 0 \qquad \text{(a)}$$

Equilibrium of the vertical forces gives the following equation:

$$\sum F_v = 0 = q \times dz - \alpha \times v \times dz + P\frac{dv}{dz} - P\frac{dv}{dz} - V + (V + dV)$$
$$\frac{dV}{dz} = \alpha v - q \qquad \text{(b)}$$

Differentiating equation (a) with respect to z and substituting in equation (b) results in the following differential equation:

$$-\frac{d^2M}{dz^2} + P\frac{d^2v}{dz^2} + \alpha v = q \qquad (2.5)$$

The curvature of the deflected shape is equal to, from elementary calculus,

$$\phi = \frac{-\frac{d^2v}{dz^2}}{\left[1 + \left(\frac{dv}{dz}\right)^2\right]^{\frac{3}{2}}} \approx -\frac{d^2v}{dz^2} \qquad (2.6)$$

because $\frac{dv}{dz} << 1$.

The stress–strain relationship is shown in Figure 2.3c to be $\sigma = E_t\varepsilon$; E_t is the slope of the nonlinear stress–strain curve, the *Tangent modulus*. We assume that there is no elastic unloading of the stress, as indicated in Figure 2.3c. (This assumption is discussed further in Chapter 3.)

The stress and strain distributions on a cross-section of the member are illustrated in Figure 2.3d. From the assumption that the strain is proportional to the distance from the neutral axis, we find that the strain equals $\varepsilon = \phi y$, where ϕ is the curvature and y is the vertical coordinate of the cross-section. The equilibrium of external and internal moments on the cross-section is equal to:

$$M = \int_{\text{Area}} (\sigma y) dA = \phi \int_{\text{Area}} (E_t y^2) dA \tag{2.7}$$

With equations 2.6 and 2.7, the differential equation 2.5 can now be expressed as

$$\frac{d}{dz^2}\left[\frac{d^2 v}{dz^2}\int_{Area} (E_t y^2) dA\right] + P\frac{d^2 v}{dz^2} + \alpha v = q \tag{2.8}$$

If the material is elastic, that is, $E_t = E$, and the column is prismatic, that is, $\int_{\text{Area}} y^2 dA = I_x$, where I_x is the moment of inertia of the cross-section about its x-axis, then equation 2.8 becomes equal to

$$EI_x v^{iv} + Pv'' + \alpha v = q \tag{2.9}$$

In this equation, $v^{iv} = \frac{d^4 v}{dz^4}$ and $v'' = \frac{d^2 v}{dz^2}$. Following we will consider a number of applications of this differential equation.

2.4 THE BASIC CASE: PIN-ENDED COLUMN

In section 2.2, we derived the Euler buckling equation using a derivation based on equilibrium of the deformed cross-section and small displacement theory. In this section, we will use the general equation of planar flexure and discuss the results in greater detail. Columns with other boundary conditions will be measured against this fundamental element through the device of the effective length concept that is familiar to structural engineering students and professionals.

The column under consideration is shown in Figure 2.4. Equilibrium is again formulated on the deformed deflection configuration. The differential equation is a special form of equation 2.9 when $\alpha = 0$ and $q = 0$, namely, when there is no elastic foundation and no distributed load. The boundary conditions at each end specify that there is no deflection and there is zero moment.

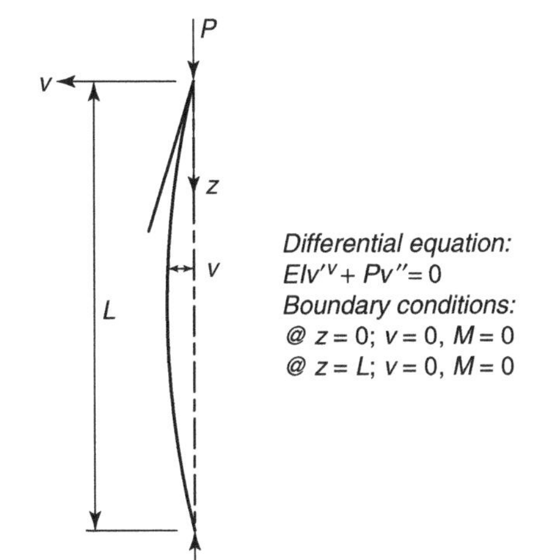

Fig. 2.4 The pin-ended column and its boundary conditions.

It was shown in the previous section that the curvature $\phi = -v''$ and $M = EI\phi$, and thus the boundary conditions are $v(0) = v''(0) = v(L) = v''(L) = 0$. The differential equation is rearranged as follows:

$$v^{iv} + k^2 v'' = 0 \tag{2.10}$$

where

$$k^2 = \frac{P}{EI} \tag{2.11}$$

Note that the subscript x was omitted for the moment of inertia I_x, and the axial force symbol P represents the critical load P_{cr}. The solution of the differential equation 2.10 is illustrated in a detailed step-by-step process. The deflection v equals

$$v = C_1 e^{r_1} + C_2 e^{r_2} + C_3 e^{r_3} + C_4 e^{r_4}$$

The coefficients C_i depend on the four boundary conditions, and the exponents r are the roots of the fourth-order differential equation:

$$r^2(r^2 + k^2) = 0$$
$$r_1 = r_2 = 0;\ r_3 = ik;\ r_4 = -ik$$
$$v = C_1 e^0 + C_2 z e^0 + C_3 e^{ikz} + C_4 e^{-ikz}$$
$$v = C_1 + C_2 z + C_3 e^{ikz} + C_4 e^{-ikz}$$

In the previous equations, $i = \sqrt{-1}$. A more convenient form of the solution can be obtained by making use of the relationships

$$\begin{aligned} e^{ikz} &= i \sin kz + \cos kz \\ e^{-ikz} &= -i \sin kz + \cos kz \end{aligned}$$

The form that will be used in the further discussions is thus

$$v = A + Bz + C \sin kz + D \cos kz \tag{2.12}$$

where $A = C_1$; $B = C_2$; $C = C_3 i - C_4 i$; and $D = C_3 + C_4$ are the coefficients dependent on the boundary conditions. The second derivative of equation 2.12 equals

$$v'' = -Ck^2 \sin kx - Dk^2 \cos kz$$

If we substitute the boundary conditions into equation 2.12, we obtain the following four simultaneous equations:

$$\begin{aligned} v(0) &= 0 = A(1) + B(0) + C(0) + D(1) \\ v''(0) &= 0 = A(0) + B(0) + C(0) + D(-k^2) \\ v(L) &= 0 = A(1) + B(L) + C(\sin kL) + D(\cos kL) \\ v''(L) &= 0 = A(0) + B(0) + C(-k^2 \sin kL) + D(-k^2 \cos kL) \end{aligned}$$

These equations can now be presented in matrix form as

$$\begin{bmatrix} 1 & 0 & 0 & 1 \\ 0 & 0 & 0 & -k^2 \\ 1 & L & \sin kL & \cos kL \\ 0 & 0 & -k^2 \sin kL & -k^2 \cos kL \end{bmatrix} \begin{bmatrix} A \\ B \\ C \\ D \end{bmatrix} = 0 \tag{2.13}$$

A, B, C, D define the deflection of the buckled bar and so at least one, if not all, have values other than zero. Thus, the determinant of the coefficients must be equal to zero. The equations are homogeneous simultaneous equations, and the value of k, and thus the critical load P_{cr}, are found by setting the determinant equal to zero. Mathematical names for the types of problems described here are *eigenvalue problems*, or *characteristic value problems*. The determinant is

$$\begin{vmatrix} 1 & 0 & 0 & 1 \\ 0 & 0 & 0 & -k^2 \\ 1 & L & \sin kL & \cos kL \\ 0 & 0 & -k^2 \sin kL & -k^2 \cos kL \end{vmatrix} = 0 \tag{2.14}$$

The decomposition of the 4×4 determinant into four 3×3 determinants is shown next:

$$1 \times \begin{vmatrix} 0 & 0 & -k^2 \\ L & \sin kL & \cos kL \\ 0 & -k^2 \sin kL & -k^2 \cos kL \end{vmatrix} - 0 \times \begin{vmatrix} 0 & 0 & -k^2 \\ 1 & \sin kL & \cos kL \\ 0 & -k^2 \sin kL & -k^2 \cos kL \end{vmatrix}$$

$$+0 \times \begin{vmatrix} 0 & 0 & -k^2 \\ 1 & L & \cos kL \\ 0 & 0 & -k^2 \cos kL \end{vmatrix} - 1 \times \begin{vmatrix} 0 & 0 & 0 \\ 1 & L & \sin kL \\ 0 & 0 & -k^2 \sin kL \end{vmatrix} = 0$$

Only the first determinant is not equal to zero, and its decomposition leads to $Lk^4 \sin kL = 0$. Since $Lk^4 \neq 0$, solution to the critical load is contained in

$$\sin kL = 0 \tag{2.15}$$

This is the *characteristic equation*, or the *eigenfunction*. It has an infinite number of roots that give an infinite number of critical loads:

$$kL = \sqrt{\frac{PL^2}{EI}} = n\pi \rightarrow P_{\text{cr}} = \frac{n\pi^2 EI}{L^2} \qquad n = 1, 2, 3 \ldots\ldots \tag{2.16}$$

Substitution of kL into the original simultaneous equations (equation 2.14) results in $A = B = D = 0$ and

$$v = C \sin kL = C \sin \frac{n\pi z}{L} \tag{2.17}$$

The value of C cannot be determined from this analysis. It is simply the unknown amplitude of the sinusoidal deflected shape. The mathematical name of the shape is the *eigenvector*. The three first shapes are shown in Figure 2.5.

Given the solution for the basic case of a pinned-end column, we can now consider the effects of different parameters on the critical column load, including boundary conditions (section 2.5) and imperfection (section 2.6).

2.5 FIVE FUNDAMENTAL CASES

The following five cases of column buckling are presented here in Table 2.1 as mile-posts so that the structural designers can compare the reasonableness of their answers that are obtained for other cases. The five stability cases are shown in Figure 2.6. They are distinct from each other by having different boundary conditions:

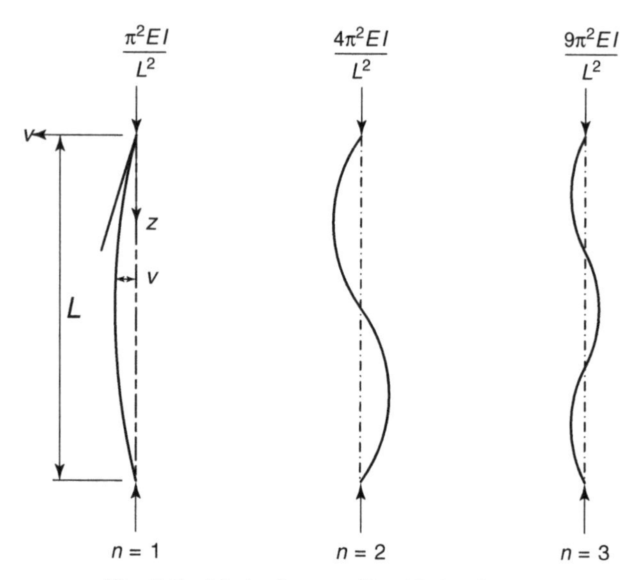

Fig. 2.5 Mode shapes of buckled column.

I. Both ends are pinned.
II. One end is pinned, the other end is fixed.
III. Both ends are fixed.
IV. One end is free, the other one is fixed.
V. Both ends are fixed, but one end is not restrained from moving sideways.

The definitions of the boundary conditions are as follows:

A. A pinned end has zero deflection and zero moment. That is, $v = v'' = 0$.
B. A fixed end has zero deflection and zero slope. That is, $v = v' = 0$.
C. A free end has zero moment $(v'' = 0)$ and zero shear. In section 2.3 where the differential equation of planar deflection is derived, it was shown that

$$V + Pv' - \frac{dM}{dz} = V + Pv' - EI\frac{d(v'')}{dz}$$
$$V = -EIv''' - Pv' = 0$$

These boundary conditions are shown in the second column of Table 2.1 for the five cases. The third column shows the determinants of the coefficients

TABLE 2.1 Five Fundamental Cases of Column Buckling

Case	Boundary Conditions	Buckling Determinant	Eigenfunction Eigenvalue Buckling Load	Effective Length Factor
I	$v(0) = v''(0) = 0$ $v(L) = v''(L) = 0$	$\begin{vmatrix} 1 & 0 & 0 & 1 \\ 0 & 0 & 0 & -k^2 \\ 1 & L & \sin kL & \cos kL \\ 0 & 0 & -k^2 \sin kL & -k^2 \cos kL \end{vmatrix}$	$\sin kL = 0$ $kL = \pi$ $P_{cr} = P_E$	1.0
II	$v(0) = v''(0) = 0$ $v(L) = v'(L) = 0$	$\begin{vmatrix} 1 & 0 & 0 & 1 \\ 0 & 0 & 0 & -k^2 \\ 1 & L & \sin kL & \cos kL \\ 0 & 1 & k \cos kL & -k \sin kL \end{vmatrix}$	$\tan kl = kl$ $kl = 4.493$ $P_{cr} = 2.045\, P_E$	0.7
III	$v(0) = v'(0) = 0$ $v(L) = v'(L) = 0$	$\begin{vmatrix} 1 & 0 & 0 & 1 \\ 0 & 1 & k & 0 \\ 1 & L & \sin kL & \cos kL \\ 0 & 1 & k \cos kL & -k \sin kL \end{vmatrix}$	$\sin \frac{kL}{2} = 0$ $kL = 2\pi$ $P_{cr} = 4 P_E$	0.5
IV	$v'''(0) + k^2 v' = v''(0) = 0$ $v(L) = v'(L) = 0$	$\begin{vmatrix} 0 & 0 & 0 & -k^2 \\ 0 & k^2 & 0 & 0 \\ 1 & L & \sin kL & \cos kL \\ 0 & 1 & k \cos kL & -k \sin kL \end{vmatrix}$	$\cos kL = 0$ $kL = \frac{\pi}{2}$ $P_{cr} = \frac{P_E}{4}$	2.0
V	$v'''(0) + k^2 v' = v'(0) = 0$ $v(L) = v'(L) = 0$	$\begin{vmatrix} 0 & 1 & k & 0 \\ 0 & k^2 & 0 & 0 \\ 1 & L & \sin kL & \cos kL \\ 0 & 1 & k \cos kL & -k \sin kL \end{vmatrix}$	$\sin kL = 0$ $kL = \pi$ $P_{cr} = P_E$	1.0

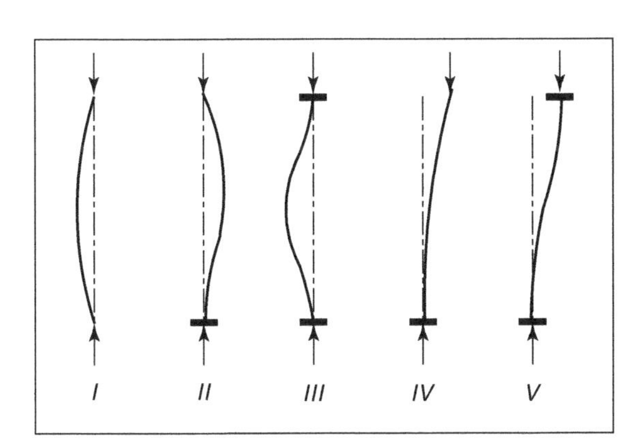

Fig. 2.6 Elementary buckling cases.

of the undetermined integration constant A, B, C, and D in equation 2.12, repeated next together with the first through third derivatives:

$$\begin{aligned} v &= A + Bz + C \sin kz + D \cos kx \\ v' &= B + Ck \cos kz - Dk \sin kz \\ v'' &= -Ck^2 \sin kz - Dk^2 \cos kz \\ v''' &= -Ck^3 \cos kz + Dk^3 \sin kz \end{aligned}$$

In these equations $k = \sqrt{P/EI}$. Substitution of the respective boundary conditions results in four simultaneous homogeneous equations. Nontrivial solutions exist only when the determinant of the coefficients is set to zero, finally giving $P = P_{cr}$, the buckling load. The equation resulting from the decomposition of the determinant is the *eigenfunction*. The *eigenvalue* is the nondimensional parameter kL that is obtained from solving the eigenfunction. All three quantities are listed in the fourth column of Table 2.1. The eigenvalue shown is the smallest value, because higher modes are not relevant on account of the presence of imperfections that will result in very large deflections as P approaches the critical value. Except for *Case II*, the pin-fix column, the solution of kL from the eigenfunction is straightforward. For *Case II* the value of $kL = 4.493$ is obtained by trial and error, or by using any one of a number of numerical equation solvers. The buckling load is expressed as a multiple of the basic pin-pin condition, $P_E = \pi^2 EI/L^2$. The fifth and final

Summary of Important Points—Elastic Column Buckling

- The Euler buckling equation, which should be familiar to structural engineers and students, is given by

$$P_{cr} = \frac{\pi^2 EI}{L^2}$$

This equation can be derived from the differential equation governing the behavior of a perfectly straight, pinned-end column loaded through the centroid.

- Higher solutions to the basic differential equation are associated with buckling modes.
- The derivation of the differential equation of planar flexure, the most general case, allows the investigation of the impact of many parameters on the buckling behavior of a member.
- The five basic cases illustrate the effects of boundary conditions and the use of the general equation.

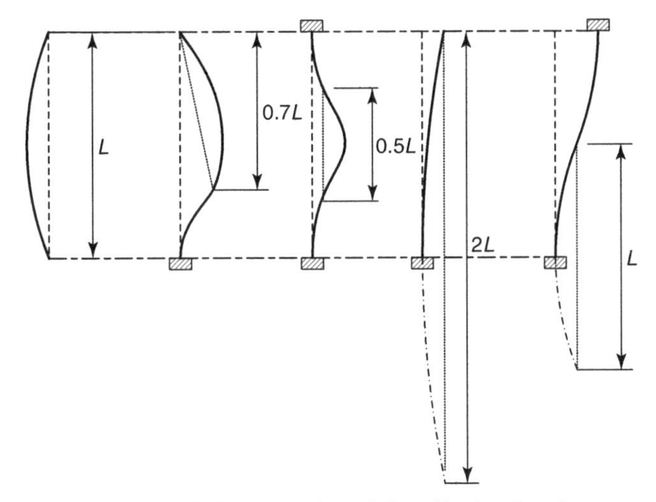

Fig. 2.7 Geometric interpretation of the effective length concept.

column in Table 2.1 lists K, the *effective length factor*. This is a popular artifice that connects any buckling load to the basic pin-pin case:

$$P_{\mathrm{cr}} = \frac{P_{\mathrm{E}}}{K^2} = \frac{\pi^2 EI}{(KL)^2} \tag{2.18}$$

For these elementary cases, one can visualize the effective length as the distance between points of inflection on the buckled shape of the column, as illustrated in Figure 2.7. This visualization is sometimes misunderstood when the effective length is taken as the distance between points of zero moment on a moment diagram from a first-order elastic analysis.

2.6 THE EFFECT OF IMPERFECTIONS

In any real column there are *imperfections* that affect the behavior near the theoretical critical load. These imperfections are small, and their occurrence is unavoidable. We consider the effects of three such phenomena:

1. Small initial crookedness (or out-of-straightness) of the column axis
2. Small load eccentricity
3. Small lateral load

These imperfections are illustrated in Figure 2.8.

2.6.1 Column with Initial Out-of-straightness

The column is shown in Figure 2.8a. The initial out-of-straightness is assumed to be sinusoidal, and its amplitude at the center of the member is v_o.

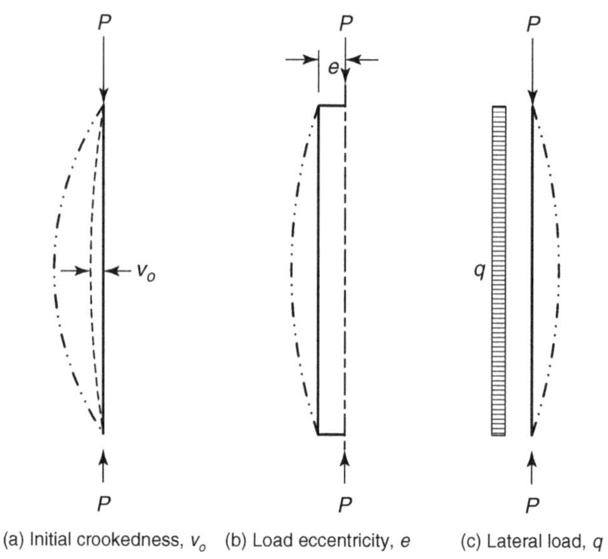

Fig. 2.8 Initial imperfections.

The assumed initial shape is expressed by the following equation:

$$v_i = v_o \sin\frac{\pi z}{L} \tag{2.19}$$

The internal moment at location z along the axis of the column is $M_{\text{int}} = -EIv''$, and the external moment equals $M_{\text{ext}} = P(v_i + v)$. Equating the internal and the external moment results in the following differential equation:

$$EIv'' + Pv = -Pv_i$$

Substitution of equation 2.19 and introducing $k^2 = \frac{P}{EI}$ leads to the equation for the deflection due to P:

$$v'' + k^2 v = -k^2 v_o \sin\frac{\pi z}{L} \tag{2.20}$$

The homogeneous solution, v_H, is $A\sin kz + B\cos kz$. The particular solution, v_P, is derived as follows:

$$v_P = C\sin\frac{\pi z}{L} + D\cos\frac{\pi z}{L}$$

Substitution into Eq. 2.20 gives

$$-C\frac{\pi^2}{L^2}\sin\frac{\pi z}{L} - D\frac{\pi^2}{L^2}\cos\frac{\pi z}{L} + Ck^2\sin\frac{\pi z}{L} + Dk^2\cos\frac{\pi z}{L} = -k^2 v_o \sin\frac{\pi z}{L}$$

From which

$$C\left[k^2 - \frac{\pi^2}{L^2}\right] = -v_o;\ D = 0$$

After some algebra and the introduction of the lowest buckling load P_E (i.e., $n = 1$) in equation 2.16,

$$P_E = \frac{\pi^2 EI}{L^2} \tag{2.21}$$

we obtain the following equation for the deflection:

$$v = v_H + v_P = A \sin kz + B \cos kz + \frac{P/P_E}{1 - P/P_E} v_o \sin \frac{\pi z}{L} \tag{2.22}$$

Substitution of the boundary conditions $v(0) = v(L) = 0$ gives $A = 0$ and $B = 0$, and thus

$$v = \frac{P/P_E}{1 - P/P_E} v_o \sin \frac{\pi z}{L} \tag{2.22}$$

The total deflection is the sum of the original initial deflection (equation 2.19) and the additional deflection due to P

$$v_{\text{total}} = v_i + v = \frac{v_o \sin \frac{\pi z}{L}}{1 - P/P_E} \tag{2.23}$$

The total deflection at the middle of the column (at $z = L/2$) becomes then equal to

$$v_{\text{Total}}/L = \frac{v_o/L}{1 - P/P_E} \tag{2.24}$$

The initial out-of-straightness v_o is the fabrication tolerance for straightness in the rolling mill, and in North American practice it is usually 1/1,000 of the length. This is a small amount, and it is not detectable by eye. However, as the ratio P/P_E approaches unity, the deflection becomes intolerably high, as can be observed in the curve shown in Figure 2.9. Since it is practically impossible to construct a column that has no imperfections, it is not possible to exceed the first mode of the buckling load; thus, $n = 1$ gives the governing buckling load for the pinned-end column that buckles elastically. This critical load, given as equation 2.21, will be the basic case against which critical loads of columns with other pinned boundary conditions will be compared. The subscript E thus denotes *elastic*, although often in the literature it also stands for *Euler* after the Swiss mathematician who first derived the formula in 1754.

In the previous discussion, it was assumed that the initial deformation is a one-half sine wave. The question is now: What happens if the initial shape

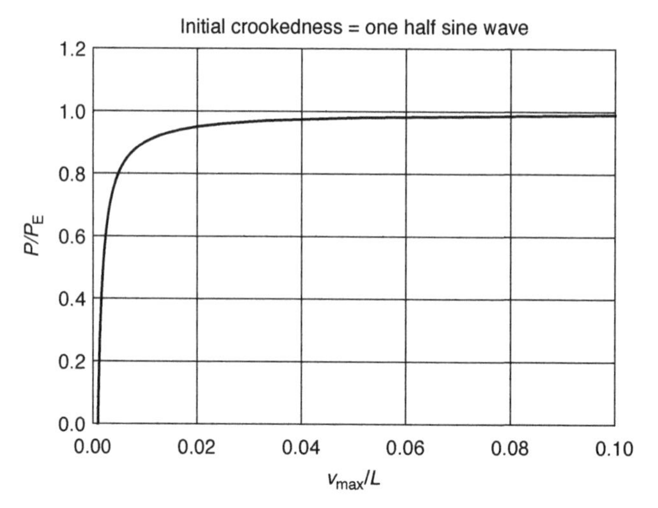

Fig. 2.9 Initial deflected shape is sinusoidal, $v_o = L/1,000$.

has some other form? Assume that the shape is a full sine wave, as shown in the side-sketch in Figure 2.10. In this case, the initial shape is

$$v_i = v_o \sin \frac{2\pi z}{L} \tag{2.25}$$

For this case the differential equation becomes equal to

$$v'' + k^2 v = -k^2 v_o \sin \frac{2\pi z}{L} \tag{2.26}$$

After differentiating equation 2.26 twice, the differential equation to be solved, the homogeneous solution and the particular solution are the following equations:

$$v^{iv} + k^2 v'' = k^2 v_o \left(\frac{4\pi^2}{L^2} \right) \sin \frac{2\pi z}{L}$$

$$v_H = A + Bz + C \sin kz + D \cos kz$$

$$v_P = E \sin \frac{2\pi z}{L} + F \cos \frac{2\pi z}{L}$$

The solution is, therefore:

$$v = A + Bz + C \sin kz + D \cos kz + \frac{v_o (P/P_{\mathrm{E}})}{4 - P/P_{\mathrm{E}}} \sin \frac{2\pi z}{L} \tag{2.27}$$

For the boundary conditions $v(0) = v(L) = v''(0) = v''(L) = 0$ one arrives at four homogeneous simultaneous equation that are identical to equation 2.13:

$$\begin{bmatrix} 1 & 0 & 0 & 1 \\ 0 & 0 & 0 & -k^2 \\ 1 & L & \sin kL & \cos kL \\ 0 & 0 & -k^2 \sin kL & -k^2 \cos kL \end{bmatrix} \begin{bmatrix} A \\ B \\ C \\ D \end{bmatrix} = 0$$

When the constants A, B, C, D are not zero, the determinant of the coefficients is zero and the resulting equation then leads to the first buckling mode with P_{E} as the critical value. When $A = B = C = D = 0$, then the deflected shape is

$$v = v_o \frac{P/P_{\mathrm{E}}}{4 - P/P_{\mathrm{E}}} \sin \frac{2\pi z}{L}$$

In this problem, it is more convenient to work with the end-slope $dv(0)/dz = \theta_o$ of the column. The total of the initial and added end-slope is equal to the following expression:

$$\theta_{o\,\mathrm{Total}} = \frac{2\pi v_o}{L} \left(\frac{1}{1 - \frac{P}{4P_{\mathrm{E}}}} \right) \tag{2.28}$$

The relationship between the axial load and the end-slope is shown in Figure 2.10. Equation 2.28 becomes very large when the axial load approaches

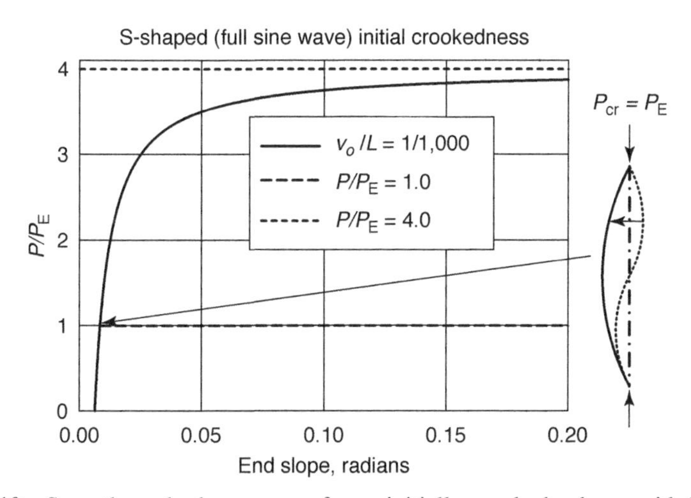

Fig. 2.10 Snap-through phenomenon for an initially crooked column with $L/1{,}000$ maximum out-of-straightness.

$4P_E$. However, the load has no chance of getting there because at $P = P_E$ the deflection of the column snaps from an S-shape into a half sine wave. A similar problem was also encountered in Chapter 1. The unexpected occurrence of such a snap-through event during the test of a full-sized column in the laboratory is something not easily forgotten.

2.6.2 Column with Eccentric Load (Figure 2.8b)

The differential equation is determined by equating the internal moment $-EIv''$ and the external moment $P(e + v)$, where e is the eccentricity of the axial load (see Figure 2.8b) and v is the deflection at the point of interest:

$$-EIv'' = P(e + v)$$
$$v'' + k^2 v = -k^2 e$$

e is the load eccentricity (Figure 2.8b), v is the deflection at location e, P is the axial force, and EI is the product of the elastic modulus and the moment of inertia, respectively. The second line is a rearrangement after introducing

$$k^2 = \frac{P}{EI}$$

The deflection is found from the general solution $v = A \sin kx + B \cos kx - e$ with the boundary conditions $v(0) = v(L) = 0$:

$$v = e\left(\cos kz + \frac{1 - \cos kL}{\sin kL} \sin kz - 1\right) \tag{2.29}$$

The maximum deflection occurs at the center at $z = L/2$ and it is equal to:

$$v(L/2) = e\left(\frac{1 - \cos \frac{kL}{2}}{\cos \frac{kL}{2}}\right) \tag{2.30}$$

When

$$P = P_E = \frac{\pi^2 EI}{L^2}; \; k = \frac{\pi}{L}; \; \cos\frac{kL}{2} = \cos\frac{\pi}{2} = 0; \; v(L/2) = \frac{e}{0} \rightarrow \infty$$

Thus, the deflection is infinite when P equals the elastic buckling load, as expected.

The center deflection of a beam with a moment M at each end is equal to (from linear structural analysis):

$$v_1(L/2) = \frac{ML^2}{8EI} = \frac{PeL^2}{8EI} = \frac{\pi^2}{8}(P/P_E)e$$

If we divide the deflection of the column (equation 2.30) by the linear beam deflection above we obtain a *magnification factor, MF,* that defines the effect of the reduction of stiffness, and thus the increase of deflection, due to the axial force:

$$MF = \frac{8}{\pi^2 (P/P_E)} \left[\frac{1 - \cos \frac{\pi}{2} \sqrt{\frac{P}{P_E}}}{\cos \frac{\pi}{2} \sqrt{\frac{P}{P_E}}} \right] \tag{2.31}$$

2.6.3 Column with Distributed Load (Figure 2.8c)

A distributed load is not specifically an imperfection; however, the transverse loading on the column creates deflection due to bending and is thus similar to the imperfection cases described in the previous sections. The differential equation for this case is equation 2.9 with the foundation modulus $\alpha = 0$:

$$EIv^{iv} + Pv'' = q \tag{2.32}$$

This equation is then rearranged and solved for the deflection:

$$v^{Iv} + k^2 v'' = \frac{q}{EI}$$

$$v = A + Bz + C \sin kz + D \cos kz + \frac{qz^2}{2P}$$

With the boundary conditions $v(0) = v(L) = v''(0) = v''(L) = 0$, the deflection at any location z and at the center of the member is, respectively:

$$\begin{aligned} v &= \frac{q}{Pk^2} \left[\left(\frac{1 - \cos, kL}{\sin kL} \right) \sin kz + \cos kz + \frac{(kz)^2}{2} - \frac{k^2 Lz}{2} - 1 \right] \\ v(L/2) &= \frac{q}{Pk^2} \left(\frac{1}{\cos \frac{kL}{2}} - \frac{(kL)^2}{8} - 1 \right) \end{aligned} \tag{2.33}$$

The first-order deflection is $\frac{5qL^4}{384EI}$, and therefore the magnification factor MF for this case is equal to

$$MF = \frac{384}{5k^4 L^4} \left(\frac{1}{\cos \frac{kL}{2}} - \frac{(kL)^2}{8} - 1 \right) \tag{2.33}$$

The variations of the magnification factors of equations 2.31 (for the eccentric axial load) and 2.33 (for the column with a distributed load) are

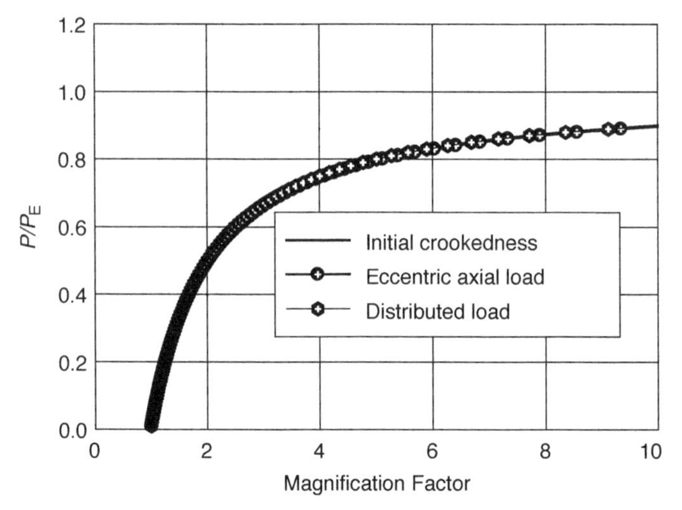

Fig. 2.11 Comparison of magnification factors.

plotted in Figure 2.11 against P/P_E. Also plotted is the magnification factor for the case of initial curvature (equations 2.24):

$$MF = \frac{1}{1 - P/P_E} \tag{2.34}$$

It is evident from comparing the curves in Figure 2.11 that they are essentially on top of each other, and thus the simpler equation 2.34 can be used for all three cases. This fact has been used for many years in design standards all over the world.

Historically, the idea of assuming an initial out-of-straightness or an accidental load eccentricity was used to arrive at formulas for the design of columns. They are named, variously, the *Rankine* formula, or the *Perry-Robertson* formula. Since such formulas were used extensively throughout the last 150 years, it is useful to give a brief derivation at this point.

The common feature of these formulas is the assumption that the maximum strength of the column is reached when the sum of the axial normal stress and the flexural normal stress equals the yield stress of the material. This is a very impractical assumption, of course, as we will show in Chapter 3, but since the initial deflection is calibrated so that the formula predicts the actual strength obtained by column tests, a useful and simple method of design is achieved. The derivation of the Rankine formula is given first:

$$\frac{P}{A} + \frac{Pv}{S} = \frac{P}{A}\left(1 + \frac{vA}{S}\right) = \sigma_y$$

P is the axial load, A is the area and S is the elastic section modulus of the cross-section. σ_y is the yield stress and v is the initial deflection at the center of the column. Note that $\frac{Av}{S} = \frac{Acv}{I} = \frac{cv}{r^2} = \frac{cv}{L^2}\left(\frac{L}{r}\right)^2$, where I is the moment of inertia, c is the distance from the neutral axis to the outer fiber of the cross-section, r is the radius of gyration, and L is the column length. The formula is then in the following form:

$$\sigma_{cr} = \frac{P}{A} = \frac{\sigma_y}{1 + \left(\frac{cv}{L^2}\right)\left(\frac{L}{r}\right)^2} = \frac{\sigma_y}{1 + \alpha\left(\frac{L}{r}\right)^2} \tag{2.35}$$

The coefficient α is obtained by calibration to test data. For example, the column formula in the 1923 Specification for steel building structures of the American Institute of Steel Construction (AISC) was of this form. There were many variants in use, and there probably are still extant some codes that use it. In the 1923 AISC Specification, the formula is as follows for a yield stress of 33 ksi and a factor of safety of 33/18:

$$\frac{\sigma_{cr}}{FS} = \frac{18\,\sigma_{cr}}{33} = \frac{18}{1 + \frac{1}{18{,}000}\left(\frac{L}{r}\right)^2} \leq 15\,\text{ksi} \tag{2.36}$$

The Perry-Robertson formula has been used in many countries, and it is still in the current Australian steel design standard. It is derived as follows:

$$v = \frac{v_o}{1 - P/P_E} = \frac{v_o}{1 - \frac{\sigma_{cr}}{\sigma_E}} = \frac{v_o \sigma_E}{\sigma_E - \sigma_{cr}}$$

$$\sigma_y = \sigma_{cr}\left[1 + \eta\left(\frac{\sigma_E}{\sigma_E - \sigma_{cr}}\right)\right]$$

Solving for the critical stress, one obtains

$$\sigma_{cr} = \frac{1}{2}[\sigma_y + \sigma_E(1 + \eta)] - \sqrt{\left\{\frac{1}{2}[\sigma_y + \sigma_E(1 + \eta)]\right\}^2 - \sigma_y \sigma_E} \tag{2.37}$$

Empirically in the Australian code, $\eta = 0.003\frac{L}{r}$.

2.7 STABILITY OF A RIGID FRAME

Next we consider the effect of elastic end-restraint on the critical load of a column. The structure is shown in Figure 2.12. The column has a pin at its bottom, and it is restrained at the top by an elastic beam that has a fixed end at its far end.

The boundary conditions at the bottom of the column at $z = L$ are equal to $v(L) = v''(L) = 0$. At the top there is no deflection, and the slope at the top of the column equals the slope at the end of the beam. The bending moments oppose each other. From structural analysis it can be determined that at the end of the beam

$$M_{AB} = \frac{4EI_{\mathrm{B}}}{L_{\mathrm{B}}}\theta_A = \alpha\theta_A = \alpha v'(0)$$

The symbol α is a spring constant that is $\alpha = \frac{4EI_{\mathrm{B}}}{L_{\mathrm{B}}}$ when the far end is fixed, and $\alpha = \frac{3EI_{\mathrm{B}}}{L_{\mathrm{B}}}$ when the far end is pinned. The moment at the top end of the

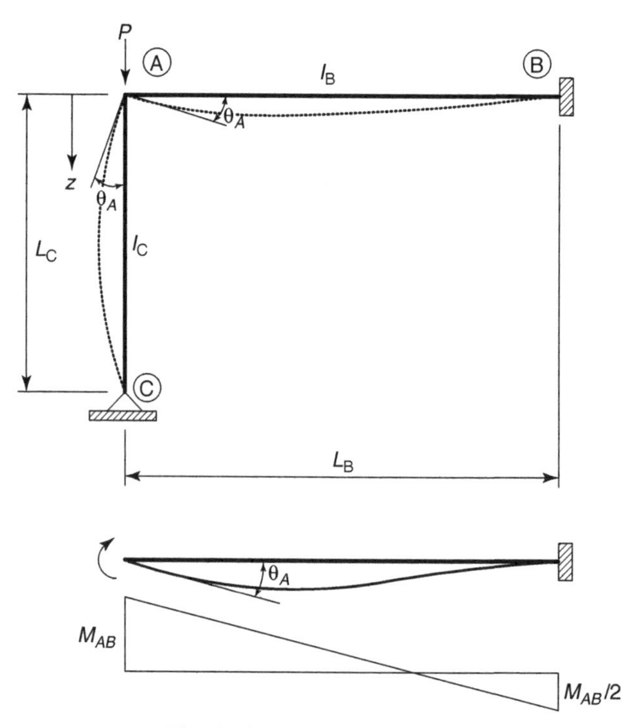

Fig. 2.12 Restrained column.

column equals

$$M_{AC} = -EI_C v''(0)$$

From the equilibrium condition $M_{AB} + M_{AC} = 0$, we then get the fourth boundary condition. The four boundary conditions are summarized next:

$$\begin{aligned}
&v(0) = 0 \\
&\alpha v'(0) - EI_C v''(0) = 0 \\
&v(L_C) = 0 \\
&v''(L_C) = 0
\end{aligned}$$

Substitution of the deflection equation $v = A + Bz + C\sin kz + D\cos kz$ and its derivatives gives four homogeneous simultaneous equations. Setting the determinant of the coefficients equal to zero

$$\begin{vmatrix} 1 & 0 & 0 & 1 \\ 0 & \alpha & \alpha k & P \\ 1 & L & \sin kL_C & \cos kL_C \\ 0 & 0 & -k^2 \sin kL_C & -k^2 \cos kL_C \end{vmatrix} = 0$$

leads to the following eigenfunction:

$$\begin{aligned}
tan\, kL_C &= \frac{\alpha\, kL_C}{PL_C + \alpha} = \frac{\gamma\, kL_C}{(kL_C)^2 + \gamma} \\
\gamma &= \frac{\alpha\, L_C}{EI_C} \\
kL_C &= \sqrt{\frac{PL_C^2}{EI_C}}
\end{aligned} \tag{2.39}$$

Equation 2.39 is the buckling equation for a column with a pinned end at one end and an elastic spring at the other end. When $I_B = \alpha = \gamma = 0$, the end restraint vanishes and we have a pinned-end column; that is, $\sin kL_C = 0 \rightarrow P_{cr} = \pi^2 EI_C / L_C^2$. When the top end is fixed, the following holds:

$$I_B = \alpha = \gamma = \infty$$

$$\tan kL_C = \frac{kL_C}{\frac{(kL_C)^2}{\gamma} + 1} = kL_C$$

$$P_{cr} = 20.19 \frac{EI_C}{L_C^2}$$

Comparing these extremes to the corresponding cases in Table 2.1 it is seen that the same answers are obtained. The variation of the critical load with the spring constant between the two extremes is shown in Figure 2.13.

The upper graph shows that the critical load is P_E when $\gamma = 0$, and it approaches $2.045P_E$ as the value of the spring constant approaches infinity. The lower graph illustrates the variation of the effective length factor K from 1.0 (pinned end) to 0.7 (fixed end). There is an important trend that can be deduced from these curves: On the one hand, when the restraint is small, large increases of the buckling load result from small increases of the spring constant α. On the other hand, when α becomes very large, a very small change in the buckling load results from very large changes in the spring

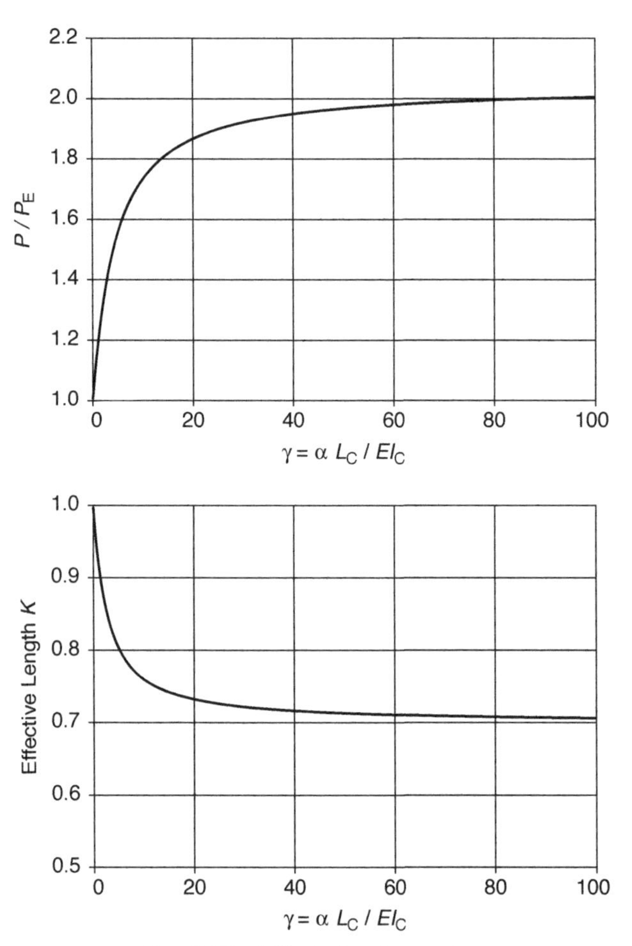

Fig. 2.13 Effect of end-restraint.

constant. *A little restraint goes a long way, but complete restraint is not worth attaining.* This principle is a general characteristic of buckling solutions discussed in more detail in Chapter 5.

2.8 END-RESTRAINED COLUMNS

In this section we consider the general case of prismatic columns that are restrained by elastic springs at their ends. By considering restrained ends, we can develop a feel for the impact of end restraint on the buckling load of the column. This situation is similar to a column restrained by beams of finite stiffness, which is discussed in depth in Chapter 5. We start the discussion with a compression member whose ends do not translate with respect to each other (often called a *non-sway* case) and that have elastic springs at each end. The column and its end boundary conditions are shown in Figure 2.14.

Substitution of the four boundary conditions into equation 2.12 results in four homogeneous simultaneous equations. The determinant of the coefficients of the constants A, B, C, D is equal to

$$\begin{vmatrix} 1 & 0 & 0 & 1 \\ 1 & L & \sin kL & \cos kL \\ 0 & \alpha_T & \alpha_T k & EIk^2 \\ 0 & -\alpha_B & -\alpha_B k \cos kL + EIk^2 \sin kL & \alpha_B k \sin kL + EIk^2 \cos kL \end{vmatrix} = 0$$

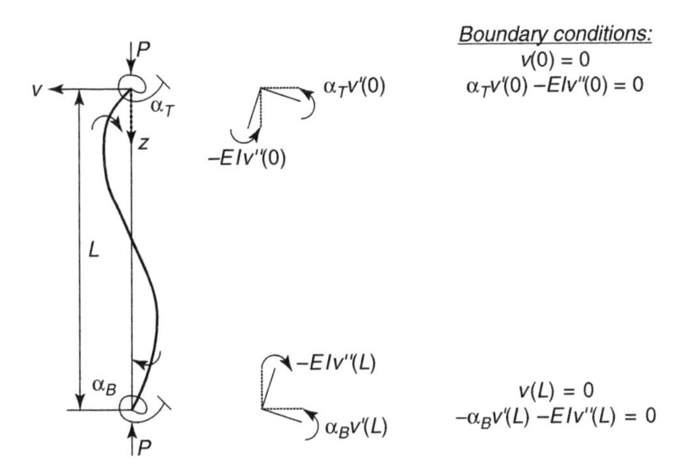

Fig. 2.14 The nonsway restrained column.

Remembering that $k = \sqrt{\frac{P}{EI}}$ and introducing the nondimensional spring constant ratios

$$\begin{aligned} R_T &= \frac{\alpha_T L}{EI} \\ R_B &= \frac{\alpha_B L}{EI} \end{aligned} \tag{2.40}$$

the algebraic decomposition of the determinant results in the following eigenfunction:

$$\begin{aligned} &-2R_T R_B + \sin kL[R_T R_B kL - kL(R_T + R_B) - (kL)^3] \\ &+\cos kL[2R_T R_B + (kL)^2(R_T + R_B)] = 0 \end{aligned} \tag{2.41}$$

Solving equation 2.41 numerically for the smallest kL gives the critical buckling load. The limiting cases of this equation are the cases of both ends pinned ($\alpha_T = \alpha_B = 0 \rightarrow R_T = R_B = 0$), and of both ends fixed ($\alpha_T = \alpha_B = \infty \rightarrow R_T = R_B = \infty$). After some algebraic and trigonometric manipulations it can be demonstrated the eigenfunction for the pinned end column is equal to $\sin kL = 0$ and for the fixed end column it equals $\sin \frac{kL}{2} = 0$. These are indeed the same functions as are shown in Table 2.1 for Cases I and III. Thus, equation 2.41 encloses all the intermediate conditions between the totally pinned ends and the totally fixed ends. The critical load thus varies from $P_{cr} = P_E$ to $4P_E$, and the effective length varies from $K = 1.0$ to 0.5.

The buckling condition of equation 2.41 is directly applicable for the situation where the elastic rotational spring constants α_T and α_B are known. Following, we consider the specialization of the expression for the case of a planar rigid frame. Such an application is within the everyday task of structural design engineers. An example is illustrated in Figure 2.15. We assume that the far ends of the top and bottom beams have the same slope as the near ends. This is not the correct situation for this given problem, but it is the assumption that governs the effective length determination in the AISC Specification (AISC 2005).

The top and bottom spring constants are:

$$\alpha_T = \frac{2EI_{gT}}{L_{gT}} \rightarrow R_T = \frac{2(I_{gT}/L_{gT})}{(I_C/L_C)} \tag{2.42}$$

$$\alpha_B = \frac{2EI_{gB}}{L_{gB}} \rightarrow R_B = \frac{2(I_{gB}/L_{gB})}{(I_C/L_C)} \tag{2.43}$$

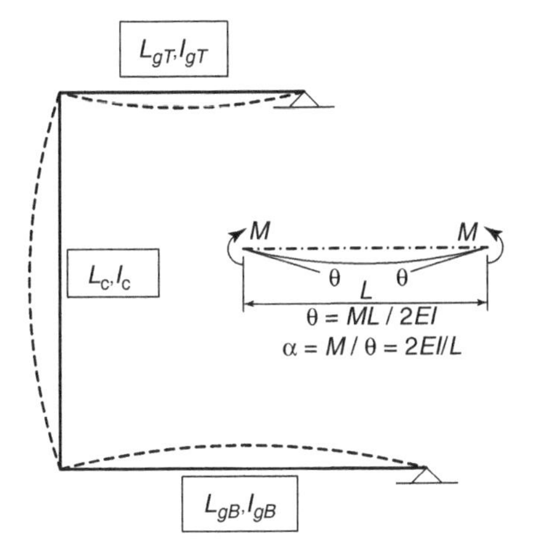

Fig. 2.15 Planar rigid-frame subassembly.

After considerable trigonometric and algebraic manipulation we arrive at the following equation:

$$\frac{(kL)^2 G_T G_B}{4} - 1 + \frac{G_T + G_B}{2}\left(1 - \frac{kL}{\tan kL}\right) + \frac{2\tan\frac{kL}{2}}{kL} = 0 \tag{2.44}$$

In this equation

$$G_T = \frac{I_C/L_C}{I_{gT}/L_{gT}}$$
$$G_B = \frac{I_C/L_C}{I_{gB}/L_{gB}} \tag{2.45}$$

Equation 2.44 can be rewritten in a nomenclature more familiar to structural engineers by introducing the effective length factor K, noting that $K = \frac{\pi}{kL}$:

$$\frac{\left(\frac{\pi}{K}\right)^2 G_T G_B}{4} - 1 + \frac{G_T + G_B}{2}\left(1 - \frac{\frac{\pi}{K}}{\tan\frac{\pi}{K}}\right) + \frac{2\tan\frac{\pi}{2K}}{\frac{\pi}{K}} = 0 \tag{2.46}$$

Equation 2.46 is the basis of the nonsway *nomograph*, also called *alignment chart* in the *Manual of the American Institute of Steel Construction* (AISC 2005). This nomograph is a graphical artifice of connecting G_T and G_B on graduated vertical lines by a straight line to read the corresponding effective length factor K on a third graduated scaled vertical line. Equation 2.46 can also be solved by numerical equation solvers in computer software.

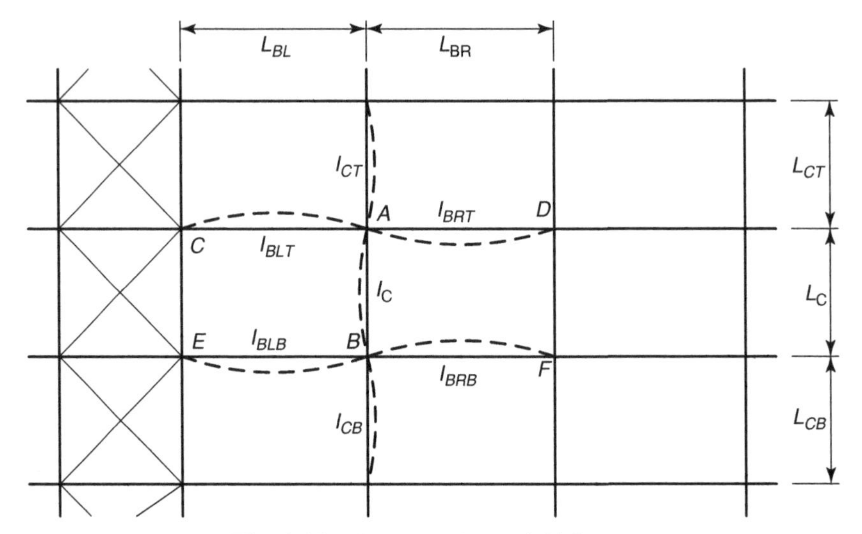

Fig. 2.16 Nonsway planar rigid frame.

The frame of Figure 2.15 can be imagined to be a subassemblage in a rigid frame, as shown in Figure 2.16. In this figure, the column under investigation is the segment *AB*. It is assumed that the rotation at the far ends of the restraining beams are of the same magnitude but of opposite direction as the rotations at the joint at the top and the bottom of column *AB*, respectively. It is further assumed that the restraining moments at these joints are distributed between the two adjoining columns in proportion to the I/L of the respective columns. The effective length of column *AB* is then determined from equation 2.46, with the following new value of the restraint parameter *G*:

$$G = \frac{\sum \frac{I_{\text{column}}}{L_{\text{column}}}}{\sum \frac{I_{\text{beam}}}{L_{\text{beam}}}} \tag{2.47}$$

The next expansion of the stability problem will be the study of the general case of the prismatic column with a rotational and a translational spring at each end, as shown in Figure 2.17. The following derivation follows from p. 80 of the text by Simitses (2005).

The boundary conditions are

$$\begin{aligned}
&\underline{@\, z = 0:} \quad -EIv''' - Pv' = \beta_T v \\
&\qquad\qquad\; -EIv'' = -\alpha_T v' \\
&\underline{@z = \text{L}:} \quad -EIv''' - Pv' = -\beta_B v \\
&\qquad\qquad\; -EIv'' = \alpha_B v'
\end{aligned} \tag{2.48}$$

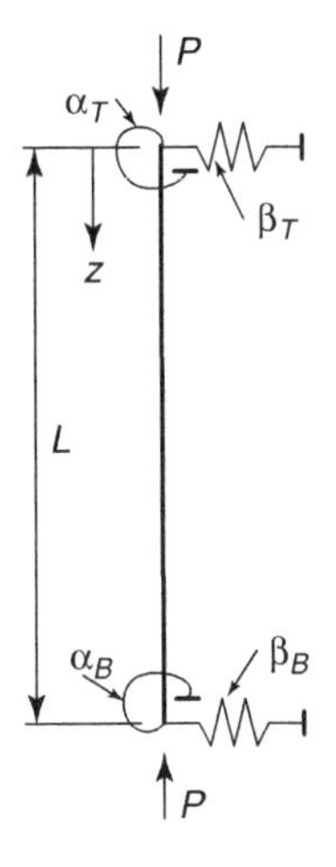

Fig. 2.17 Column with rotational and translational springs.

The following variables are now introduced:

$$R_T = \frac{\alpha_T L}{EI}; \; R_B = \frac{\alpha_B L}{EI}$$
$$T_T = \frac{\beta_T L^3}{EI}; \; T_B = \frac{\beta_B L^3}{EI} \tag{2.49}$$
$$k = \sqrt{\frac{P}{EI}}$$

The determinant of the coefficients of the undetermined unknowns *A, B, C, D* in the deflection equation $v = A + Bz + C \sin kL + D \cos kL$ is equal to the following:

$$\begin{vmatrix} T_T & (kL)^2 & 0 & T_T \\ 0 & R_T & R_T kL & (kL)^2 \\ T_B & [T_B - (kL)^2] & T_B \sin kL & T_B \cos kL \\ 0 & R_B & \begin{matrix}[R_B kL \cos kL \\ - (kL)^2 \sin kL]\end{matrix} & \begin{matrix}[-R_B kL \sin kL \\ - (kL)^2 \cos kL]\end{matrix} \end{vmatrix} = 0 \tag{2.50}$$

The eigenvalue kL can be obtained by using a suitable automatic equation solver.

For the AISC no-sway case (Figures 2.15 and 2.16) the translational spring-constants are assumed to be equal to infinity because both ends of the column are prevented from moving sideways with respect to each other. If the first and the third row of equation 2.50 are divided by T_T and T_B, respectively, and if the spring-constants are made to approach infinity, then the determinant of equation 2.50 becomes equal to equation 2.51:

$$\begin{vmatrix} 1 & 0 & 0 & 1 \\ 0 & R_T & R_T kL & (kL)^2 \\ 1 & 1 & \sin kL & \cos kL \\ 0 & R_B & \begin{matrix}[R_B kL \cos kL \\ -(kL)^2 \sin kL]\end{matrix} & \begin{matrix}[-R_B kL \sin kL \\ -(kL)^2 \cos kL]\end{matrix} \end{vmatrix} = 0 \tag{2.51}$$

Performing the decomposition algebraically results in equation 2.41, which was derived previously.

The AISC Specification recognizes a type of frame subassembly where the top of the column is able to translate with respect to the bottom. It assumed that the bottom cannot translate, i.e., $T_B = \infty$, and there is a zero translational restraint at the top, $T_T = 0$. The restraining beams are assumed to be bent in double curvature so that the slope at each far end equals the slope at the end of the respective column joint, top or bottom. This subassembly is shown in Figure 2.18.

Substituting $T_T = 0$ into the first row of equation 2.50, and dividing each term in the third row by T_B and then equating T_B to infinity results in the following buckling determinant:

$$\begin{vmatrix} 0 & (kL)^2 & 0 & 0 \\ 0 & R_T & R_T kL & (kL)^2 \\ 1 & 1 & \sin kL & \cos kL \\ 0 & R_B & \begin{matrix}[R_B kL \cos kL \\ -(kL)^2 \sin kL]\end{matrix} & \begin{matrix}[-R_B kL \sin kL \\ -(kL)^2 \cos kL]\end{matrix} \end{vmatrix} = 0 \tag{2.52}$$

From Figure 2.18 we find the following relationships:

$$\alpha_T = \frac{6EI_{BT}}{L_{BT}}; \ \alpha_B = \frac{6EI_{BB}}{L_{BB}} \tag{2.53}$$

$$\begin{aligned} R_T &= \frac{\alpha_T L_C}{EI_C} = \frac{6EI_{BT}}{L_{BT}} \times \frac{L_C}{EI_C} = 6\left(\frac{I_{BT}/L_{BT}}{I_C/L_C}\right) = \frac{6}{G_T} \\ R_B &= \frac{\alpha_B L_C}{EI_C} = \frac{6EI_{BB}}{L_{BB}} \times \frac{L_C}{EI_C} = 6\left(\frac{I_{BB}/L_{BB}}{I_C/L_C}\right) = \frac{6}{G_B} \end{aligned} \tag{2.54}$$

$$G_T = \frac{I_C/L_C}{I_{BT}/L_{BT}}; \ G_B = \frac{I_C/L_C}{I_{BB}/L_{BB}} \tag{2.55}$$

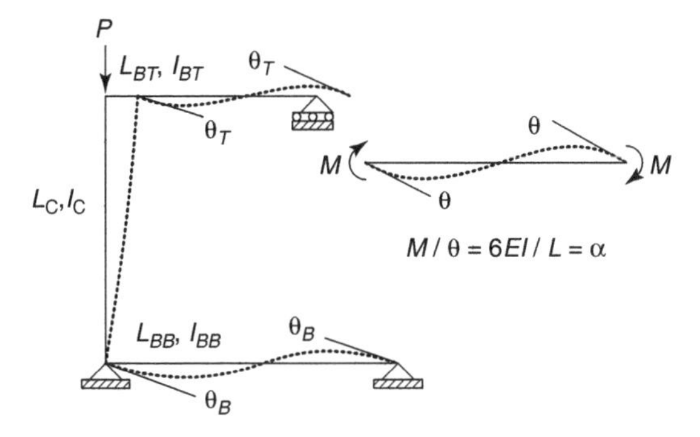

Fig. 2.18 Sway permitted subassembly.

Substitution of R_T and R_B into the decomposed determinant of equation 2.52, setting the resulting equation equal to zero, will lead, after some algebraic manipulation, to the following equation:

$$\frac{kL}{\tan KL} - \frac{(kL)^2 G_T G_B - 36}{6(G_T + G_B)} = 0 \tag{2.56}$$

This equation is the basis of the nonsway alignment chart in the AISC Specification (AISC 2005). This three-bar nomograph relates the flexibility parameters

$$G_T = \frac{\sum I_C / L_C}{\sum I_{BT} / L_{BT}}; \; G_B = \frac{\sum I_C / L_C}{\sum I_{BB} / L_{BB}} \tag{2.57}$$

with the effective length $K = \frac{\pi}{kL}$ of the column under investigation. The definitions of the geometric terms in equation 2.57 are given in Figure 2.16, and the sway-permitted frame is illustrated in Figure 2.19. The area

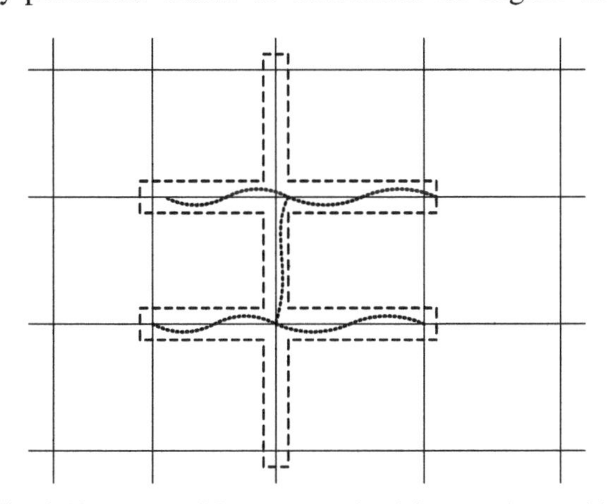

Fig. 2.19 The AISC sway-permitted frame subassembly.

enclosed by the dashed lines defines the members involved in the determination of the effective length factor. It should be noted that the strict assumption of the AISC nomograph method assumes that all columns in a story buckle simultaneously and that, therefore, there is no lateral restraint available. This assumption, as well as others, is discussed further in Chapter 5.

2.9 RESTRAINED COLUMN EXAMPLES

Following are four examples that illustrate various aspects of the stability of elastically restrained columns. These problems are also introduced to show various fallacies one can commit if one does not have a clear picture of the assumptions of buckling theory.

2.9.1 Example 2.1: Geometry and Loading

The column and its restraining members are shown in Figure 2.20a. The left column in the rigid frame is loaded by an axial compressive force P. The other two members of the frame are the restraining elements. The sketch in Figure 2.20b is the restraining subassembly. If this structure is subjected to a bending moment M and a lateral force H at the top joint of the column, then the rotational and translational spring constants are equal to

$$\alpha_T = \frac{M}{\theta} = \frac{40}{17}\frac{EI}{L}$$

$$\beta_T = \frac{H}{\Delta} = \frac{90}{7}\frac{EI}{L^3}$$

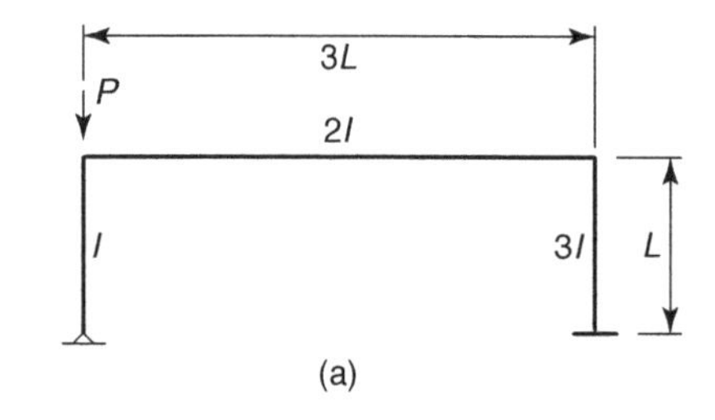

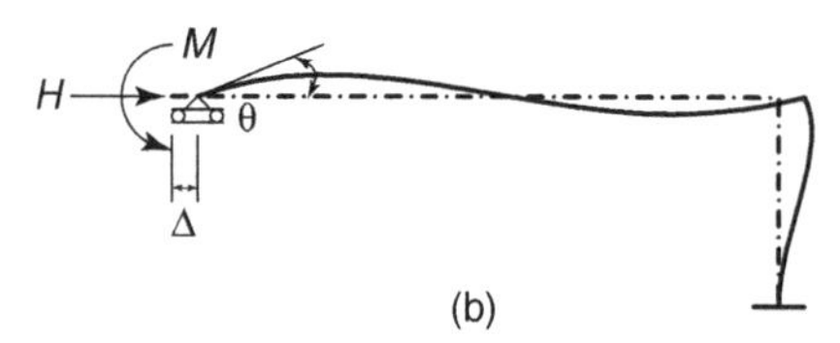

Fig. 2.20 Example 2.1 geometry and loading.

The statically indeterminate structure of Figure 2.20b was analyzed using the classical slope-deflection method.

The nondimensional restraint factors are then determined using equation 2.49:

$$R_T = \frac{\alpha_T L_{\text{Column}}}{EI_{\text{Column}}} = \frac{40}{17}\frac{EI}{L}\frac{L}{EI} = \frac{40}{17}$$

$$T_T = \frac{\beta_T L^3_{\text{column}}}{EI_{\text{column}}} = \frac{90}{7}\frac{EI}{L^3}\frac{L^3}{EI} = \frac{90}{7}$$

The restraint factors at the bottom of the column are those of a pinned end: $R_B = 0$ and $T_B = \infty$. The eigenfunction of a column with rotational and translational restraints at each end, as shown in Figure 2.17, is equation 2.50. If the third line in the matrix is divided by T_B and then T_B is set equal to infinity, the following set of homogeneous simultaneous equations are obtained:

$$\begin{vmatrix} T_T & (kL)^2 & 0 & T_T \\ 0 & R_T & R_T kL & (kL)^2 \\ 1 & 1 & \sin kL & \cos kL \\ 0 & 0 & -(kL)^2 \sin kL & (kL)^2 \cos kL \end{vmatrix} = 0 \qquad (2.58)$$

This determinant is now solved numerically in an automatic equation solver to obtain the eigenvalue $kL = 3.381$. The effective length factor $K = \frac{\pi}{kL} = 0.93$, and the ratio of the critical load to the basic pinned-end buckling load equals $\frac{P}{P_E} = 1.16$.

Following, we examine the consequences of making various assumptions with regard to the boundary conditions at the top end of the column. Table 2.2 lists six stability analyses that were made using equation 2.58. The solution just presented, using the complete beam and the column as restraints, is case No. 4. Case 1 is the buckling solution that would result if the strict assumptions of the AISC sway permitted nomograph method were followed: The lateral restraint is ignored, and rotational restraint is provided only by the beam that has, additionally, the assumption of equal end slopes. As can be seen the difference between the two solutions gives an order of magnitude difference, 750 percent. In this particular example, the lateral restraint overpowers the rotational restraint, and ignoring it results in very conservative answers. Lateral restraint is also neglected in case 2 and 3, with the resulting low values of the critical load. In case 2 the far end of the beam is fixed against rotation, and in case 3 it is pinned. In case 5 there is no

TABLE 2.2 Critical Loads for Example 2.1

	Boundary Conditions	T_T	R_T	k_L	K	P_{cr}/P_E
1	P; 2I,3L; θ; θ; I,L; AISC Model	0	4	1.265	2.48	0.16
2	P; 2I, 3L; I,L	0	8/3	1.159	2.71	0.14
3	P; 2I, 3L; I,L; 3I, L	0	2	1.077	2.92	0.12
4	P; 2I, 3L; I,L; 3I, L	90/7	40/17	3.381	0.93	1.16
5	P; 2I, 3L; I,L; 3I, L	90/7	0	3.142	1.00	1.00
6	P; 2I, 3L; I,L; 3I, L	9	0	3.000	1.05	0.91

assumed rotational restraint, but lateral restraint is provided by both the beam and the unloaded vertical member on the right. In case 6 only the right column provides lateral restraint. This problem shows that it is important to assess the relative importance of the types of available restraint.

2.9.2 Example 2.2: Boundary Conditions for Bracing Structures

This example explores some aspects of the bracing of structures. The subject is important enough to deserve a separate chapter in this book (Chapter 7), but this present problem will introduce the topic. The braced column and the respective boundary conditions at its top and bottom are shown in Figure 2.21.

The column of length L and stiffness EI is pinned at its base and is restrained from lateral deflection by a linear spring at the top. At the top the shear is equal to the deflection times the spring constant β. Upon substitution of the four boundary conditions into the general deflection expression $v = A + Bz + C\sin kz + D\cos kz$, we can derive the following buckling determinant:

$$\begin{vmatrix} 0 & 0 & 0 & -k^2 \\ -\beta & -P & 0 & -\beta \\ 1 & L & \sin kL & \cos kL \\ 0 & 0 & -k^2 \sin kL & -k^2 \cos kL \end{vmatrix} = 0 \tag{2.59}$$

In this equation, $k^2 = \frac{P}{EI}$. Decomposition of the determinant results in the following eigenfunction:

$$k^4 (P - \beta L) \sin kL = 0 \tag{2.60}$$

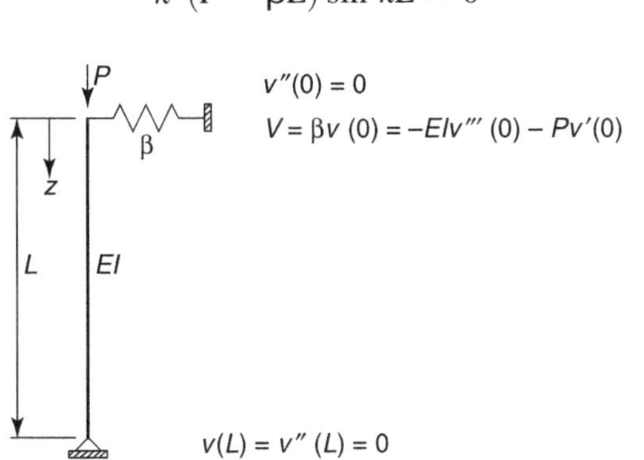

Fig. 2.21 Example 2.2: Geometry, loading, and boundary conditions.

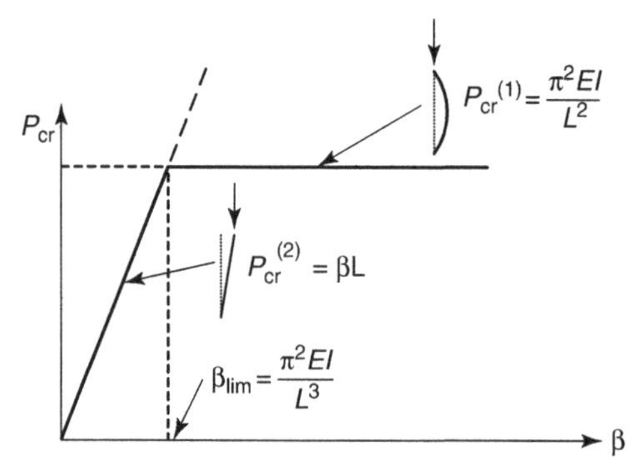

Fig. 2.22 The two buckling modes of the column in Figure 2.21.

Since k^4 is not equal to zero, we get two buckling conditions, and thus two critical loads:

$$\begin{aligned} P - \beta L = 0 \rightarrow P_{cr}^{(1)} &= \beta L \\ \sin kL = 0 \rightarrow P_{cr}^{(2)} &= \frac{\pi^2 EI}{L^2} \end{aligned} \tag{2.61}$$

The lowest of these critical loads is the controlling buckling strength. The relationship between these two buckling loads is given in the graphs in Figure 2.22, where the ordinate represents the buckling load and the abscissa is the magnitude of the spring constant β:

The diagonal straight line corresponds to the rigid-body rotation of the column, when the spring constant is relatively small. However, when the second buckling load is reached, it governs the strength. Thus, it is not necessary to have a spring stiffness larger than the limiting value when the two critical loads are equal to each other: $\beta_{lim} = \frac{\pi^2 EI}{L^3}$.

The following example illustrates the concept of minimum lateral bracing stiffness required to permit the column to reach the basic simply supported Euler load, P_E. Figure 2.23 shows a pinned-end column that is braced by being laterally connected to a vertical cantilever. When this restraining member is subjected to a lateral force H, it deflects an amount $\Delta = \frac{HL^3}{3EI_S} \rightarrow \beta = \frac{H}{\Delta} = \frac{3EI_S}{L^3}$. The required stiffness is obtained from equating $\frac{3EI_S}{L^3}$ to $\beta_{lim} = \frac{\pi^2 EI}{L^3}$. This results in the minimum required moment of inertia $I_S = \frac{\pi^2 I}{3} = 3.29I$.

Another application is the determination of the required bracing stiffness of the two-column braced frame in Figure 2.24a.

The frame consists of two pinned-end columns that are connected by a beam and by diagonal braces. These braces are assumed to be acting in

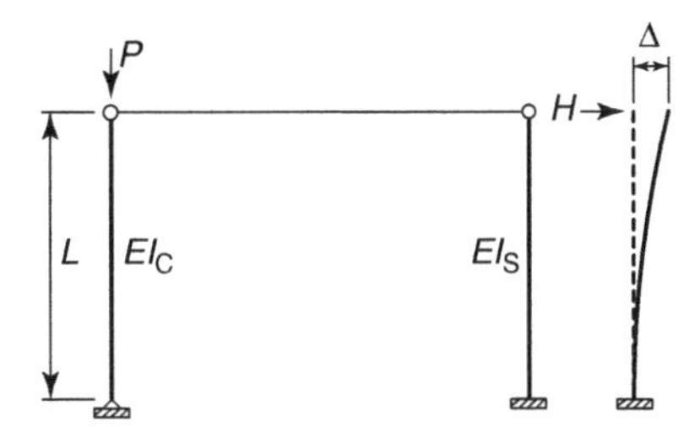

Fig. 2.23 Bracing of a pinned-end column.

tension only. That is, the brace that is in compression, while the top of the frame moves laterally, will buckle and will not participate in providing stiffness. The spring constant is determined by subjecting the frame to a force F and calculating the resulting deflection Δ. The spring constant is then $\beta = \frac{F}{\Delta}$. The tensile force T in the diagonal (see Figure 2.24b) is obtained by elementary statical equilibrium considerations under the assumption of small deflections and rotations:

$$\text{Tensile force in brace} \quad T = \frac{F}{\cos\theta}$$

$$\text{Length of brace: } L_{BR} = \frac{L_B}{\cos\theta}$$

$$\text{Bar elongation: } e = \frac{TL_{BR}}{EA_{BR}} = \frac{FL_B}{EA_{BR}\cos^2\theta}$$

$$\Delta = \frac{e}{\cos\theta} = \frac{FL_B}{EA_{BR}\cos^3\theta} = \frac{F}{EA_{BR}} \frac{(L_B^2 + L_C^2)^{\frac{3}{2}}}{L_B^2}$$

$$\beta = \frac{EA_{BR}L_B^2}{(L_B^2 + L_C^2)^{\frac{3}{2}}}$$

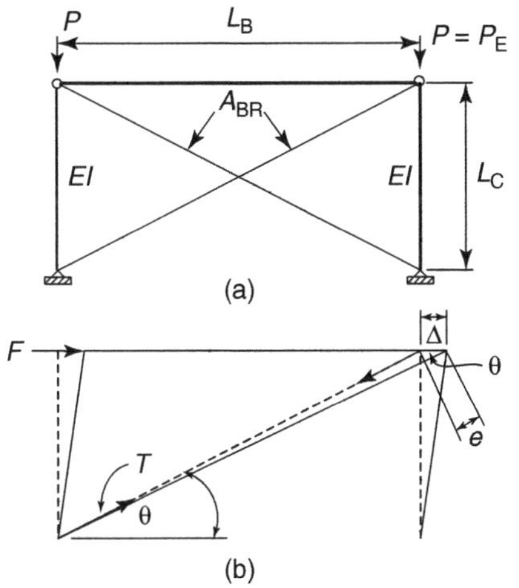

Fig. 2.24 Diagonally braced frame.

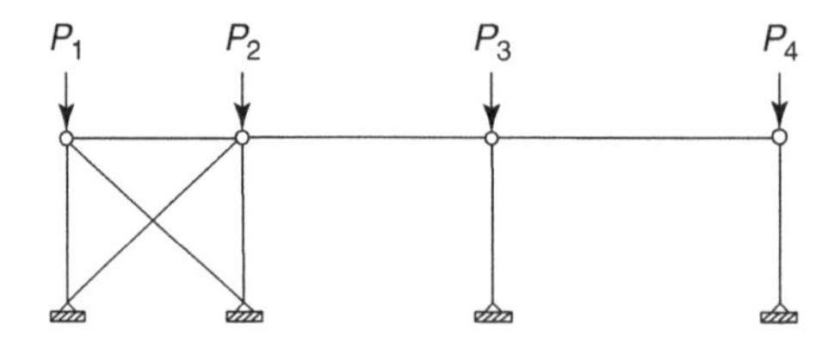

Fig. 2.25 Braced frame with multiple columns.

The required brace area needed to provide minimum stiffness so that the two columns can support the Euler load is obtained by equating the two critical loads:

$$\sum P_{\mathrm{E}} = \frac{2\pi^2 EI}{L_{\mathrm{C}}^2} = \beta L_{\mathrm{C}} = \frac{EA_{\mathrm{BR}} L_{\mathrm{B}}^2 L_{\mathrm{C}}}{(L_{\mathrm{B}}^2 + L_{\mathrm{C}}^2)^{\frac{3}{2}}}$$
$$A_{\mathrm{BR}} \geq 2\pi^2 I \left[\frac{(L_{\mathrm{B}}^2 + L_{\mathrm{C}}^2)^{\frac{3}{2}}}{L_{\mathrm{B}}^2 L_{\mathrm{C}}^3} \right] \tag{2.62}$$

For a frame with several columns in a story, as shown in Figure 2.25, the required brace area is

$$A_{\mathrm{BR}} \geq \sum P_{\mathrm{E}} \left[\frac{(L_{\mathrm{B}}^2 + L_{\mathrm{C}}^2)^{\frac{3}{2}}}{EL_{\mathrm{B}}^2 L_{\mathrm{C}}} \right] \tag{2.63}$$

2.9.3 Example 2.3: Left Column Acting as Restraining Lateral Support for Right Column

The next example examines another case of a structure where two types of buckling modes are possible. The whole frame is shown in Figure 2.26a. The left column is pinned at the top and fixed at the bottom. The right column is pinned at both ends. Thus, the left column acts as a restraining lateral support for the right column. The two columns are connected at their tops by a horizontal beam. The applied load is located at a distance of aL from the left corner of the frame, where a is a ratio varying from zero to one. The axial load in the left column is $P(1 - a)$, and in the right column it is Pa. The moment of inertia in the left column is I, and in the critical right column it is bI, where b is the ratio of the two moments of inertia.

The sketch Figure 2.26b shows the free-body of the right, pinned-end column. It is a member with a hinge at its bottom and a restraining spring at the top. The spring constant β is provided by the left column. The drawing Figure 2.26c shows the supporting column. The spring constant is determined by the ratio $\beta = \frac{H}{\Delta}$. What is different from Example 2.1 here is that

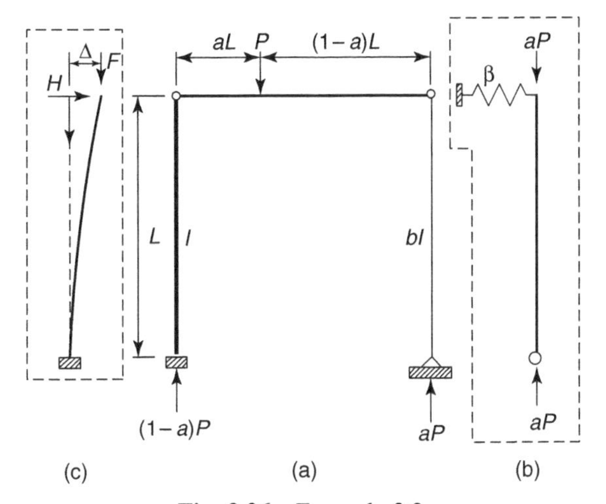

Fig. 2.26 Example 2.3.

the restraining member has an axial load on it. Because of this, the stiffness is reduced, and a nonlinear analysis must be performed. The deflection is $v = A + Bz + C \sin kz + D \cos kz$, and the boundary conditions are

$$\begin{aligned} v''(0) &= 0 \\ -H &= -EIv'''(0) - Fv'(0) \\ v(L) &= 0 \\ v'(L) &= 0 \end{aligned}$$

The second boundary condition is the shear equilibrium condition at the top of the member. Note that these are not homogeneous boundary conditions. From the substitution of the boundary conditions into deflection equation we obtain (after some algebraic manipulation) the following expression for the deflection:

$$v = \frac{HL}{F}\left(-1 + \frac{\sin kL}{kL \cos kL} + \frac{z}{L} - \frac{\sin kz}{kL \cos kL}\right)$$

The deflection at $z = 0$, and the spring constant β, are, therefore, equal to:

$$\Delta = v(0) = \frac{HL}{F}\left(\frac{\sin kL - kL \cos kL}{kL \cos kL}\right)$$

$$\beta = \frac{H}{\Delta} = \frac{F}{L}\left(\frac{kL \cos kL}{\sin kL - kL \cos kL}\right)$$

From Figure 2.26a it is seen that $F = (1 - a)P$, and thus

$$(kL)^2 = \frac{PL^2}{EI}(1 - a) \tag{2.64}$$

From equation 2.61, in Example 2.2 we know that there are two critical loads for the column that is restrained by a spring at the top: the rigid-body rotation mode, and the Euler buckling mode. The axial load in the right column is aP, as shown in Figure 2.26b. The first buckling load is obtained from the first of equation 2.61:

$$(aP_{\text{cr}})^{(1)} = \beta L = (1 - a)P\left(\frac{kL\cos kL}{\sin kL - kL\cos kL}\right) \tag{2.65}$$

The left and right sides of equation 2.65 are rearranged and kL from equation 2.64 is substituted to result in the following two equations:

$$\frac{aPL^2}{EI} = \frac{aPL^2}{EI}\frac{1-a}{1-a} = \frac{a}{1-a}(kL)^2$$

$$\frac{(1-a)PL^2}{EI}\left(\frac{kL\cos kL}{\sin kL - kL\cos kL}\right) = (kL)^2\left(\frac{kL\cos kL}{\sin kL - kL\cos kL}\right)$$

If these are again substituted into equation 2.65, the following buckling equation results:

$$\frac{a}{1-a} - \frac{kL\cos kL}{\sin kL - kL\cos kL} = 0 \tag{2.66}$$

This equation can then be solved for kL by an automatic equation solver routine. The critical load is then obtained from equation 2.64:

$$P_{\text{cr}}^{(1)} = \frac{(kL)^2 EI(1 - a)}{L^2} \tag{2.67}$$

The second critical load is when the critical right column fails as a pinned-end member according to the second of equation 2.61:

$$aP_{\text{cr}}^{(2)} = \frac{\pi^2 EbI}{L^2}$$

The symbol b is the ratio of the moments of inertia in the right column to that in the left member. Rearrangement of this equation results in the following equation for the second critical load:

$$P_{cr}^{(2)} = \frac{\pi^2 EbI}{aL^2} \tag{2.68}$$

The limiting values of $P_{cr}^{(1)}$ are when $a = 0$ (i.e., when the force P sits on top of the left column) and when $a = 1$ (i. e., when the column sits on top of the right column). These loads are equal to, respectively:

$$P_{cr,a=0}^{(1)} = \frac{\pi^2 EI}{4L^2}$$

$$P_{cr,a=1}^{(1)} = \frac{3EI}{L^2}$$

The curves in Figure 2.27 are computed for $b = 0.2$. Rigid body buckling occurs for approximately $a < 0.7$, while column buckling governs for $0.7a < 1.0$.

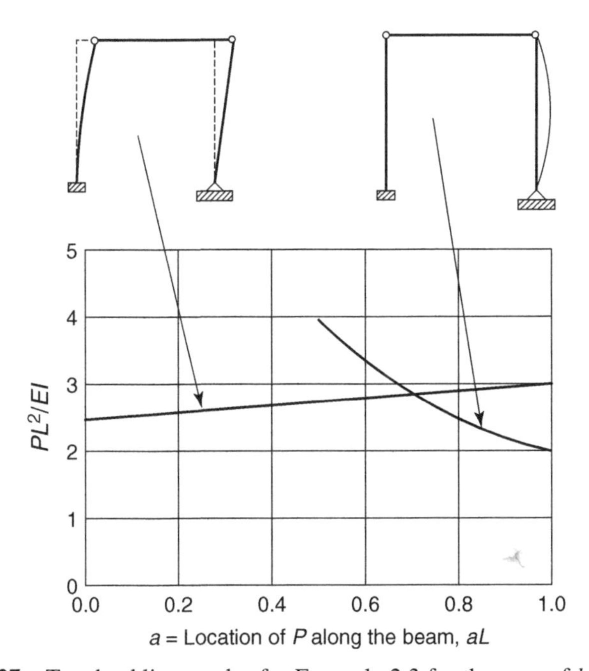

Fig. 2.27 Two buckling modes for Example 2.3 for the case of $b = 0.2$.

2.9.4 Example 2.4: Buckling Load for a Stepped Column

In this example, we determine the buckling load of a stepped column. The column is pinned at the top and fixed at the bottom, and it is composed of two cross-sections (Figure 2.28): The top third of the length has a moment of inertia of I and the bottom two thirds has a moment of inertia of $2I$.

Buckled deflection of the top part (equation 2.12):

$$y_1 = A_1 + B_1 x + C_1 \sin k_1 x + D_1 \cos k_1 x$$

Buckled deflection of the bottom part:

$$y_2 = A_2 + B_2 z + C_2 \sin k_2 z + D_2 \cos k_2 z$$

In these equations $k_1 = \sqrt{\frac{P}{EI}} = k, \quad k_2 = \sqrt{\frac{2P}{2EI}} = k$, (equation 2.11).

There are eight unknown coefficients, and thus eight boundary conditions are required: Each end furnishes two conditions, and compatibility of deflection and slope, plus the equilibrium of moment and shear at the location where the cross-section changes stiffness gives the remaining four:

1. At the top of the column: $y_1(0) = 0$ and $y_1''(0) = 0$.
2. At the bottom of the column $y_2(2L/3) = 0$, and $y_2''(2L/3) = 0$.
3. Deflection and slope compatibility at $x = L/3$ and $z = 0$: $y_1(L/3) = y_2(0)$, and $y_1'(L/3) = y_2'(0)$.
4. Moment equilibrium at $x = L/3$ and $z = 0$: $EIy_1''(L/3) = 2EIy_2''(0)$.

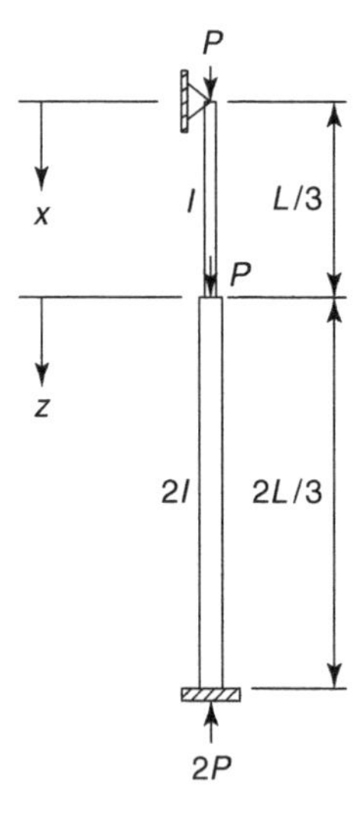

Fig. 2.28 Stepped column stability.

Shear equilibrium at $x = L/3$ and $z = 0$: $Py_1'(L/3) + EIy_1'''(L/3) = 2Py_2'(0) + 2EIy_2'''(0)$. This relationship can be derived from equation 2.5.

From the two boundary conditions at the top of the column it can be demonstrated that $D_1 = -A_1$, and thus $y_1 = B_1 x + C_1 \sin kx$. Performing the operations in the remaining six conditions leads to the following homogeneous simultaneous equations:

$$A_2 + B_2 \frac{2L}{3} + C_2 \sin\left(\frac{2kL}{3}\right) + D_2 \cos\left(\frac{2kL}{3}\right) = 0$$

$$B_2 L + C_2 kL \cos\left(\frac{2kL}{3}\right) - D_2 kL \sin\left(\frac{2kL}{3}\right) = 0$$

$$B_1 \frac{L}{3} + C_1 \sin\left(\frac{kL}{3}\right) - A_2 - D_2 = 0$$

$$B_1 L + C_1 kL \cos\left(\frac{kL}{3}\right) - B_2 L - C_2 kL = 0$$

$$-C_1 \sin\left(\frac{kL}{3}\right) + D_2 = 0$$

$$B_1 L - 2B_2 L = 0$$

Setting the determinant of the coefficients of the unknown deflection amplitudes equal to zero, shown on the following MATHCAD solver equation, gives the following eigenvalue for the unknown parameter $kL = 4.712$. Since $k^2L^2 = \frac{PL^2}{EI} = 22.20$, the critical buckling load becomes equal to $P_{cr} = 22.20 \frac{EI}{L^2} = \frac{22.20}{\pi^2} \frac{\pi^2 EI}{L^2} = 2.25 P_E$. P_E is the Euler load.

$$\text{Ans} = \text{root}\left(\left|\begin{bmatrix} 0 & 0 & 1 & \frac{2}{3} & \sin\left(\frac{2 \cdot kL}{3}\right) & \cos\left(\frac{2 \cdot kL}{3}\right) \\ 0 & 0 & 0 & 1 & kL \cdot \cos\left(\frac{2 \cdot kL}{3}\right) & -\left(kL \cdot \sin\left(\frac{2 \cdot kL}{3}\right)\right) \\ \frac{1}{3} & \sin\left(\frac{kL}{3}\right) & -1 & 0 & 0 & -1 \\ 1 & kL \cdot \cos\left(\frac{kL}{3}\right) & 0 & -1 & -kL & 0 \\ 0 & -\sin\left(\frac{kL}{3}\right) & 0 & 0 & 0 & 1 \\ 1 & 0 & 0 & -2 & 0 & 0 \end{bmatrix}\right|, kL\right)$$

2.10 CONTINUOUSLY RESTRAINED COLUMNS

The subject of this section is the stability of a column on an elastic foundation. The solution follows from the book *Beams on Elastic Foundation* (Hetenyi 1946). The differential equation for this case is given in equation 2.69:

$$EIv^{iv} + Pv'' + \alpha v = 0 \tag{2.69}$$

In this equation α is the foundation modulus, having units of force/length^2. The column is shown in Figure 2.29.

By introducing the following two definitions, the differential equation can be rearranged.

$$k^2 = \frac{P}{EI}$$
$$\lambda^4 = \frac{\alpha}{4EI} \tag{2.70}$$

$$v^{iv} + k^2 v'' + 4\lambda^4 v = 0 \tag{2.71}$$

The four roots of this equation are $r_{1,2,3,4} = \pm\sqrt{\frac{-k^2}{2} \pm \sqrt{\frac{k^4}{4} - 4\lambda^4}}$, and therefore the deflection is equal to

$$v = C_1 e^{r_1 z} + C_2 e^{r_2 z} + C_3 e^{r_3 z} + C_4 e^{r_4 z} \tag{2.72}$$

There are three possible cases that must be considered:

1. $\frac{k^4}{4} - 4\lambda^4 < 0$ or $P < 2\sqrt{\alpha EI}$
2. $\frac{k^4}{4} - 4\lambda^4 = 0$ or $P = 2\sqrt{\alpha EI}$
3. $\frac{k^4}{4} - 4\lambda^4 > 0$ or $P > 2\sqrt{\alpha EI}$

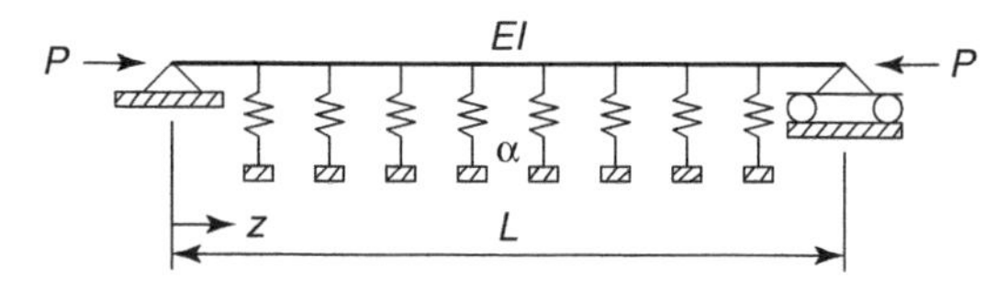

Fig. 2.29 Column on an elastic foundation.

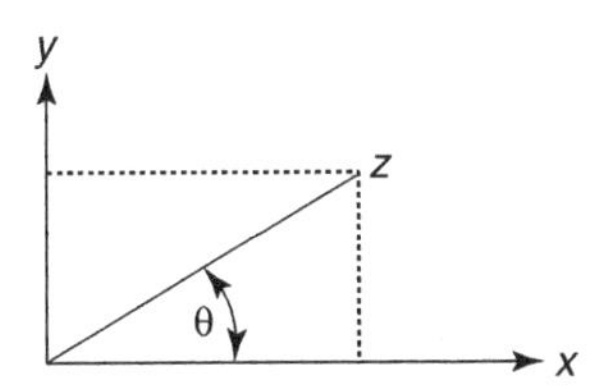

Fig. 2.30 Complex space.

We consider Case 1 first. The four roots are $r_{1,2,3,4} = \pm\sqrt{\frac{-k^2}{2} \pm i\sqrt{\frac{k^4}{4} - 4\lambda^4}}$, where $i = \sqrt{-1}$. The following derivation was taken from a text on functions of complex variables (Churchill 1948). A complex space is shown in Figure 2.30, where for the first root $x = \frac{-k^2}{2}; y = \sqrt{4\lambda^4 - \frac{k^4}{4}}$

$$r_1 = \sqrt{x + iy}$$

$$z = x + iy$$

$$|z| = r = \sqrt{x^2 + y^2}$$

$$\sqrt{z} = \sqrt{r}\left(\cos\frac{\theta}{2} + i\sin\frac{\theta}{2}\right)$$

$$\cos\frac{\theta}{2} = \sqrt{\frac{1 + \cos\theta}{2}}; \quad \sin\frac{\theta}{2} = \sqrt{\frac{1 - \cos\theta}{2}}; \quad \cos\theta = \frac{x}{r}$$

$$\sqrt{r}\cos\frac{\theta}{2} = \sqrt{r}\sqrt{\frac{r + x}{2r}} = \sqrt{\frac{r + x}{2}}; \quad \sqrt{r}\sin\frac{\theta}{2} = \sqrt{r}\sqrt{\frac{r - x}{2r}} = \sqrt{\frac{r - x}{2}}$$

$$r_1 = \sqrt{z} = \sqrt{\frac{r + x}{2}} + i\sqrt{\frac{r - x}{2}} = \sqrt{\frac{\sqrt{x^2 + y^2} + x}{2}} + i\sqrt{\frac{\sqrt{x^2 + y^2} - x}{2}}$$

Substitution of x and y from this yields, after some algebra,

$$r_1 = \sqrt{\lambda^2 - \frac{k^2}{4}} + i\sqrt{\lambda^2 + \frac{k^2}{4}} \tag{2.73}$$

Finally,

$$r_{1,2,3,4} = \pm(b \pm ia)$$
$$a = \sqrt{\lambda^2 + \frac{k^2}{4}}; \quad b = \sqrt{\lambda^2 - \frac{k^2}{4}} \tag{2.74}$$

The final equation for the deflection is attained by the steps shown next.

$$v = C_1 e^{bz+aiz} + C_2 e^{-bz+aiz} + C_3 e^{bz-aiz} + C_4 e^{-bz-aiz}$$

$$v = e^{aiz}(C_1 e^{bz} + C_2 e^{-bz}) + e^{-aiz}(C_3 e^{bz} + C_4 e^{-bz})$$

$$e^{aiz} = i \sin az + \cos az; \quad e^{-aiz} = -i \sin az + \cos az$$

$$v = \cos az[e^{bz}(C_1 + C_3) + e^{-bz}(C_2 + C_4)] + \sin az[e^{bz}(iC_1 - iC_3) + e^{-bz}(iC_2 - iC_4)]$$

$$v = (A_1 e^{bz} + A_2 e^{-bz})\cos az + (A_3 e^{bz} + A_4 e^{-bz})\sin az \qquad (2.75)$$

The boundary conditions of the simply supported column are equal to

$$v(0) = v(L) = v''(0) = v''(L) = 0$$

The second derivative of equation 2.75 is the following expression:

$$v'' = [(-a^2 + b^2)(A_1 e^{bz} + A_2 e^{-bz}) + 2ab(A_3 e^{bz} - A_4 e^{-bz})]\cos az + [2ab(-A_1 e^{bz} + A_2 e^{-bz}) + (-a^2 + b^2)(A_3 e^{bz} + A_4 e^{-bz})]\sin az$$

Substitution of the boundary conditions at $z = 0$ leads to

$$v(0) = A_1 + A_2 = 0 \rightarrow A_2 = A_1$$

$$v''(0) = (-a^2 + b^2)(A_1 - A_1) + 2ab(A_3 - A_4) = 0 \rightarrow A_3 = -A_4$$

The remaining two boundary conditions at $z = L$ lead to the following two homogeneous simultaneous equations:

$$[2 \sin h(bL)\cos(aL)]A_1 + [2 \cos h(bL) \sin(aL)]A_3 = 0$$
$$[-2(a^2 - b^2) \sin h(bL) \cos(aL) - 4ab \cos h(bL)\sin(aL)] A_1 + [-2(a^2 - b^2) \cos h(bL) \sin(aL) + 4ab \sin h(bL)\cos(aL)] A_3 = 0$$

The following substitutions were made in the derivation of the two equations above:

$$e^{bL} - e^{-bL} = 2 \sin h(bL)$$
$$e^{bL} + e^{-bL} = 2 \cos h(bL)$$

where sinh, cosh, and tanh are hyperbolic sine, cosine, and tangent functions.

If we set the determinant of the coefficients of A_1 and A_3 equal to zero, we obtain the following buckling condition after performing some algebra:

$$\tan h^2(bL) + \tan^2(aL) = 0 \qquad (2.76)$$

Since $\tan h(bL) = \frac{\sin h(bL)}{\cos h(bL)} = 0$ if, and only if, $b = 0$ then, $b = 0 = \sqrt{\lambda^2 - \frac{k^2}{4}}$. With equations 2.70 and 2.74 it can be demonstrated that this can only occur if $P > 2\sqrt{\alpha EI}$. However, our derivation started with the assumption that $P < 2\sqrt{\alpha EI}$. Thus, when this is the case there is no possibility of buckling: the foundation is too stiff and buckling is prevented. To investigate the situation when buckling does occur, it is necessary to consider Case 3, that is, $P > 2\sqrt{\alpha EI}$. The limiting case when $P = 2\sqrt{\alpha EI}$ corresponds to the buckling of an infinitely long column on an elastic foundation (Hetenyi 1946, p. 114).

The roots of the differential equation of buckling, equation 2.71, are

$$r_{1,2,3,4} = \pm\sqrt{\frac{-k^2}{2} \pm \sqrt{\frac{k^4}{4} - 4\lambda^4}}$$

Assume that

$$r_o = \pm(i\bar{b} \pm ia)$$

where

$$\bar{b} = \sqrt{\frac{k^2}{4} - \lambda^2}; \quad a = \sqrt{\frac{k^2}{4} + \lambda^2} \tag{2.77}$$

Squaring the expression for r_o leads to $r_o^2 = -\frac{k^2}{2} \pm \sqrt{\frac{k^4}{4} - 4\lambda^4}$. Thus, $r_o = r$ and, therefore, we can use the results of the previous derivation by replacing b with $\bar{b}$. The buckling condition of equation 2.76 becomes, then, equal to

$$\tan h^2(\bar{b}L) + \tan^2(aL) = 0$$

or

$$\sin h^2(i\bar{b}L)\cos^2(aL) + \cos h^2(i\bar{b}L)\sin^2(aL) = 0 \tag{2.78}$$

The following algebraic and trigonometric manipulations lead to the final buckling equation (equation 2.78):

$$\sin h(i\bar{b}L) = -\frac{\sin(\bar{b}L)}{i}; \quad \cos h(i\bar{b}L) = \cos(\bar{b}L)$$

$$\frac{\sin^2(\bar{b}L)\cos^2(aL)}{i^2} + \cos(\bar{b}L)\sin^2(aL) = 0$$

$$-\sin^2(\bar{b}L)\cos^2(aL) + \cos^2(\bar{b}L)\sin^2(aL) = 0$$

$$\tan^2(aL) = \tan^2(\bar{b}L)$$

$$\tan(aL) = \tan(\bar{b}L)$$

$$\sin(aL)\cos(\bar{b}L) = \cos(aL)\sin(\bar{b}L)$$

$$\sin(aL - \bar{b}L) = 0 \tag{2.78}$$

This condition is possible only when

$$aL - \overline{b}L = n\pi \tag{2.79}$$

Substitution of the expressions for a and $\overline{b}$ from equation 2.77 gives, after some algebra, the following equation:

$$k^2L^2 = \frac{4\lambda^4 L^4}{n^2\pi^2} + n^2\pi^2$$

Note that $n = 1, 2, 3\ldots$ and therefore it is not possible to determine beforehand which value of n will furnish the lowest buckling load. If we now substitute the formulas for k and λ from equation 2.70, we arrive at the following expression for the critical load:

$$\frac{PL^2}{EI} = \frac{\alpha L^4}{n^2\pi^2 EI} + n^2\pi^2 \tag{2.80}$$

Dividing both sides by the Euler load $P_E = \frac{\pi^2 EI}{L^2}$ and introducing the nondimensional foundation modulus ratio

$$\overline{\alpha} = \left(\frac{\alpha L^4}{\pi^4 EI}\right)^{\frac{1}{4}} \tag{2.81}$$

we arrive at the critical elastic buckling load for a pinned-end column supported by an elastic foundation (note that the units of the foundation modulus α are force/unit area, psi or pascal):

$$P_{cr} = P_E\left(\frac{\overline{\alpha}^4}{n^2} + n^2\right) \tag{2.82}$$

This critical load depends on n, the number of waves that the buckling deflection develops. The first three modes are presented in Table 2.3.

The relationship between the critical load and the foundation modulus is also shown in Figure 2.30 for the first three modes $n = 1, 2, 3$. The curves represent the variation of P/P_E as the abscissa and the nondimensional foundation modulus $\overline{\alpha}$ (equation 2.81) as the ordinate. For $\overline{\alpha} \leq \sqrt{2}$ the first mode, $n = 1$, controls, and so forth. The envelope of the lowest critical loads is shown as a heavy line. As long as $\overline{\alpha}$ is less than about 0.6, the effect of the foundation is negligible. However, as the foundation stiffness increases, the buckling load rises rapidly. For the first mode ($n = 1$) the buckling load is equal to $P_{cr} = P_E(1 + \overline{\alpha}^4)$ (equation 2.82 with $n = 1$). For

TABLE 2.3 Buckling Loads and Modes

n	P_{cr}/P_E	Mode	Limits of Foundation Modulus	Limits of critical Load
1	$1+\overline{\alpha}^4$		$0 \leq \overline{\alpha}^4 \leq 4$	$1 \leq P/P_E \leq 5$
2	$4+\frac{\overline{\alpha}^4}{4}$		$4 \leq \overline{\alpha}^4 \leq 36$	$5 \leq P/P_E \leq 13$
3	$9+\frac{\overline{\alpha}^4}{9}$		$36 \leq \overline{\alpha}^4 \leq 144$	$13 \leq P/P_E \leq 45$
etc.	etc.	etc.	etc.	etc.

higher modes $(n > 1)$ it is slightly conservative to assume that the member has infinite length. The buckling load for this case is

$$P_{cr} = 2\sqrt{\alpha EI} = 2P_E\overline{\alpha}^2 \tag{2.83}$$

This curve is shown as a thin dash-dot-dot line in Figure 2.30. This simplification can be utilized in developing the required minimum brace stiffness in design standards (Winter 1960).

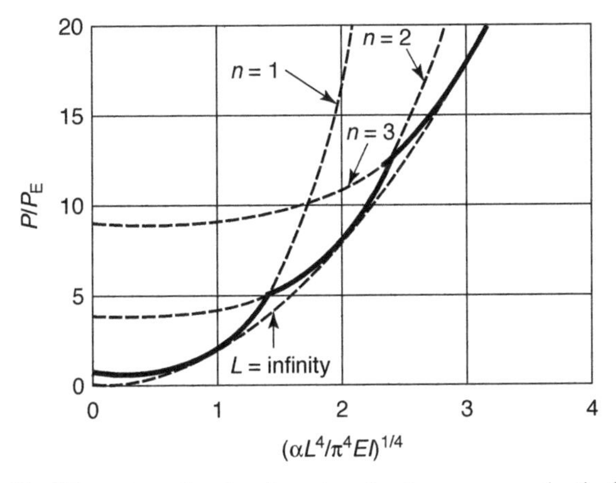

Fig. 2.31 Buckling strength of a pinned-end column on an elastic foundation.

2.11 SUMMARY

The in-plane buckling of continuous axially loaded members can be described by deriving the differential equation of planar flexure. The simplest application of this is the simply-supported column with no transverse loading or imperfections. The critical load determined for this problem is the familiar form of the Euler buckling load. For simple columns, this critical load is affected by end restraint (boundary conditions) and imperfections as shown in the five fundamental cases of section 2.5 and the three imperfection cases discussed in section 2.6. End restraint may also be provided by spring supports or adjoining members; these effects will be covered in more detail in Chapter 5, which discusses frame behavior. One method of handling the effects of end restraint in design practice is by using effective length, or K-factors. The development of the alignment charts used in the AISC specification is based on the theoretical derivation of critical loads given end restraint as a function of the relative stiffness of the columns being restrained and the members of the subassembly that restrain them. A more detailed discussion of effective length factors is provided in Chapter 8.

PROBLEMS

2.1. Calculate the multiplier α for $I_B = I/2$, I, $2I$. Examine the results and write down a one-sentence observation.

$$P_{cr} = \frac{\pi^2 EI\alpha}{L^2}$$

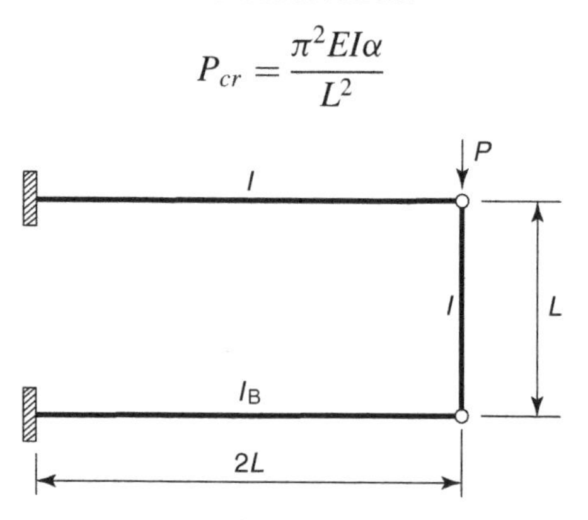

Fig. p2.1

2.2. The elastic buckling load of a pinned-end column is $P_E = \frac{\pi^2 EI}{L^2}$ at temperature $t = t_o$. Assuming the length L cannot vary, what is the elastic buckling load at $t = t_1$? Calculate also the temperature for which no axial load can be applied. Given: $E = 29000\,\text{ksi}$; $t_o = 60^\circ$ F; $t_1 = 120^\circ$ F; $L/r = 140$; $A = 10\,\text{in}^2$; $F_y = 60\,\text{ksi}$ and the coefficient of thermal expansion is $\alpha = 6.5 \times 10^{-6}$.

2.3. Determine the elastic buckling load of this frame, Q_{cr}. The frame is an equilateral triangle with rigid joints.

The critical load is $Q_{cr} = \sqrt{3}C^2 \frac{\pi^2 EI}{L^2}$. Before making any derivations and calculations, estimate C to be in one of the ranges of values 0–0.5, 0.5–1.0, 1.0–1.5, 1.5–2.0, 2.0–3.0, 3.0–5.0, 5.0–10, $C > 10$. What was the reason for selecting the answer? Determine the exact answer after becoming acquainted with the slope-deflection method in Chapter 4.

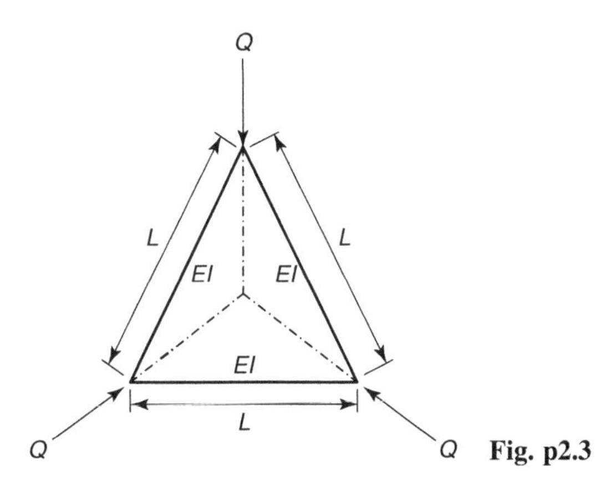

Fig. p2.3

2.4. During buckling the direction of P must pass through the point C. Before making any derivations, sketch a curve relating $\frac{P_{cr}L^2}{\pi^2 EI}$ as the ordinate and a as the abscissa as a varies from -1 to 5. Calculate numerical values for $a = -1$, 0, $+1$, and $+5$.

2.5. Determine the elastic buckling load of this column for $a = 0$, $a = b = L/2$, and $a = 10b$. Express the answer in the form $P_{cr} = \frac{\pi^2 EI}{(KL^2)}$. Note that $L = a + b$.

2.6. A pinned-end beam-column is a roof member and it supports an axial load and ponded water. The water level at the support is h, and inside the span it is $h + y$. The distributed load due to the water is $w = S\gamma(h + y)$, where S = the spacing of parallel beam-columns and $\gamma = 62.4\,\text{lb./ft}^3$, the specific weight of water. Show that when the axial load is zero, the ponded water will result in runaway deflection (instability) if $\frac{S\gamma L^4}{\pi^4 EI} \rightarrow 1.0$. Show also that instability in the presence

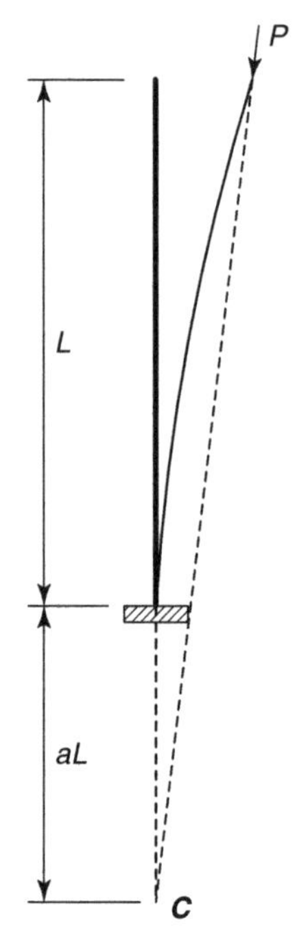

Fig. p2.4

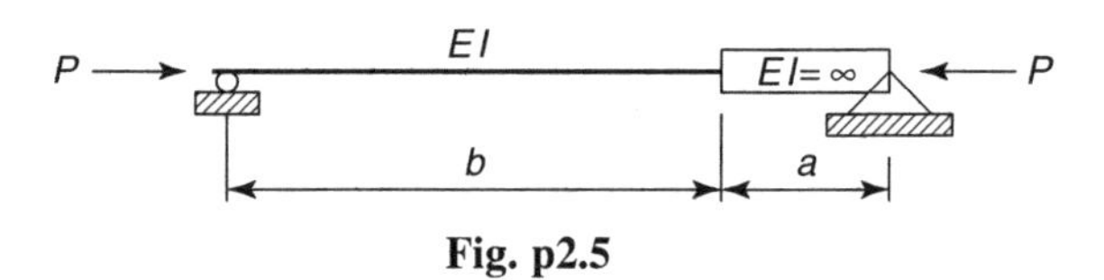

Fig. p2.5

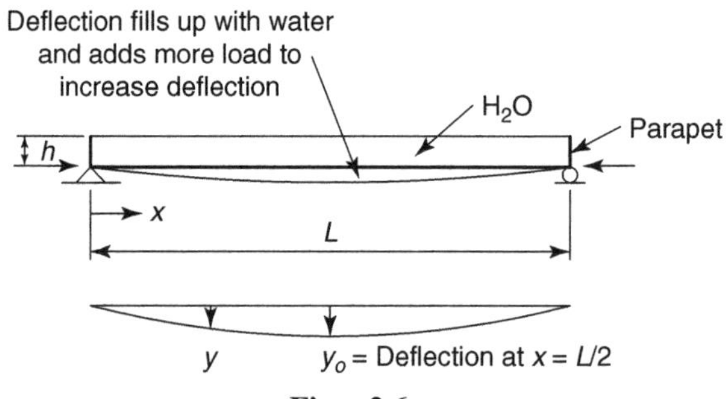

Fig. p2.6

of ponding and axial compression results when the interaction equation is satisfied: $\frac{S\gamma L^4}{\pi^4 EI} + \frac{PL^2}{\pi^2 EI} = 1$.
Hints:

The differential equation equals $EIy^{iv} + Py'' - S\gamma y = S\gamma h$.
The deflection is equal to $y = A\sinh\alpha x + B\cosh\alpha x + C\sin\beta x + D\cos\beta y - h$, where

$$\alpha = \sqrt{\frac{1}{2}\left(-k^2 + \sqrt{k^4 + 4\overline{w}^4}\right)}$$

$$\beta = \sqrt{\frac{1}{2}\left(k^2 + \sqrt{k^4 + 4\overline{w}^4}\right)}$$

$$k^2 = \frac{P}{EI}; \; \overline{w}^4 = \frac{S\gamma}{EI}$$

2.7. **a.** Derive all the formulas in Figure 2.6 for Cases II, III, IV, and V.
b. Derive equation 2.50.
c. Solve all six problems in Table 2.1.

2.8. Determine the elastic buckling load of this frame. The columns are continuous over the two-story height, and they have stiffness *EI*. The top story is laterally braced by the diagonal members so that point *C* does not translate laterally with respect to point *B*.

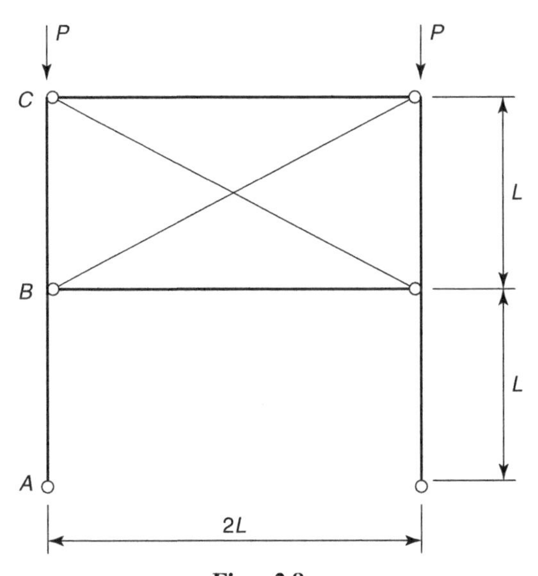

Fig. p2.8

2.9. Investigate the critical load on this pile. Find P_{cr} if $\lambda L = 10$. Plot kL versus λL for the range $0 \leq \lambda L \leq 100$. Discuss the results.

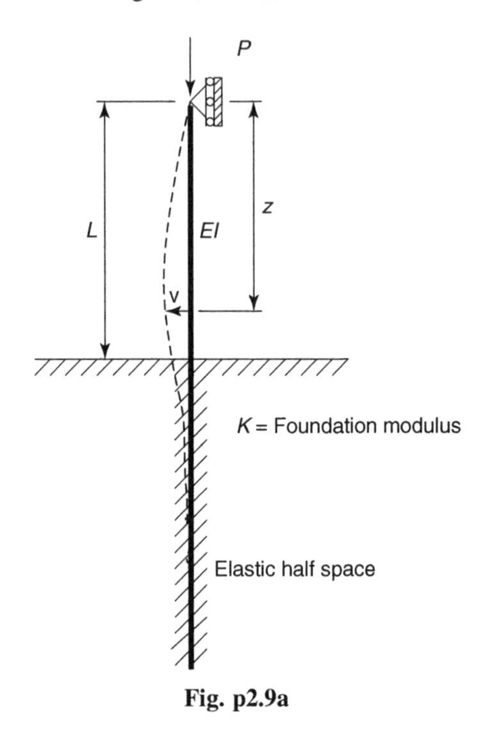

Fig. p2.9a

From "Beams on Elastic Foundations" (Hetenyi 1961, p. 130) for an elastic half-space:

$$\alpha = \sqrt{\lambda^2 - \frac{Q}{EI}}; \quad \beta = \sqrt{\lambda^2 + \frac{Q}{EI}}; \quad \lambda = \left(\frac{K}{4EI}\right)^{\frac{1}{4}}$$

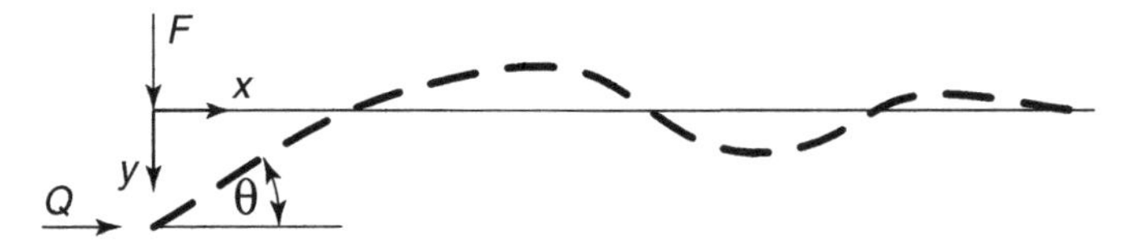

Fig. p2.9b

$$y(0) = \frac{F}{\beta K}\left[\frac{2\lambda^2(2\alpha\beta)}{3\alpha^2 - \beta^2}\right]; \quad \theta(0) = -\frac{F}{EI}\left(\frac{1}{3\alpha^2 - \beta^2}\right);$$
$$M(0) = 0; V(0) = -F$$

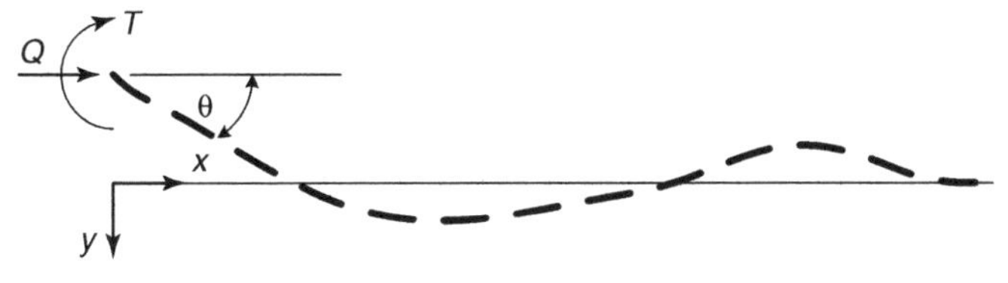

Fig. p2.9c

$$y(0) = -\frac{T}{EI}\left(\frac{1}{3\alpha^2 - \beta^2}\right); \quad \theta(0) = \frac{T}{EI}\left(\frac{2\alpha}{3\alpha^2 - \beta^2}\right); \quad M(0) = T;$$

$$V(0) = T\left(\frac{-4\alpha(\alpha^2 - \beta^2)}{3\alpha^2 - \beta^2}\right)$$

APPENDIX 2.1

Equation 2.2: $\frac{d^2\theta}{ds^2} + K^2 \sin\theta = 0$

Multiply by $d\theta$ and integrate: $\int \frac{d^2\theta}{ds^2} d\theta + K^2 \int \sin\theta d\theta = 0$

$$\int \frac{d^2\theta}{ds^2}\frac{d\theta}{ds} ds + K^2 \int \sin\theta d\theta = 0$$

$$\frac{1}{2}\int_0^s \frac{d}{ds}\left(\frac{d\theta}{ds}\right)^2 ds + K^2 \int_{\theta_o}^{\theta} \sin\theta d\theta = 0 = \left[\frac{1}{2}\left(\frac{d\theta}{ds}\right)^2\right]_0^s - K^2[\cos\theta]_{\theta_o}^{\theta}$$

at $s = 0$, the curvature $\phi(0) = 0 = \frac{d\theta}{ds}(0)$

$$\therefore \left(\frac{d\theta}{ds}\right)^2 = 2K^2(\cos\theta - \cos\theta_o) = 4K^2\left(\sin^2\frac{\theta_o}{2} - \sin^2\frac{\theta}{2}\right)$$

$$\frac{d\theta}{ds} = \pm 2K\sqrt{\sin^2\frac{\theta_o}{2} - \sin^2\frac{\theta}{2}}$$

Use the negative sign because θ decreases from $z = 0$ to $z = L/2$

$$\therefore -2Kds = \frac{d\theta}{\sqrt{\sin^2\frac{\theta_o}{2} - \sin^2\frac{\theta}{2}}}$$

At $z = L/2$, $\theta = 0$; Integration from $s = L/2$ to $s = 0$ leads to

$$\int_0^{\theta_o} \frac{d\theta}{\sqrt{\sin^2\frac{\theta_o}{2} - \sin^2\frac{\theta}{2}}} = \int_{L/2}^{0} -2K ds = 2K\int_0^{L/2} ds = KL$$

Let $\sin\frac{\theta}{2} = p\sin\alpha$, where $p = \sin\frac{\theta_o}{2}$
when θ varies from 1 to 0, $\sin\alpha$ varies from $\frac{\pi}{2}$ to 0 and α varies from $\frac{\pi}{2}$ to 0.

$$\theta = 2\sin^{-1}(p\sin\alpha); \quad d\theta = \frac{2p\cos\alpha\, d\alpha}{\sqrt{1 - p^2\sin^2\alpha}}$$

then $KL = \displaystyle\int_0^{\theta_o} \frac{d\theta}{\sqrt{\sin^2\frac{\theta_o}{2} - \sin^2\frac{\theta}{2}}} = \int_0^{\frac{\pi}{2}} \frac{2p\cos\alpha\, d\alpha}{\sqrt{1 - p^2\sin^2\alpha}\sqrt{p^2 - p^2\sin^2\alpha}}$

Finally $\dfrac{KL}{2} = \displaystyle\int_0^{\frac{\pi}{2}} \frac{d\alpha}{\sqrt{1 - p^2\sin^2\alpha}};$

complete elliptic integral of the first kind

CHAPTER THREE

INELASTIC COLUMN BUCKLING

Previously, the discussions of buckling loads have focused on columns exhibiting elastic behavior. In this chapter, we discuss the behavior of columns that buckle inelastically, methods of modeling that behavior, and the development of column strength curves that establish the critical loads of both elastic and inelastic columns.

3.1 TANGENT AND REDUCED MODULUS CONCEPTS

In Chapter 2 we showed that a simply supported, symmetric, perfectly straight, elastic column will buckle when the axial load becomes equal to the critical load defined by

$$P_{cr} = \frac{\pi^2 EI}{L^2} \tag{3.1}$$

This equation can also be expressed in terms of a critical stress as

$$\sigma_{cr} = \frac{P}{A} = \frac{\pi^2 E}{(L/r)^2} \tag{3.2}$$

As discussed in Chapter 2, the elastic critical load, P_{cr}, is the load at which the column begins to deflect laterally. In the elastic range, no change in load

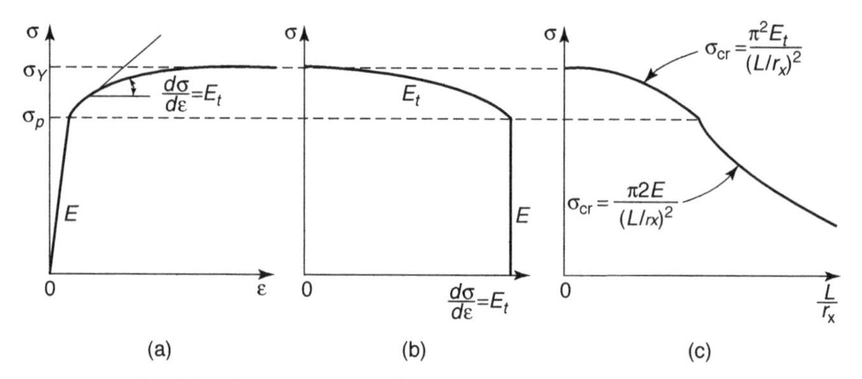

Fig. 3.1 Construction of the tangent modulus column curve.

is required to obtain this deflection, and the column can be perfectly straight and slightly deflected under the same load P_{cr}. We call this phenomenon the bifurcation of the equilibrium.

This same reasoning was expanded into the inelastic range in 1889 by Engesser, who postulated that a homogeneous column made of a material having a stress-strain curve as shown in Figure 3.1a will experience bifurcation of the equilibrium in the range above the proportional limit when the average stress P/A is equal to

$$\sigma_{cr} = \frac{\pi^2 E_t}{(L/r)^2} \tag{3.3}$$

In this equation, E_t is the slope $d\sigma/d\varepsilon$ of the stress-strain curve at the stress σ_{cr}. The axial load corresponding to this stress is called the *tangent modulus load* P_T and is equal to:

$$P_T = \frac{\pi^2 E_t I}{L^2} \tag{3.4}$$

In equation 3.4 the stress can no longer be computed directly, since E_t is also a function of the stress. Therefore, the computational process is reversed, that is,

$$\left(\frac{L}{r}\right)_{cr} = \pi\sqrt{\frac{E_t}{\sigma}} \tag{3.5}$$

Since it may not be convenient to express the σ–ε relationship analytically, a *column curve* (σ_{cr} versus L/r curve) can be constructed graphically, as shown in Figure 3.1. From an experimentally determined σ–ε curve (Figure 3.1a) we can construct a σ–$(d\sigma/d\varepsilon)$ curve graphically (Figure 3.1b), and the column curve is then constructed by choosing a value of σ and computing the critical slenderness ratio from equation 3.5 (Figure 3.1c). For any given type of material we can thus construct column curves that can be used directly in design.

Even though Engesser's tangent-modulus concept is simple to apply, and the critical loads computed by it usually correspond closely to experimental results, the theoretical reasoning behind it is incorrect. Engesser stated that bifurcation occurs without a change in load (Figure 3.2a), and therefore the sum of the stresses introduced by the bending moment Pv_o (Figure 3.2b), which exist after buckling, must be equal to zero across any section. This bending moment causes further compression on one side of the cross-section, and it reduces the stress on the other side (Figure 3.2c). This is no problem if buckling takes place in the elastic range, since both loading and unloading of the strains is governed by the elastic modulus E. By extending the elastic concept into the inelastic range, Engesser in fact assumed that both loading and unloading are governed by E_t in this range also (Figure 3.3a). In reality, however, materials unload according to the elastic modulus E, and therefore *the real column should be stronger than its strength predicted by the tangent modulus concept.*

The error in Engesser's reasoning was pointed out in 1895 by Jasinsky, and in 1898 Engesser corrected his theory to include the effect of elastic unloading. At the same time, Considere proposed the reduced modulus theory independently. The theory is referred to as the *reduced modulus concept*; it is also referred to as the *double modulus concept* (Timoshenko 1953).

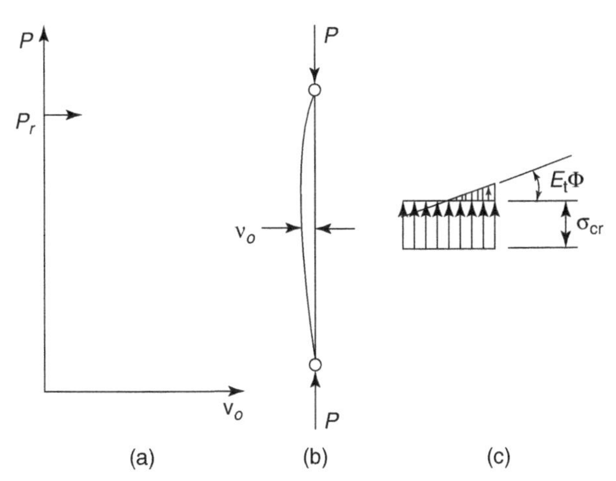

Fig. 3.2 Engesser's concept of inelastic column buckling.

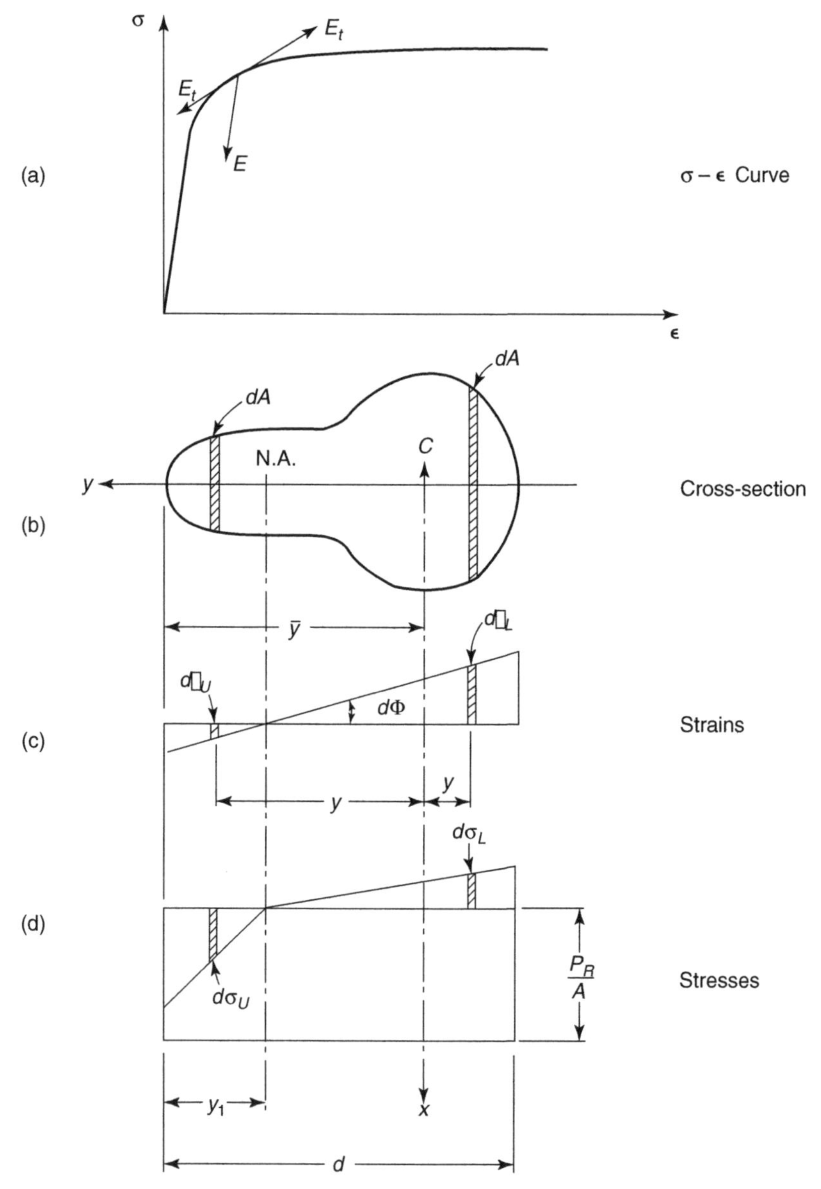

Fig. 3.3 The reduced modulus concept.

The reduced modulus concept (as it applies to in-plane buckling) is based on these assumptions:

- The stress-strain curve of the material is known.
- The displacements are small.

- Plane sections before bending are plane after bending.
- No change of load is associated with bifurcation.

Figure 3.3b shows a cross-section that is symmetric about the y-axis and that is part of a column that is assumed to buckle at a stress $\sigma_{cr} = P_R/A$ where P_R is the reduced modulus load.

In the loading (L) portion

$$d\sigma_L = E_t d\varepsilon_L \tag{3.6}$$

In the unloading (U) portion

$$d\sigma_U = E d\varepsilon_U \tag{3.7}$$

that is, the stress–strain relationship is governed by the elastic modulus E.

From the geometry of the strain distribution we find (Figure 3.3c) that

$$d\varepsilon_L = [y + (\bar{y} - y_1)]d\phi \quad \text{and} \quad d\varepsilon_U = [y - (\bar{y} - y_1)]d\phi \tag{3.8}$$

where $\bar{y}$ and y_1 are defined in Figure 3.3.The curvature $d\phi$ is related to the deflection of the column v by the formula $d\phi = -v''$, and therefore the stresses are

$$d\sigma_L = -v'' E_t(\bar{y} - y_1 + y) \quad \text{and} \quad d\sigma_U = -v'' E(-\bar{y} + y_1 + y) \tag{3.9}$$

According to our assumption, the change in P due to buckling is zero, and so

$$dP = 0 = \int_{\bar{y}-y_1}^{\bar{y}} d\sigma_U dA - \int_{-(d-\bar{y})}^{\bar{y}-y_1} d\sigma_L dA \tag{3.10}$$

If we substitute equation 3.9 into equation 3.10, we find that

$$ES_1 = E_t S_2 \tag{3.11}$$

where

$$S_1 = \int_{\bar{y}-y_1}^{\bar{y}} (y - \bar{y} + y_1) dA \tag{3.12}$$

$$S_2 = \int_{-(d-\bar{y})}^{\bar{y}-y_1} (\bar{y} - y_1 + y) dA \tag{3.13}$$

are the statical moments of the areas to the left and to the right of the neutral axis (N. A. in Figure 3.3) about this axis, respectively. The expression in equation 3.11 permits the determination of the location of the neutral axis of the bending stresses.

Equilibrium of the moments due to the bending stresses (Figure 3.3d) about the neutral axis is expressed as follows:

$$M = Pv = \int_{\bar{y}-y_1}^{\bar{y}} d\sigma_U (y - \bar{y} + y_1) dA + \int_{-(d-\bar{y})}^{\bar{y}-y_1} d\sigma_L (\bar{y} - y_1 + y_1) dA \quad (3.14)$$

From equation 3.14 we can obtain, after substitution of $d\sigma_L$ and $d\sigma_U$ from equation 3.9, the following formula:

$$Pv = -v''(EI_1 + E_t I_2) \quad (3.15)$$

where

$$I_1 = \int_{\bar{y}-y_1}^{\bar{y}} (y - \bar{y} + y_1)^2 dA \quad (3.16)$$

$$I_2 = \int_{-(d-\bar{y})}^{\bar{y}-y_1} (\bar{y} - y_1 + y)^2 dA \quad (3.17)$$

are the moments of inertia about the neutral axis of the area to the left and to the right of this axis, respectively. Rearrangement and differentiation of equation 3.15 gives the following differential equation:

$$v^{iv} + \frac{Pv''}{\bar{E}I_x} = 0 \quad (3.18)$$

where

$$\bar{E} = E\left(\frac{I_1}{I_x}\right) + E_t\left(\frac{I_2}{I_x}\right) \quad (3.19)$$

This latter term is the *reduced modulus* and it is a *function of both the material properties E* and E_t and the *cross-section geometry.* The reduced modulus buckling load follows from the differential equation:

$$P_R = \frac{\pi^2 \bar{E} I_x}{L^2} \quad (3.20)$$

The load P_R is the *reduced modulus load.* Since $\bar{E} > E_t$, P_R will always be larger than P_T.

3.2 SHANLEY'S CONTRIBUTION

For almost 50 years engineers were faced with a dilemma: They were convinced that the reduced modulus concept was correct, but the test results had an uncomfortable tendency to lie near the tangent modulus load. For this reason, the tangent modulus concept was used in column design, and the discrepancy was ascribed to initial out-of-straightness and end eccentricities of the load that could not be avoided when testing columns.

In order to determine the cause of the discrepancy, Shanley conducted very careful tests on small aluminum columns (Shanley 1947). He found that lateral deflection started very near the theoretical tangent modulus load, but that additional load could be carried until the column began to unload. At no time could Shanley's columns support as much load as that predicted by the reduced modulus theory.

In order to explain the behavior, Shanley developed the model shown in Figure 3.4. The model is composed of a column made up of two inextensible bars connected in the center of the column by a deformable cell. Upon buckling, all deformations take place in the cell, which consists of two flanges of area $A/2$; these two flanges are connected by a web of zero area.

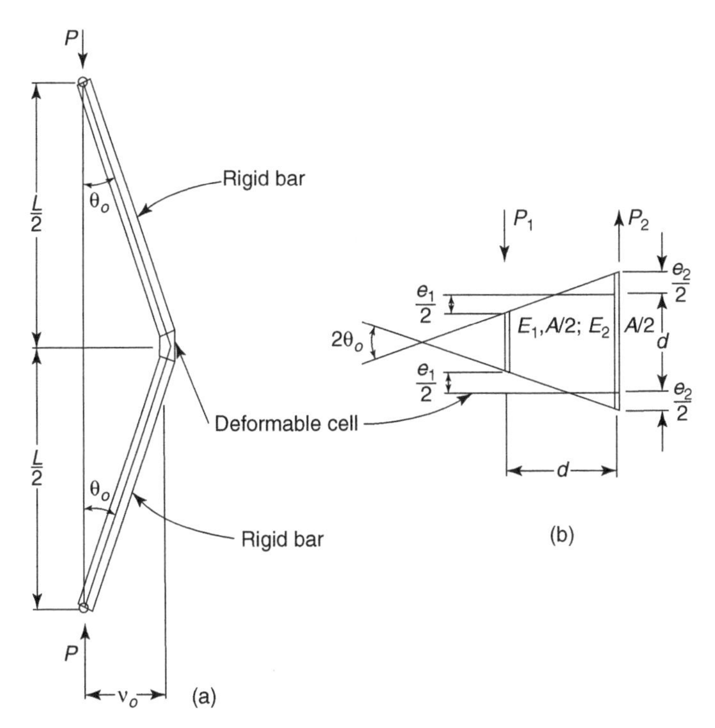

Fig. 3.4 Shanley's column model.

One flange consists of a material with an elastic modulus E_1, and the other of a material with a modulus E_2. The length of the column L is much larger than the dimensions d of the cell.

In the buckled configuration (Figure 3.4a) the following relationship exists between the end slope θ_o, the lateral deflection v_o (where $v_o \ll d$), and the strains e_1 and e_2 of the flanges (Figure 3.4b):

$$v_o = \frac{\theta_o L}{2} \quad \text{and} \quad \theta_o = \frac{1}{2d}(e_1 + e_2) \tag{3.21}$$

By combining these two equations we eliminate θ_o, and thus

$$v_o = \frac{L}{4d}(e_1 + e_2) \tag{3.22}$$

The external moment (e) at the mid-height of the column is

$$M_e = P v_o = \frac{PL}{4d}(e_1 + e_2) \tag{3.23}$$

The forces in the two flanges due to buckling are

$$P_1 = \frac{E_1 e_1 A}{2d} \quad \text{and} \quad P_2 = \frac{E_2 e_2 A}{2d} \tag{3.24}$$

The internal moment (i) is then

$$M_i = \frac{d}{2}(P_1 + P_2) = \frac{A}{4}(E_1 e_1 + E_2 e_2) \tag{3.25}$$

With $M_e = M_i$ we get an expression for the axial load P:

$$P = \frac{Ad}{L}\left(\frac{E_1 e_1 + E_2 e_2}{e_1 + e_2}\right) \tag{3.26}$$

In case the cell is elastic $E_1 = E_2 = E$, and so

$$P_{\mathrm{E}} = \frac{AEd}{L} \tag{3.27}$$

For the tangent modulus load $E_1 = E_2 = E_T$, thus

$$P_T = \frac{AE_t d}{L} \tag{3.28}$$

When we consider the elastic unloading of the *tension flange*, the $E_1 = E_t$ and $E_2 = E$. Thus

$$P = \frac{Ad}{L}\left(\frac{E_t e_1 + E e_2}{e_1 + e_2}\right) \tag{3.29}$$

Upon substitution of e_1 from equation 3.22 and P_T from equation 3.28 and using the abbreviation

$$\tau = \frac{E_t}{E} \tag{3.30}$$

we find that

$$P = P_T\left[1 + \frac{Le_2}{4dv_o}\left(\frac{1}{\tau} - 1\right)\right] \tag{3.31}$$

There are forces P_1 and P_2 acting on the two flanges if the member is deflected, and the difference of these two loads is the amount by which P is increased above the tangent modulus load. That is,

$$P = P_T + (P_1 - P_2) \tag{3.32}$$

By employing equations 3.22, 3.24, and 3.30 it can be shown that

$$P_1 - P_2 = \frac{AE_t}{2d}\left[\frac{4v_o d}{L} - \left(1 + \frac{1}{\tau}\right)e_2\right] \tag{3.33}$$

and

$$P = P_T\left[1 + \frac{2v_o}{d} - \frac{Le_2}{2d^2}\left(1 + \frac{1}{\tau}\right)\right] \tag{3.34}$$

By eliminating e_2 from equations 3.31 and 3.34, we obtain the load P to be:

$$P = P_T\left[1 + \frac{1}{d/2v_o + \frac{(1+\tau)}{(1-\tau)}}\right] \tag{3.35}$$

If, for example, $\tau = 0.5$ and it remains constant after P_T is exceeded, then

$$\frac{P}{P_T} = 1 + \frac{1}{d/2v_o + 3} \tag{3.36}$$

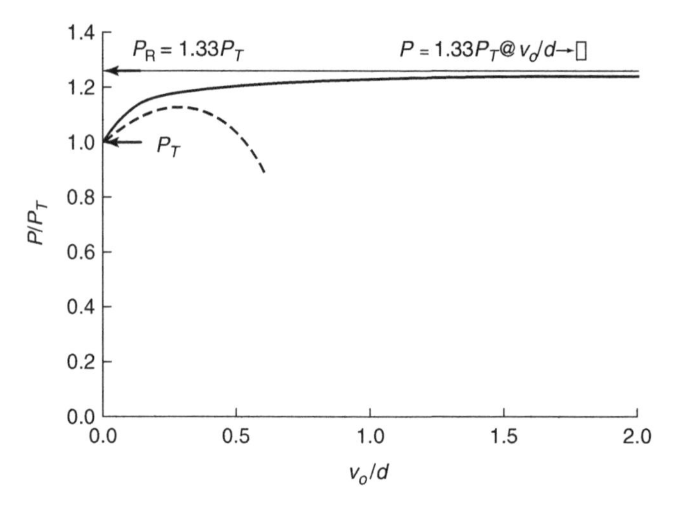

Fig. 3.5 Post-buckling behavior in the inelastic range.

and the curve relating P/P_T and v_o/d is the solid line in Figure 3.5. This curve approaches $P = 1.333P_T$ asymptotically as $v_o/d \to \infty$. In reality, τ will vary with the strain, and as P is increased, it will become progressively smaller and the curve shown by the dashed line in Figure 3.5 results. This relationship will have a maximum between P_T and $1.333P_T$.

The reduced modulus concept defines the load at which deflection occurs without a change in load. Thus in equation 3.33 $P_1 - P_2 = 0$, and therefore

$$e_2 = \frac{4v_o d}{L}\left(\frac{1}{1 + 1/\tau}\right) \tag{3.37}$$

Substituting equation 3.37 into equation 3.34 gives the reduced modulus load P_R

$$P_R = P_T\left(1 + \frac{1-\tau}{1+\tau}\right) \tag{3.38}$$

that becomes equal to $1.333P_T$ for $\tau = 0.5$. P_R is the maximum load that can theoretically be reached if the deflection v_o approaches infinity.

Equation 3.35 defines the relationship between the axial load P and the resulting deflection of the mid-height of the column v_o for any constant value τ (Figure 3.5). When $v_o = 0$, this equation gives $P = P_T$, the tangent modulus load. For any value of v_o the axial load is larger than P_T, or conversely, if $P > P_T$, a deflection v_o is necessary for maintaining equilibrium.

The maximum value of $P = P_R$ is reached when the deflection v_o becomes infinitely large. This load represents the maximum possible theoretical load. Because bending after P_T is accompanied by further straining of the compression flange of the cell, τ is reduced from its initial value and the $P - v_o$ curve therefore reaches a peak below the reduced modulus load (dashed curve, Figure 3.5).

Although the Shanley model (Figure 3.4) has no resemblance to an actual column, the conclusions from its analysis apply also to such real members. These conclusions are

1. The tangent modulus concept gives the maximum load up to which an initially straight column will remain straight.
2. The actual maximum load exceeds the tangent modulus load, but it cannot be as large as the reduced modulus load.
3. Any load above P_T will cause the column to be laterally deflected.
4. In the load range $P_T < P < P_{max}$ there is always strain reversal present.

Shanley resolved the paradox that had existed for half a century, and he defined the concepts on which a rational inelastic column theory could be based. Following Shanley, many investigators refined and expanded these concepts (e.g., Johnston 1961 and 1964). This research has further proved Shanley's model, and it has brought out the following additional facts about column behavior:

1. Theoretically, it is possible to have an infinite number of "critical" loads (i.e., loads at which a column may deflect) between the tangent modulus load P_T and the elastic buckling load (see Figure 3.6). Upon deflection, the gradient of the load-deflection curve is positive when $P_T < P < P_R$, zero when $P = P_R$, and negative (i.e., deflection is accompanied by unloading) when $P_R < P < P_E$.
2. In reality, there are no initially straight columns. If the behavior of a straight column is regarded as the limiting case of an initially crooked column as the crookedness vanishes (Figure 3.6), then the only significant critical load among the many theoretical possibilities is the tangent modulus load.

Shanley's theory also led to the development of a rational column theory for ideally straight metal columns, which is discussed in sections 3.4 and 3.5 (Beedle and Tall 1960). The effect of initial out-of-straightness will be discussed in section 3.6.

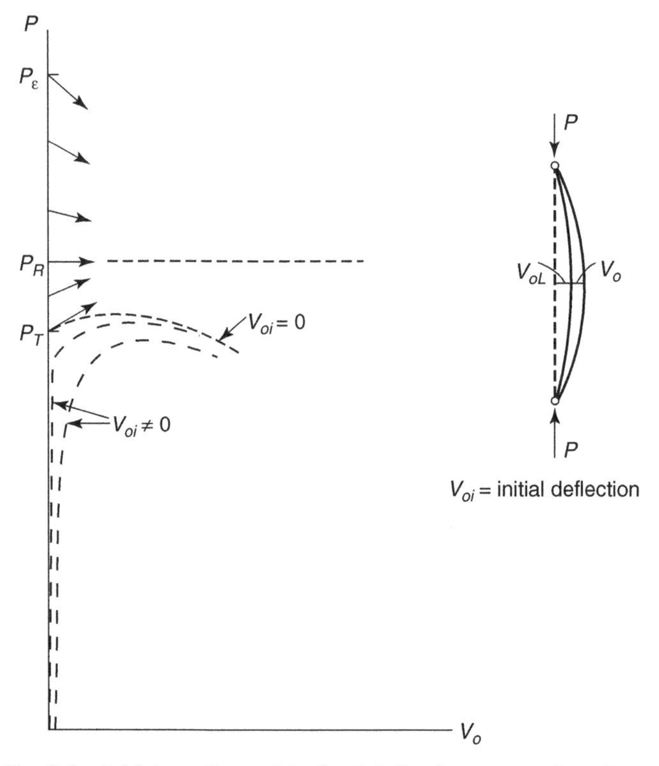

Fig. 3.6 Initial gradient of the load-deflection curve of a column.

3.3 EXAMPLE ILLUSTRATING THE TANGENT MODULUS AND THE REDUCED MODULUS CONCEPTS

To illustrate the tangent modulus and the reduced modulus concepts, we determine these two loads for a pinned-end column having a rectangular cross-section (Figure 3.7a) when buckling is about the x-axis. Let the stress–strain relationship be given by the formula

$$\varepsilon = \frac{\sigma}{E} + \left(\frac{3\sigma_y}{7E}\right)\left(\frac{\sigma}{\sigma_y}\right)^{10} \tag{3.39}$$

This is a special case of a generalized formula suggested for material with a nonlinear stress–strain curve (Ramberg and Osgood 1943).

Nondimensionalizing this equation by the yield strain $\varepsilon_y = \sigma_y/E$, we get

$$\frac{\varepsilon}{\varepsilon_y} = \frac{\sigma}{\sigma_y} + \frac{3}{7}\left(\frac{\sigma}{\sigma_y}\right)^{10} \tag{3.40}$$

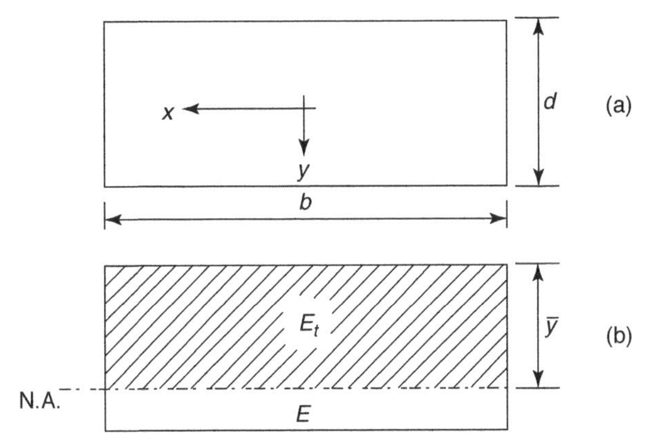

Fig. 3.7 The yielded rectangular cross-section.

The curve representing this relationship is shown in the right corner of Figure 3.8. It is nearly linear almost up to 90 percent of the yield stress. Beyond this point, the curve bends sharply, but it continues to rise. Such a stress–strain curve is typical of aluminum and stainless steel alloys.

The tangent modulus E_t is obtained by differentiating equation 3.39:

$$\frac{d\varepsilon}{d\sigma} = \frac{1}{E}\left[1 + \frac{30}{7}\left(\frac{\sigma}{\sigma_y}\right)^9\right] \tag{3.41}$$

$$\tau = \frac{E_t}{E} = \frac{d\sigma/d\varepsilon}{E} = \frac{1}{1 + (30/7)(\sigma/\sigma_y)^9} \tag{3.42}$$

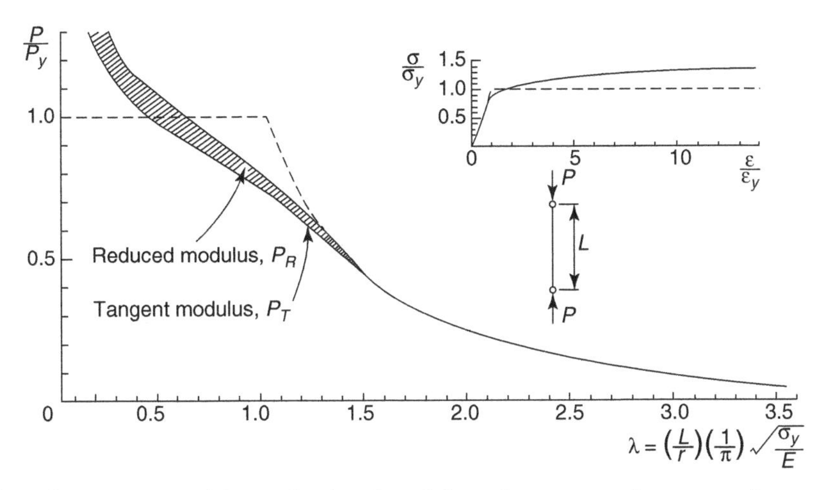

Fig. 3.8 Tangent modulus and reduced modulus column curves for rectangular columns.

The tangent modulus buckling load is equal to equation 3.4: $P_T = \pi^2 E_t I_x / L^2$. In a nondimensionalized form

$$\frac{P}{P_y} = \frac{\tau \pi^2 E/\sigma_y}{(L/r_x)^2} \tag{3.43}$$

With the definition of the slenderness parameter,

$$\lambda_x = \frac{(L/r_x)\sqrt{\sigma_Y/E}}{\pi} \tag{3.44}$$

Equation 3.43 can be expressed as

$$(\lambda_x)_T = \sqrt{\frac{\tau}{P/P_y}} \tag{3.45}$$

This equation can be solved for the critical nondimensional slenderness parameter by specifying $P/P_y = \sigma/\sigma_y$ and computing τ by using equation 3.42. The resulting curve is the lower of the two solid line curves in Figure 3.8.

The reduced modulus critical slenderness parameter is similarly obtained (from equation 3.20) as

$$(\lambda_x)_R = \sqrt{\frac{\bar{E}/E}{P/P_y}} \tag{3.46}$$

where (from equation 3.19)

$$\frac{\bar{E}}{E} = \frac{I_1}{I_x} + \tau\left(\frac{I_2}{I_x}\right) \tag{3.47}$$

I_1 and I_2 are the moments of inertia of the elastic and inelastic zones about the neutral axis (Figure 3.7b), respectively. The location of the neutral axis is determined from equation 3.11 as

$$\frac{E_t}{E} = \tau = \frac{S_1}{S_2} \tag{3.48}$$

where $S_1$and S_2 are the statical moments of the elastic and inelastic zones about the neutral axis, respectively. From 3.7b:

$$S_1 = \frac{b}{2}(d - \bar{y})^2 \quad \text{and} \quad S_2 = \frac{b\bar{y}^2}{2} \tag{3.49}$$

From equations 3.48 and 3.49, we can solve for $\bar{y}$:

$$\bar{y} = d\left(\frac{1 - \sqrt{\tau}}{1 - \tau}\right) \tag{3.50}$$

From Figure 3.7 we also find that

$$I_x = \frac{bd^3}{12}, \quad I_1 = \frac{d(d - \bar{y})^3}{3} \quad \text{and} \quad I_2 = \frac{b\bar{y}^3}{3} \tag{3.51}$$

With equations 3.51, 3.50, and 3.47, we finally get from equation 3.46 the reduced modulus relationship

$$(\lambda_x)_R = 2\sqrt{\frac{\tau(1 - \sqrt{\tau})^3 + (\sqrt{\tau} - \tau)^3}{(P/P_Y)(1 - \tau)^3}} \tag{3.52}$$

The resulting curve is shown as the upper solid line curve in Figure 3.8.

The curves in Figure 3.8 show that indeed the reduced modulus load is above the tangent modulus load. The theoretical maximum strength of the geometrically perfect column lies between the narrow band bounded by the two curves. The dashed line represents the ideal elastic Euler buckling strength that is terminated when the whole section is yielded.

Summary of Important Points from sections 3.1 to 3.3

- The tangent modulus concept assumes that the load does not change during bifurcation, and that stiffness is a function of the tangent modulus (E_t) of the stress–strain curve. This theory provides a lower bound solution of the inelastic buckling load.
- The reduced modulus concept uses a stiffness based on both material and cross-sectional properties. This theory provides an upper-bound solution for the inelastic buckling load.
- Shanley's model, which utilizes a deformable cell at the point of buckling, shows that any load above P_T will cause the column to be laterally deflected, and in the load range $P_T < P < P_{max}$ there is always strain reversal present.

3.4 BUCKLING STRENGTH OF STEEL COLUMNS

Constructional steels have stress–strain curves that are very nearly elastic-plastic before strain hardening sets in. By the strict application of the tangent modulus concept, the critical stress below σ_y is governed by the elastic

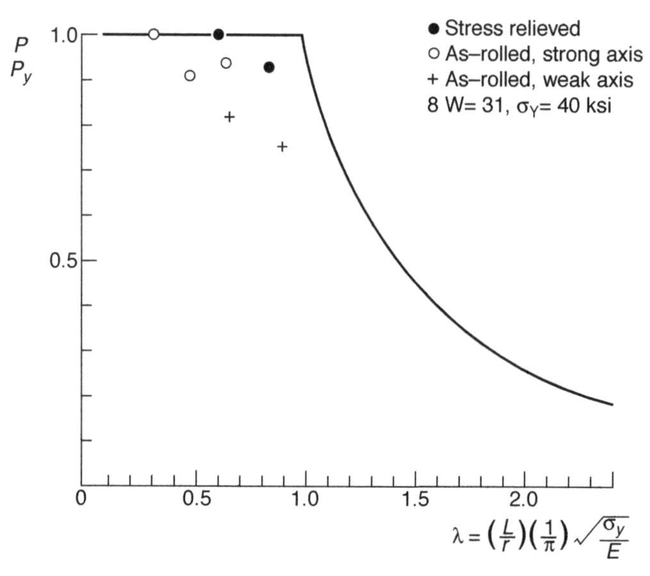

Fig. 3.9 Scatter of test points for steel columns.

formula, and the column curve would take the form shown in Figure 3.9. However, a great number of tests on steel columns have very convincingly shown that the column strength predicted by the reasoning above is usually higher than the actual strength if the columns are of intermediate length ($0.3 < \lambda < 1.4$, approximately). The test points in Figure 3.9 illustrate this. These test results were taken from a test program performed at the Fritz Engineering Laboratory of Lehigh University. (Beedle and Tall 1960). The specimens were W8 $\times$ 31 columns of A7 steel. Other tests, notably those on welded H shapes, show an even larger discrepancy.

The traditional explanation of the apparent scatter of the test points has been that initial out-of-straightness and unavoidable eccentricity of the load are the causes for the difference between theory and practice. However, the tests in Figure 3.9 were performed carefully on well-centered and essentially straight specimens, and the measured amount of initial out-of-straightness did not account for the low test strengths.

The idea that residual stresses could be held accountable for the lower strength of steel columns of intermediate length had been advanced as early as 1888, but it was not until 1952 that it was convincingly shown in a paper by Osgood (1952) that this was the case. At that time extensive research was already in progress at the Fritz Engineering Laboratory of Lehigh University, where the distribution and the magnitude of residual stresses, and their effect on column strength was extensively investigated (Beedle and Tall 1960). Rolled, riveted, and welded columns of various shapes and types of steel were tested and their column strength was compared with theoretical values computed from the measured material and residual stress properties.

This research demonstrated that residual stresses account for a large share of the deviation of the column strength from the ideal curve predicted from the stress–strain relationship (Figure 3.9). A new interpretation of the tangent modulus concept that includes the effects of premature yielding due to residual stresses was introduced (Beedle and Tall 1960). With this modification, the strength of ideally straight columns can be predicted very satisfactorily. Residual stresses do not, however, account for all effects, as can be seen in Figure 3.9, where the *as-rolled* shapes (those containing residual stresses from uneven cooling after hot rolling) have lower strengths than the stress-relieved specimens. For these latter columns the deviation can only be explained by considering initial out-of-straightness.

In the next section of this chapter we illustrate the effects of residual stresses using the example of a rectangular pinned-end column. section 3.6 discusses the effect of initial out-of-straightness, and section 3.7 pulls all of this together and discusses the design criteria of actual metal columns.

3.5 ILLUSTRATION OF THE EFFECT OF RESIDUAL STRESSES ON THE BUCKLING STRENGTH OF STEEL COLUMNS

A pinned-end column of rectangular cross section (Figure 3.10a) is made of steel with an ideal elastic-plastic stress–strain relationship and having a residual stress distribution, as shown in Figure 3.10b. This residual stress pattern is the same in any plane that passes through the cross-section parallel to the x-axis. The residual stress varies linearly from a tensile stress of $\sigma_y/2$ at the center of the section to a compressive stress of $\sigma_y/2$ at the edges.

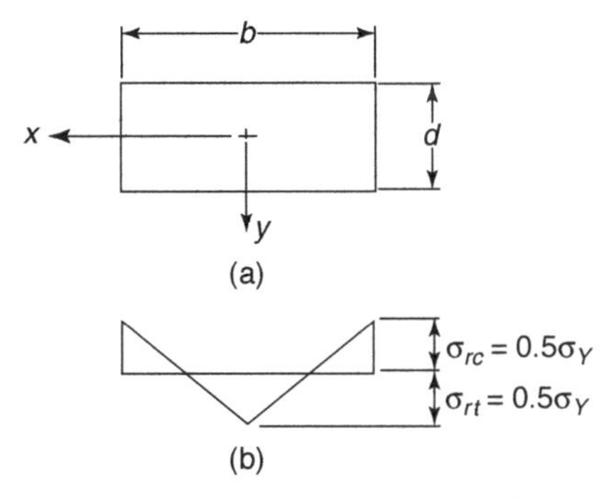

Fig. 3.10 Residual stress in a rectangular column.

Any stress, σ, due to an axial load is evenly distributed across the section until yielding starts when

$$\sigma + \sigma_r = \sigma_y.$$

This occurs at $x = \pm b/2$ when

$$\sigma + \sigma_Y/2 = \sigma_y, \quad \text{or} \quad \sigma/\sigma_y = P/P_y = 1/2.$$

For an increased amount of P, portions of the cross-section are yielded (cross-hatched area in Figure 3.11a) and the stress distribution is as shown in Figure 3.11b. From this figure we obtain the following relationship between the applied stress, σ, and the parameter α, that defines the extent of the elastic core, by using the equilibrium condition $P = \int_A \sigma dA$:

$$\frac{P}{P_y} = \frac{\sigma}{\sigma_y} - \left(\frac{\sigma}{\sigma_y} - \frac{1}{2}\right)\left(\frac{1}{2} - \alpha\right) \tag{3.53}$$

From the geometry of similar triangles in Figure 3.11b we can also show that

$$\frac{\sigma}{\sigma_y} = \frac{3}{2} - 2\alpha \tag{3.54}$$

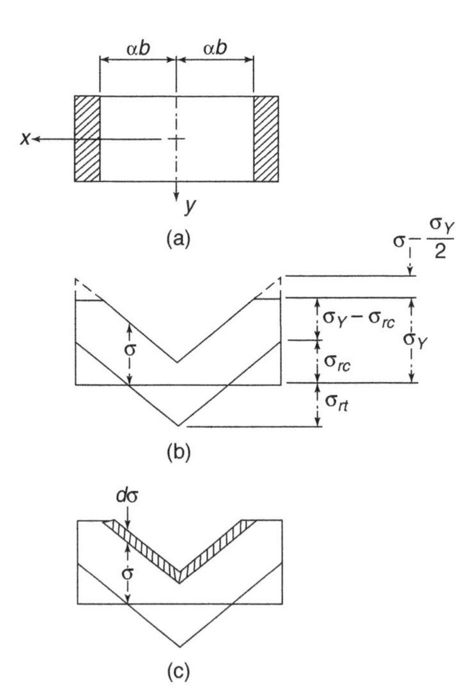

Fig. 3.11 Residual and compression stresses.

By combining equations 3.53 and 3.54, we prove that

$$\alpha = \sqrt{\frac{1}{2}\left(1 - \frac{P}{P_y}\right)} \tag{3.55}$$

Another way of establishing α is demonstrated next. If an infinitesimally small increase of stress, $d\sigma$, is added to the column (Figure 3.11c), then the increase in P is equal to

$$(dP) = (d\sigma) \times d \times 2\alpha b = A_{\mathrm{E}} \times (d\sigma) \tag{3.56}$$

where A_{E} is the area of the elastic core. The average stress increase is $(d\sigma)_{av} = \frac{(dP)}{A}$. The symbol A defines the full cross-sectional area. The increase in the strain is equal to:

$$(d\varepsilon) = \frac{(d\sigma)}{E} = \frac{(dP)}{A_{\mathrm{E}}E} \tag{3.57}$$

When a short length of this column is compressed in a testing machine and the value of P/A is plotted against ε, a so-called *stub-column stress–strain curve* is generated. The slope at any location of the stub-column stress–strain curve is

$$\frac{(d\sigma)_{av}}{(d\varepsilon)} = \frac{(dP)/A}{(dP)/A_{\mathrm{E}}E} = \frac{EA_{\mathrm{E}}}{A} \tag{3.58}$$

We can think of this slope of the stub-column stress–strain curve as the *tangent modulus of the cross-section* E_t, and so

$$\frac{E_t}{E} = \tau = \frac{A_{\mathrm{E}}}{A} \tag{3.59}$$

In the inelastic region of column behavior, the ratio of the tangent modulus to the elastic modulus is equal to the ratio of the area of the elastic core to the area of the whole cross-section. This is in contrast to the tangent modulus defined previously for the material; the tangent modulus for the cross-section includes the effect of the residual stresses. It can be calculated analytically or numerically from the known or measured residual stress pattern, or it can be obtained experimentally by compressing a short piece of the column, a *stub column*, and plotting the average stress P/A versus the strain curve. This is a routine test, and it has been standardized internationally (Galambos 1998). The stub column stress–strain curve plays the same role as the material stress-strain curve for the material in the calculation of the inelastic buckling load (see Figure 3.1).

In the case of the rectangular cross section of Figure 3.11a,

$$\tau = \frac{A_E}{A} = \frac{2\alpha bd}{bd} = 2\alpha \tag{3.60}$$

And so, from equation 3.55

$$\tau = \sqrt{2(1 - P/P_y)} \tag{3.61}$$

It is now possible to develop formulas for the buckling strength of the rectangular column that include the effects of residual stresses. In the elastic range $(0 \leq P/P_y \leq 0.5)$, this is equal to

$$P_X = \frac{\pi^2 EI_x}{L^2} \quad \text{and} \quad P_Y = \frac{\pi^2 EI_y}{L^2} \tag{3.62}$$

with the abbreviations

$$\lambda_x = \frac{(L/r_x)}{\pi}\sqrt{\frac{\sigma_y}{E}} \quad \text{and} \quad \lambda_y = \frac{(L/r_y)}{\pi}\sqrt{\frac{\sigma_y}{E}} \tag{3.63}$$

We now nondimensionalize equation 3.62 to give

$$(\lambda_x)_E = (\lambda_y)_E = \frac{1}{\sqrt{P/P_y}} \tag{3.64}$$

Where $P_y = A\sigma_y$

The tangent modulus buckling strength is defined by the elastic core, since $E = 0$ in the yielded zone; that is,

$$P_T = \frac{\pi^2 EI_E}{L^2} \tag{3.65}$$

In nondimensional form, this equation is

$$\lambda_T = \sqrt{\frac{I_E/I}{P/P_Y}} \tag{3.66}$$

For buckling about the y-axis

$$I_{yE} = \frac{d}{12}(2\alpha b)^3 \quad \text{and} \quad I_y = \frac{db^3}{12} \tag{3.67}$$

and for buckling about the x-axis

$$I_{x\mathrm{E}} = \frac{d^3}{12}(2\alpha b) \quad \text{and} \quad I_x = \frac{d^3 b}{12} \tag{3.68}$$

Setting equation 3.60 into equations 3.67 and 3.68, and substituting these, in turn, into equation 3.66, we obtain the following equations for the tangent modulus loads:

$$\begin{aligned}(\lambda_x)_T &= \sqrt{\frac{\tau}{P/P_y}} = \sqrt{\frac{2(1-P/P_y)^{1/2}}{P/P_y}} \\ (\lambda_y)_T &= \sqrt{\frac{\tau^3}{P/P_y}} = \sqrt{\frac{2(1-P/P_y)^{3/2}}{P/P_y}}\end{aligned} \tag{3.69}$$

The curves showing the P/P_y versus $(\lambda_x)_T$ and $(\lambda_y)_T$ relationship are shown as the solid-line curves in Figure 3.12. For y-axis bending, the tangent modulus is nearly a straight line between $P/P_y = 1.0$, where the whole cross-section is yielded, and $P/P_y = 0.5$, where yielding commences. The curve for buckling about the x-axis is considerably above the y-axis curve.

The reason for this is seen from equations 3.67 and 3.68; the moment of inertia of the elastic core about the y-axis varies as α^3, whereas this property varies as α about the x-axis. Since

$$\alpha < 1.0, \ \alpha^3 < \alpha,$$

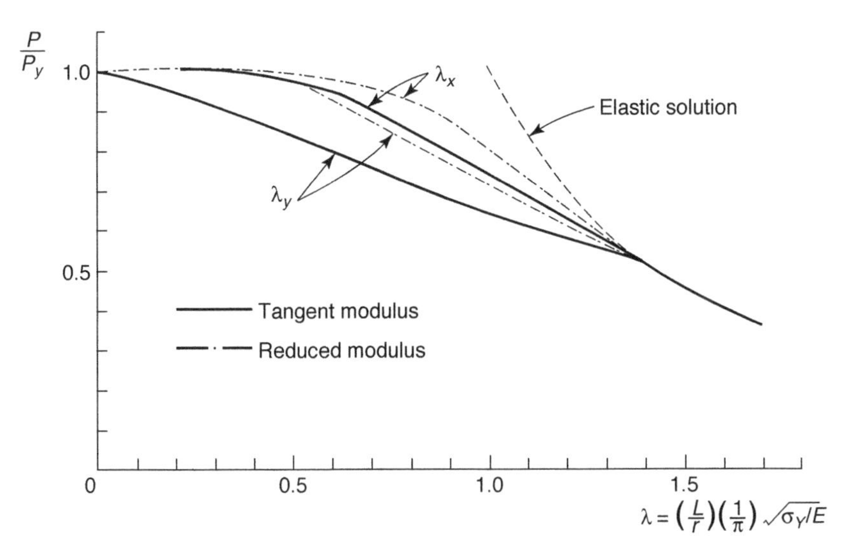

Fig. 3.12 Reduced and tangent modulus column curves for buckling about the x- and y-axes of a rectangular column.

then it follows that

$$P_{TX}/P_y > P_{TY}/P_y.$$

As the buckling strength of wide-flange shapes is governed principally by the stiffness of the flanges, and the residual stress distribution in these flanges is similar to that in our rectangular column, the nondimensional buckling strength of wide-flange shapes should be larger for strong-axis buckling than for weak-axis buckling. That this is indeed so has been verified by numerous experiments on rolled wide-flange steel columns (Galambos 1998).

Since the reduced modulus strength is only of academic interest, the development of the equations will be omitted here. The resulting curves are included in Figure 3.12 for comparison only. In the next section it will be demonstrated that for steel columns one must also consider initial out-of-straightness in determining column strength.

3.6 EFFECT OF INITIAL OUT-OF-STRAIGHTNESS AND LOAD ECCENTRICITY

The previous portions of the chapter on inelastic column buckling dealt with the theory of the buckling strength of ideally straight compression members. Examples were presented to show how strength is determined for columns made from materials that have a nonlinear stress–strain relationship, such as aluminum and stainless steel. The effect of residual stresses on the inelastic buckling strength of steel columns was also elaborated. This section introduces the additional effect on column strength that results from initial out-of-straightness of the member, and from the eccentricity of the load application. The phenomenon is illustrated on a pinned-end rectangular column made from a material with an ideal-elastic-plastic stress–strain curve. The final part of this chapter, section 3.7 pulls together the various ideas that were presented so far in order to explain the origins of column formulae used in modern design specifications for the design of metal structures.

Columns have initial deflections as a result of manufacturing and fabrication. These are small and must be within prescribed tolerances, and they are thus unavoidable. These "imperfections" may affect the strength of the compression member, as shown schematically in Figure 3.6. If the material is infinitely elastic, then the column strength will approach asymptotically the elastic buckling load of a perfect column, as was demonstrated in Chapter 2. However, if the material is inelastic, then the attainment of the maximum capacity necessarily occurs after some part of the column has yielded. Inevitably there is deflection and bending as soon as there is any load on the member. In order to determine the strength, it is necessary to consider the nonlinear moment-curvature relationship of the cross-section. In the following portions of this section, we first derive the moment-curvature equations, and then we use these relations to find the maximum strength.

3.6.1 Moment-Curvature-Thrust Relations for a Rectangular Section

It is assumed that the material has an ideal-elastic-plastic stress–strain curve (Figure 3.18). The cross-section is assumed rectangular with dimensions $b \times d$. An axial force P is acting at the centroid of the cross-section, and it is subjected to a bending moment M. It is assumed that plane sections before bending will remain plane after bending, and therefore the strain varies linearly. According to the stress–strain diagram, there are three possible stress distributions:

- The stresses are all below the yield stress.
- The top of the rectangle is yielded.
- Both top and bottom are yielded.

The cross-section, loading, strain distribution, and the three stress distributions are shown in Figure 3.13.

The relationships between the bending moment M, the curvature ϕ and the axial force P will be next developed for the three stress distributions shown in Figure 3.13.

Case I (Figure 3.14)

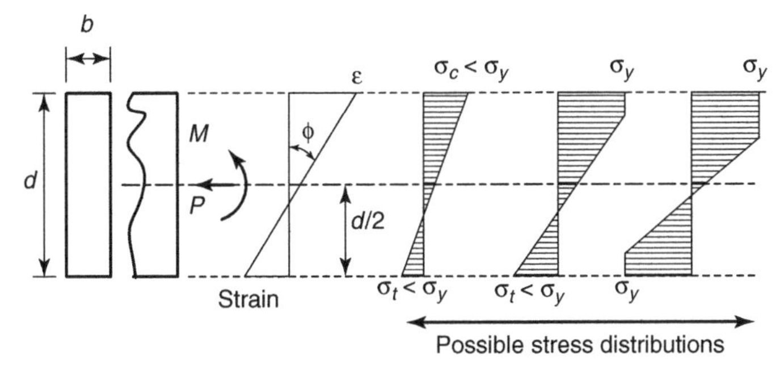

Fig. 3.13 Rectangular cross-section. Showing loading, strain distribution, and three stress distributions.

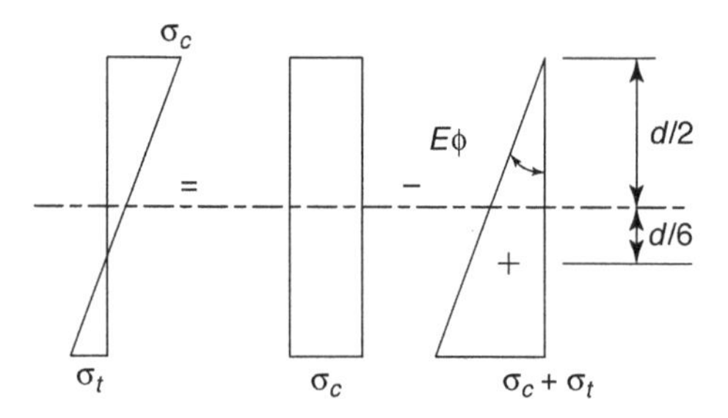

Fig. 3.14 Case I stress distribution.

The equilibrium of the applied forces with the stresses is

$$P = \int_A \sigma dA = \sigma_c bd - (\sigma_c + \sigma_t)\left(\frac{bd}{2}\right) = \left(\frac{\sigma_c - \sigma_t}{2}\right) bd \tag{3.70}$$

$$M = \int_A \sigma y dA = (\sigma_c + \sigma_t)\left(\frac{bd}{2}\right)\left(\frac{d}{6}\right) = (\sigma_c + \sigma_t)\left(\frac{bd^2}{12}\right) \tag{3.71}$$

Assuming small strains, that is $\tan\phi = \phi$, we get from the geometry of the stress block in Figure 3.14:

$$E\phi = \frac{\sigma_t + \sigma_c}{d} \tag{3.72}$$

The following quantities are introduced here for use in the derivations:

$$\textit{Yield force}(\textit{``squash load''})\text{: } P_y = bd\sigma_y$$

$$\textit{Yield moment}\text{: } M_y = bd^2\sigma_y/6$$

$$\textit{Yield curvature}\text{: } \phi_y = 2\sigma_y/Ed$$

Equation 3.70 can be rearranged as follows:

$$\frac{P}{P_y} = \frac{1}{2}\left(\frac{\sigma_c}{\sigma_y} - \frac{\sigma_t}{\sigma_y}\right) \rightarrow \frac{\sigma_c}{\sigma_y} = 2\frac{P}{P_y} + \frac{\sigma_t}{\sigma_y} \tag{3.73}$$

From equations. 3.72 and 3.73 we get:

$$\begin{aligned} \frac{\sigma_t}{\sigma_y} &= \frac{\phi}{\phi_y} - \frac{P}{P_y} \\ \frac{\sigma_c}{\sigma_y} &= \frac{\phi}{\phi_y} + \frac{P}{P_y} \end{aligned} \tag{3.74}$$

Substitution into equation 3.71 finally results in:

$$\frac{M}{M_y} = \frac{1}{2}\left(\frac{\sigma_c}{\sigma_y} + \frac{\sigma_t}{\sigma_y}\right) = \frac{\phi}{\phi_y} \tag{3.75}$$

The upper limit of applicability of this equation is when $\frac{\sigma_c}{\sigma_y} = 1$ and $\frac{\phi}{\phi_y} = 1 - \frac{P}{P_y}$

Case II (Figure 3.15)
In this case, yielding is in compression; yielding penetrates the top of the cross-section. From the geometry of the stress blocks in Figure 3.15, the curvature equals

$$E\phi = \frac{\sigma_t + \sigma_y}{d - \xi d}.$$

From this we get (after some algebra),

$$\frac{\sigma_t}{\sigma_y} = 2\frac{\phi}{\phi_y}(1 - \xi) - 1.$$

The equilibrium of the axial stresses

$$P = \sigma_y bd - \frac{1}{2}(\sigma_t + \sigma_y)(d - \xi d)b$$

leads to

$$\frac{P}{P_y} = 1 - \frac{1}{2}\left(\frac{\sigma_t}{\sigma_y} + 1\right)(1 - \xi).$$

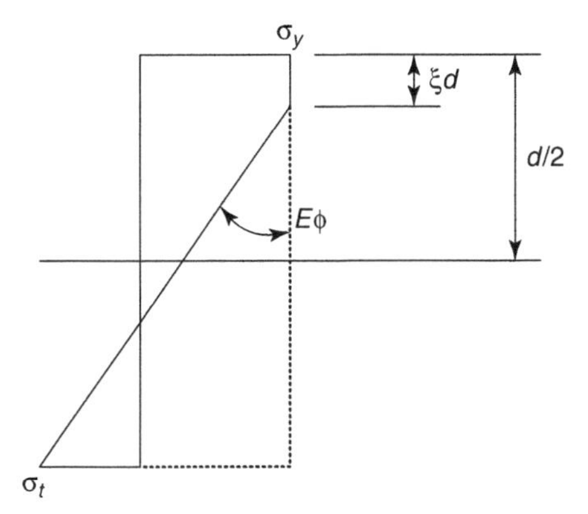

Fig. 3.15 Yielding in compression only.

Substituting $\frac{\sigma_t}{\sigma_y}$ from above we get

$$\frac{P}{P_y} = 1 - \frac{\phi}{\phi_y}(1-\xi)^2.$$

From the equation for $\frac{P}{P_y}$ we can now solve the quadratic equation for

$$\xi = 1 - \sqrt{\frac{1 - \frac{P}{P_y}}{\frac{\phi}{\phi_y}}}$$

The negative sign controls because ξ must be less than 1.0. Substituting ξ into the equation for $\frac{\sigma_t}{\sigma_y}$ gives

$$\frac{\sigma_t}{\sigma_y} = 2\frac{\phi}{\phi_y}\sqrt{\frac{1 - \frac{P}{P_y}}{\frac{\phi}{\phi_y}}} - 1$$

The moment equilibrium equation is

$$M = (\sigma_t + \sigma_y)\frac{1}{2}b(d - \xi d)\left(\frac{d}{2} - \frac{d - \xi d}{3}\right).$$

Substituting ξ and σ_t, and nondimensionalizing M, we get the following equation for the ratio $\frac{M}{M_y}$ for stress Case II:

$$\frac{M}{M_y} = 3(1 - \frac{P}{P_y}) - 2\sqrt{\frac{\left(1 - \frac{P}{P_y}\right)^3}{\frac{\phi}{\phi_y}}} \tag{3.76}$$

The lower limit of this equation is when

$$\xi = 0 = 1 - \sqrt{\frac{1 - \frac{P}{P_y}}{\frac{\phi}{\phi_y}}} \rightarrow \frac{\phi}{\phi_y} = 1 - \frac{P}{P_y}$$

The upper limit is when

$$\frac{\sigma_t}{\sigma_y} = 1 = 2\frac{\phi}{\phi_y}\sqrt{\frac{1 - \frac{P}{P_y}}{\frac{\phi}{\phi_y}}} - 1 \rightarrow \frac{\phi}{\phi_y} = \frac{1}{1 - \frac{P}{P_y}}$$

Case III (Figure 3.16)

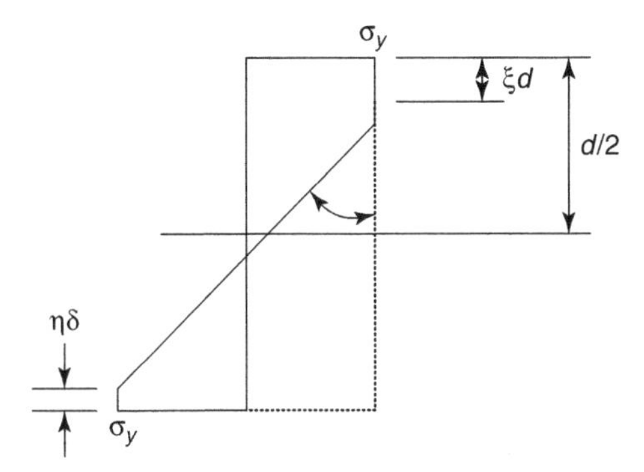

Fig. 3.16 Stress distribution for Case III.

The curvature is

$$E\phi = \frac{2\sigma_y}{d - \eta d - \xi d}$$

Or nondimensionally,

$$\frac{\phi}{\phi_y} = \frac{1}{1 - \eta - \xi}.$$

The axial stress equilibrium is

$$P = \sigma_y bd - \eta db(2\sigma_y) - \frac{1}{2}(2\sigma_y)b(d - \eta d - \xi d), \text{ or}$$

nondimensionalizing, $\frac{P}{P_y} = \xi - \eta$.

From the equations for $\frac{\phi}{\phi_y}$ and $\frac{P}{P_y}$ we can solve for η and ξ:

$$\eta = \frac{1}{2}\left(1 - \frac{1}{\frac{\phi}{\phi_y}} - \frac{P}{P_y}\right)$$

$$\xi = \frac{1}{2}\left(1 - \frac{1}{\frac{\phi}{\phi_y}} + \frac{P}{P_y}\right)$$

The moment equilibrium is expressed as follows:

$$M = 2\sigma_y b\eta d\left(\frac{d}{2} - \frac{\eta d}{2}\right) + \frac{1}{2}(2\sigma_y)b(d - \eta d - \xi d)\left(\frac{d}{2} - \eta d - \frac{d - \eta d - \xi d}{3}\right)$$

After substitution of the previously derived values of η and ξ we get the following nondimensional equation for the moment:

$$\frac{M}{M_y} = \frac{3}{2}\left[1 - \left(\frac{P}{P_y}\right)^2 - \frac{1}{3\left(\frac{\phi}{\phi_y}\right)^2}\right] \tag{3.77}$$

The lower limit is when

$$\eta = \frac{1}{2}\left(1 - \frac{1}{\frac{\phi}{\phi_y}} - \frac{P}{P_y}\right) = 0 \rightarrow \frac{\phi}{\phi_y} = \frac{1}{1 - \frac{P}{P_y}}$$

The flow-diagram of Figure 3.17 summarizes the moment-curvature-thrust relationships for the rectangular cross-section. Figure 3.18 recapitulates the definitions that were used for the loading, the cross-section and the material stress–strain relationship.

A typical moment-curvature curve is shown in Figure 3.19. When $\phi \rightarrow \infty$, practically when $\phi \approx 10\phi_y$, the moment reaches a plateau. This

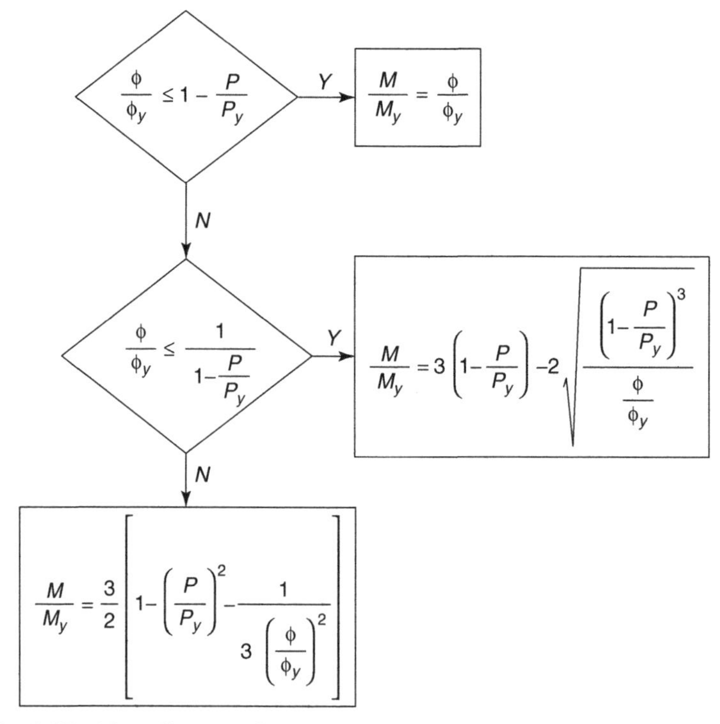

Fig. 3.17 Flow-diagram of moment-curvature thrust formulas for rectangle.

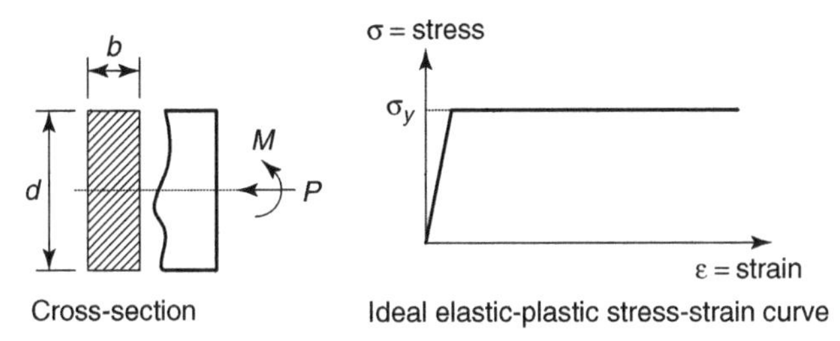

Fig. 3.18 Recapitulation of definitions.

maximum moment is defined as the plastic moment in the presence of axial force, M_{pc}.

$$\frac{M_{pc}}{M_y} = \frac{3}{2}\left[1 - \left(\frac{P}{P_y}\right)^2\right] \tag{3.78}$$

In case $P = 0$, that is, bending moment only,

$$M_p = \frac{3}{2} M_y = \frac{3}{2} \times \frac{bd^2\sigma_y}{6} = \frac{bd^2\sigma_y}{4} \tag{3.79}$$

and therefore

$$\frac{M_{pc}}{M_p} = \left[1 - \left(\frac{P}{P_y}\right)^2\right] \tag{3.80}$$

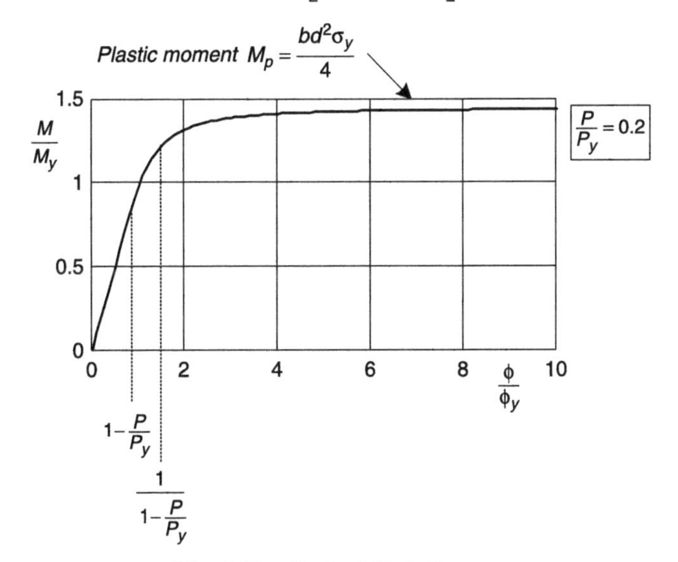

Fig. 3.19 Typical M-ϕ-P curve.

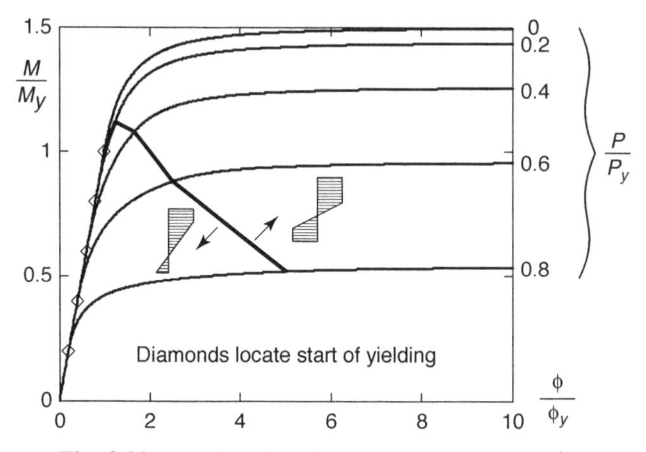

Fig. 3.20 Family of M-ϕ curves for values of P/P_y.

In the following portion of this section we use the M-ϕ-P relationships to derive the strength of initially crooked axially loaded columns. The information in Figure 3.20 delineates the regions where elastic and inelastic behavior controls the deformations of the column.

3.6.2 Maximum Strength of a Column with Initial Out-of-Straightness

The sketch in Figure 3.21a shows an ideal column that is initially perfectly straight. In contrast, Figure 3.21b illustrates an actual column that has an initial bow with an amplitude v_{oi} in the middle even before any axial load is applied. In addition, the load is applied through an eccentricity e_i. This represents the actual case of a compression member in a real structure.

It is assumed that the initial out-of-straightness v_i and the additional lateral deflection v during loading are sinusoidal in shape and given by

$$v_i = v_{oi} \sin \frac{\pi z}{L} \tag{3.81}$$

$$v = v_o \sin \frac{\pi z}{L} \tag{3.82}$$

The curvature at the center of the column (at $z = L/2$) is then

$$\phi_o = -v''(L/2) = v_o \left(\frac{\pi^2}{L^2}\right) \sin \frac{\pi}{2} = v_o \left(\frac{\pi^2}{L^2}\right) \tag{3.83}$$

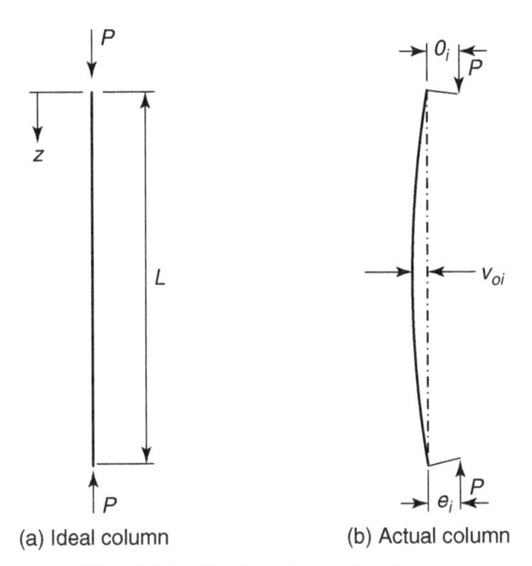

Fig. 3.21 Ideal and actual column.

Equation 3.83 is divided by the yield curvature $\phi_y = \frac{2\sigma_y}{Ed}$ to obtain the following nondimensional expression for the curvature ϕ_o:

$$\frac{\phi_o}{\phi_y} = 6\left(\frac{v_o}{d}\right)\left(\frac{1}{\lambda^2}\right) \tag{3.84}$$

where

$$\lambda = \frac{1}{\pi}\frac{L}{r}\sqrt{\frac{\sigma_y}{E}} \tag{3.85}$$

The radius of gyration for the rectangular section is $r = \frac{d}{\sqrt{12}}$.

The external moment at $z = L/2$ is equal to

$$M_e = P(e_i + v_{oi} + v_o) \tag{3.86}$$

If the expression in equation 3.86 is divided by the yield moment $M_y = \frac{bd^2\sigma_y}{6}$, then

$$\frac{M_e}{M_y} = 6\frac{P}{P_y}\left(\frac{e_i}{d} + \frac{v_{oi}}{d} + \frac{v_o}{d}\right) \tag{3.87}$$

where $P_y = bd\sigma_y$. The internal moments are determined from the formulas in the flowchart of Figure 3.17.

The maximum value of the axial load P will be attained when the change in the external moment δM_e can no longer be resisted by the change in the internal moment δM_i at the center of the member where the moment is largest. The stability criterion is thus $\delta M_e/\delta v_o = \delta M_i/\delta v_o$. Nondimensionally

$$\frac{\delta(M_e - M_i)/M_y}{\delta(v_o/d)} \tag{3.88}$$

The maximum occurs when some portion of the column is yielded. The load-deflection curve is shown schematically in Figure 3.22. There are two cases of yielding: compression only (Case II, Figure 3.15) and compression and tension (Case III, Figure 3.16).

Case II, yielding in compression

The moment-curvature-thrust relationship is equal to (from equation 3.76):

$$\frac{M_i}{M_y} = 3\left(1 - \frac{P}{P_y}\right) - \frac{2\left(1 - \frac{P}{P_y}\right)^{\frac{3}{2}}}{\left(\frac{\phi}{\phi_y}\right)^{\frac{1}{2}}} \tag{3.89}$$

With equations 3.87 and 3.84, we obtain the following equation for the difference between the external and the internal moments:

$$\frac{M_e - M_i}{M_y} = 6\frac{P}{P_y}\left(\frac{e_i}{d} + \frac{v_{oi}}{d} + \frac{v_o}{d}\right) - 3\left(1 - \frac{P}{P_y}\right) + \frac{2\left(1 - \frac{P}{P_y}\right)^{\frac{3}{2}}}{\left(\frac{\phi}{\phi_y}\right)^{\frac{1}{2}}} \tag{3.90}$$

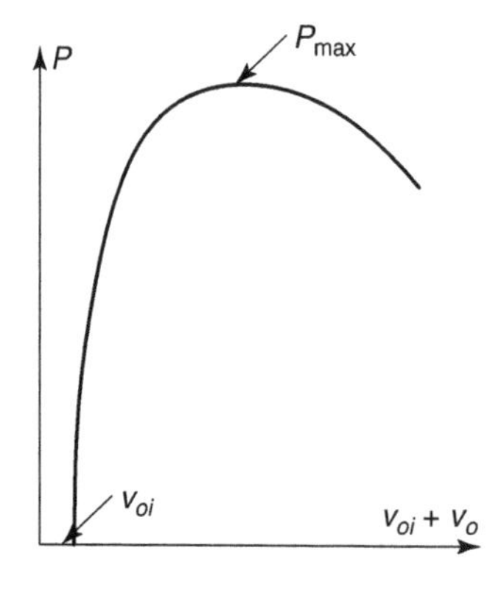

Fig. 3.22 Sketch of load-deflection curve.

Taking the derivative of this equation with respect to v_o/d, setting the derivative equal to zero, and solving for v_o/d, we get

$$\frac{v_o}{d} = \frac{1}{6}(\lambda)^{\frac{2}{3}}\left[\frac{1 - P/P_y}{(P/P_y)^{\frac{2}{3}}}\right] \tag{3.91}$$

Setting equation 3.91 into the equilibrium relationship $M_i = M_e$, the following equation is obtained for the interaction between the maximum axial force and the slenderness ratio:

$$6\left(\frac{P}{P_y}\right)\left(\frac{e_i}{d} + \frac{v_{oi}}{d}\right) = 3\left(1 - \frac{P}{P_y}\right)\left[1 - \left(\frac{P}{P_y}\lambda^2\right)^{\frac{1}{3}}\right] \tag{3.92}$$

Case III, yielding in compression and tension

In this case, from equation 3.77:

$$\frac{M_i}{M_y} = \frac{3}{2}\left[1 - \left(\frac{P}{P_y}\right)^2\right] - \frac{1}{2\left(\frac{\phi}{\phi_y}\right)^2} \tag{3.93}$$

and thus

$$\frac{M_e - M_i}{M_y} = 6\left(\frac{P}{P_y}\right)\left(\frac{e_i}{d} + \frac{v_{oi}}{d} + \frac{v_o}{d}\right) - \frac{3}{2}\left[1 - \left(\frac{P}{P_y}\right)^2\right] + \frac{\lambda^4}{72}\left(\frac{v_o}{d}\right)^{-2} \tag{3.94}$$

Setting the derivative of equation 3.94 with respect to v_o/d equal to zero and solving for v_o/d results in

$$\frac{v_o}{d} = \frac{\lambda^{\frac{4}{3}}}{6\left(\frac{P}{P_y}\right)^{\frac{1}{3}}} \tag{3.95}$$

Setting equation 3.95 into the equilibrium relationship $M_i = M_e$, the following equation is obtained for the interaction between the maximum axial force and the slenderness ratio:

$$6\left(\frac{P}{P_y}\right)\left(\frac{e_i}{d} + \frac{v_{oi}}{d}\right) = \frac{3}{2}\left[1 - \left(\frac{P}{P_y}\right)^2 - \lambda^{\frac{4}{3}}\left(\frac{P}{P_y}\right)^{\frac{2}{3}}\right] \tag{3.96}$$

Equation 3.92 applies in the range $1 - \frac{P}{P_y} \leq \frac{\phi}{\phi_y} \leq \frac{1}{1-\frac{P}{P_y}}$.
By substituting equations 3.84 and 3.91, we get the following domain of applicability:

$$\frac{\left(1 - \frac{P}{P_y}\right)^3}{\frac{P}{P_y}} \leq \lambda^2 \leq \frac{1}{\frac{P}{P_y}} \tag{3.97}$$

Equation 3.96 applies in the range $\frac{1}{1-\frac{P}{P_y}} \leq \lambda^2 < \infty$. By substituting equations 3.84 and 3.95, we get the following domain of applicability:

$$0 \leq \lambda^2 \leq \frac{\left(1 - \frac{P}{P_y}\right)^3}{\frac{P}{P_y}} \tag{3.98}$$

The interaction equations and their ranges of applicability are summarized in Figure 3.23.

We use the model for the strength of rectangular steel columns to illustrate the effects of unavoidable load eccentricity and initial out-of-straightness. In some steel design codes used during the middle of the twentieth century, it was customary to assume an unavoidable load eccentricity ratio of

$$e_i A/S = 0.25.$$

For the rectangle the area $A = bd$ and the elastic section modulus $S = bd^2/6$. The ratio $e_i/d \approx 0.04$. The initial out-of-straightness is expressed as a ratio of the length of the column. The average initial out-of-straightness of the test columns used as the basis of the development of the column design equation of the AISC (to be presented in section 3.7) is $L/1{,}500$ (Galambos 1998). Other standards use $L/1{,}000$. These are small deflections that are not noticeable by an unaided eye. Based on an $L/1{,}500$ out-of-straightness:

$$\frac{v_{oi}}{d} = \frac{L}{1{,}500} \times \frac{d}{r\sqrt{12}} \times \frac{1}{\pi}\sqrt{\frac{\sigma_y}{E}} \times \frac{\pi}{d}\sqrt{\frac{E}{\sigma_y}} = \frac{\pi}{\sqrt{12}}\sqrt{\frac{E}{\sigma_y}}\frac{\lambda}{1{,}500}.$$

For $\sigma_y = 50\,\text{ksi}$ and $E = 29{,}000\,\text{ksi}$, $v_{oi}/d = 0.015$.
The combined effect of initial out-of-straightness and end eccentricity is illustrated in Figure 3.23 for

$$\varepsilon_i = \frac{e_i}{d} = 0.04 \quad \text{and} \quad v_i = \frac{v_o}{d} = 0.02.$$

$\sqrt{\frac{(1-p)^3}{p}} \le \lambda \le \frac{1}{\sqrt{p}}$ $6p(\varepsilon_i + v_i) - 3(1-p)\left[1 - (p\lambda^2)^{\frac{1}{3}}\right] = 0$	$0 \le \lambda \le \sqrt{\frac{(1-p)^3}{p}}$ $6p(\varepsilon_i + v_i) - 1.5\left[1 - p^2 - \lambda^{\frac{4}{3}} p^{\frac{2}{3}}\right] = 0$
$p = \frac{P}{P_y}; \varepsilon_i = \frac{e_i}{d}; v_i = \frac{v_{oi}}{d}; \lambda = \frac{1}{\pi}\frac{L}{r}\sqrt{\frac{\sigma_y}{E}}$	

Fig. 3.23 Interaction between axial load, slenderness, load eccentricity, and initial out-of-straightness.

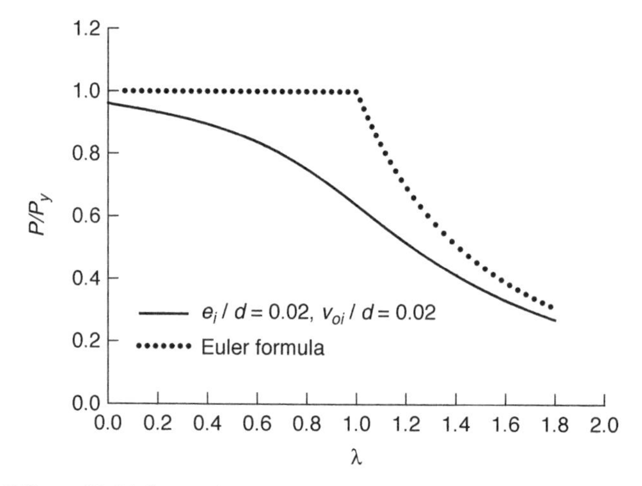

Fig. 3.24 Effect of initial out-of-straightness and load eccentricity on column strength.

The comparison with the Euler curve illustrates the reduction of strength, especially in the region $0.2 < \lambda < 1.4$, where most of the practical-length columns occur. The effect of the mean initial out-of-straightness of $L/1{,}500$ is shown in Figure 3.25. Also shown is a curve representing the strength of columns according to the specification of the AISC for steel structures. This curve is discussed in more detail in the next section. The AISC design column curve is below the curve that only includes the initial out-of-straightness because the former includes the effects of both the initial curvature and residual stresses.

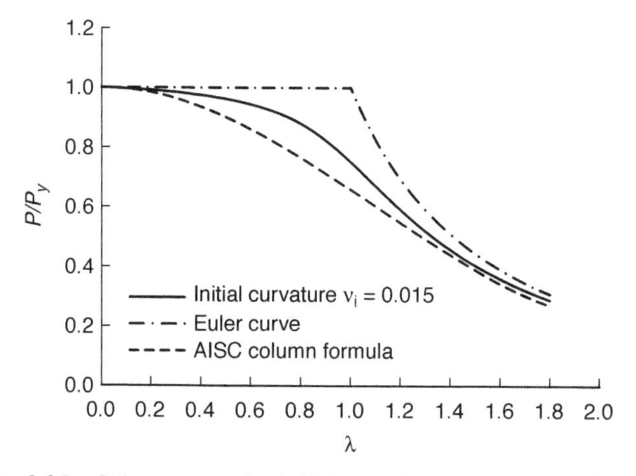

Fig. 3.25 Column curve for initial out-of-straightness of L/1,500.

3.7 DESIGN FORMULAS FOR METAL COLUMNS

In the previous sections of this chapter on the inelastic behavior of axially loaded compression members, several models of behavior were illustrated on relatively simple problems. First, the behavior of geometrically perfect, straight members was considered. The effects of a nonlinear stress–strain curve on the tangent modulus and the reduced modulus strength were examined. The paradox between the two idealized theories was resolved by Shanley, and his derivation was explained in detail. Practical tangent modulus column strengths were solved by using the Ramberg–Osgood equation for the stress–strain curve. The concept of residual stresses was then introduced for the solution of the strength of steel columns. Finally, the inelastic behavior of geometrically crooked columns was studied for the simple case of rectangular members.

Engineers in the design office need formulas that can be employed to quickly check the strength of many columns in the course of a working day. The history of column design formulas is a long and interesting study, going back some 250 years to Leonard Euler. Only a brief overview is given here. For a more detailed history of column curve development, the reader is referred to the *Guide to Design Criteria for Metal Structures* (Galambos 1998).

Engineers have always been aware that the behavior of practical-length metal columns is not governed by elastic behavior. Basically, three historic threads for column formulas run through the past 150 years, since the beginning of the extensive use of metal columns in construction:

1. Purely empirical formulas based on laboratory tests of specific types of cross sections and materials
2. Formulas that were derived theoretically by assuming a fictional or a measured unavoidable load eccentricity or initial curvature
3. Formulas that are based on the tangent modulus theory

For the past 100 years it was known that column strength is dependent on the stress–strain curve of the metal, on the locked-in initial residual stresses, on the shape of the cross-section, on the eccentricity of the applied load, and on the unavoidable initial out-of-straightness of the erected member. Many ingenious behavioral models were used to arrive at curves relating critical stress and slenderness ratio. However, prior to the computer being an indispensable engineering tool, these models had to be simple enough to allow computation by slide-rule or mechanical calculators.

In North American structural steel specifications of the 1960s and 1970s the Tangent Modulus concept was used as the basis for column design equations. The theoretical model assumed geometrically straight members, and ideal elastic-plastic stress strain diagram for the material, and idealized residual stresses due to cooling after rolling (Galambos 1998). The residual stress distribution in the flanges was assumed to vary linearly from a compressive stress at the four flange tips, σ_{rc}, to a tensile residual stress at the flange-to-web junction, σ_{rt}. The residual stress in the web was assumed to be a uniform tensile residual stress, σ_{rt}. This distribution of residual stresses is shown in Figure 3.26. The compressive residual stress was assumed to be equal to $\sigma_{rc} = 0.3F_y$, or, 30% of the yield stress. The tensile residual stress is defined by the condition that the net force on the cross section due to the residual stresses must equal zero, and thus $\sigma_{rt} = \frac{\sigma_{rc}}{1+\frac{dt_w}{b_f t_f}\left(1-\frac{2t_f}{d}\right)}$. The role of the initial curvature was acknowledged in the design standards by providing safety factors that became larger as the column length increased. As discussed below, this method of designing columns has been abandoned in modern codes.

Based on the geometry of the wide-flange shape and the distribution of the residual stress, assuming an ideal elastic-plastic stress-strain diagram, tangent modulus column equations have been derived. The method is illustrated on a rectangular shaped column with residual stress earlier in this chapter. The resulting column curves for major and minor axis buckling of wide-flange columns are shown as the dashed curves in Figure 3.27. The

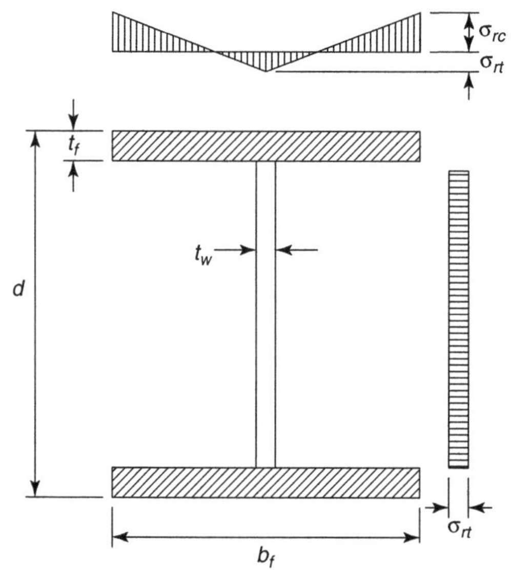

Fig. 3.26

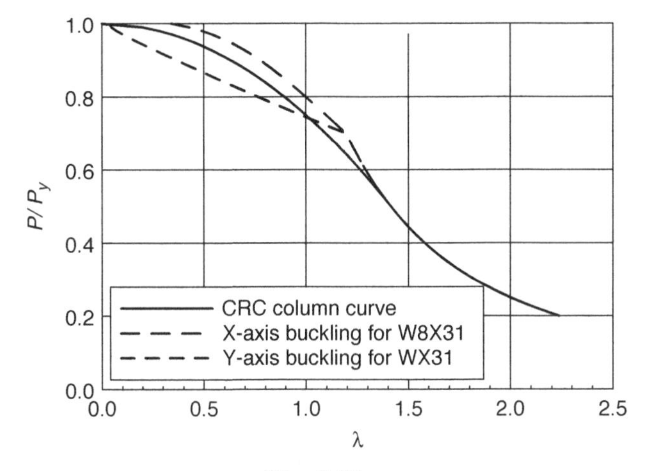

Fig. 3.27

corresponding equations were thought to be too cumbersome for design office use, and so a compromise design curve was adopted for the specifications. This curve is shown as the solid curve in Figure 3.27. It is often referred to as the Column Research Council column equation, or the CRC Column Curve (Galambos 1998). The corresponding equations are

$$\frac{P}{P_y} = 1 - \frac{\lambda^2}{4} \quad \text{for} \quad \frac{P}{P_y} > 0.5, \quad \text{or} \quad \lambda < \sqrt{2} \tag{3.99}$$

$$\frac{P}{P_y} = \frac{1}{\lambda^2} \quad \text{for} \quad \frac{P}{P_y} \leq 0.5, \quad \text{or} \quad \lambda \geq \sqrt{2} \tag{3.100}$$

where

$$P_y = AF_y \tag{3.101}$$

$$\lambda = \frac{L}{\pi r}\sqrt{\frac{F_y}{E}} \tag{3.102}$$

Even though the CRC Column Curve is no longer used directly in design codes, a tangent modulus equation that is derived from it is defined in Appendix 7 of AISC Specification (2005) for use in the stability analysis of steel framed structures. This method of design is presented in Chapter 8 of this text.

The tangent modulus implicit in equation 3.99 is derived as follows:

define $\tau = \dfrac{E_t}{E}$ and $p = \dfrac{P}{P_y}$

from equation 3.99: $\lambda^2 = 4(1 - p)$
critical tangent modulus buckling load. $p = \dfrac{\tau}{\lambda^2} = \dfrac{\tau}{4(1 - p)}$

$$\tau = 4p(1 - p) \quad \text{for } 0.5 < p \leq 1.0 \tag{3.103}$$

$$\tau = 1.0 \quad \text{for } p \leq 0.5 \tag{3.104}$$

These equations will be used later in Chapters 4 and 5 to solve inelastic column and frame examples.

Starting in 1945, when Shanley removed the intellectual barrier of the tangent and reduced model paradox, a very intense area of research commenced. On the experimental side, reliable data became available on material properties and residual stress distributions, and careful and coordinated column tests were performed by many laboratories. It was possible to connect a given column test with the material properties, residual stress distributions, and geometric imperfections of the same specimen. At the same time, researchers could use computer models that incorporated all the data of the test column to attempt to predict the total load-deflection behavior, especially also the strength.

It not unreasonable to state that one can predict with a high degree of certainty how strong a column is going to be if we know the pertinent characteristics of material and geometry. This was demonstrated in the 1972 dissertation of Bjorhovde (Bjorhovde 1978), in which 26 column tests were performed for which initial out-of-straightness, in addition to cross-sectional measurements, material property distributions across the section, and residual stresses were carefully determined and reported. The average test strength-to-predicted strength was reported as 1.03, with a coefficient of variation of 0.05. This level of accuracy provided confidence in the method of calculating strength. Of course, the individual details of every column cannot be determined prior to design, so a more general approach was required.

In addressing this problem, Bjorhovde found 112 column cross-sections in the literature for which there was available the complete information on geometry, material characteristic, and residual stress needed to perform a numerical analysis of the column strength for the range of practical column lengths. He produced 112 column curves for initial sinusoidal curvatures of $L/1{,}000$ as the maximum amplitude at the center of the column, the tolerance for straightness in the steel industry, and for the amplitude of $L/1{,}470$, the average measured out-of-straightness of tested columns.

The column curves were then grouped into three column categories. For each category, an average column curve was constructed and an equation was fitted to this average. Coefficients of variation were also determined and reported so that these statistics could be used in developing the resistance factors ϕ for the AISC Specification (AISC 2005). The column sections for each of the three categories is provided in the Column Curve Selection Table (Figure 3.27 of Galambos 1998) and is reproduced here as Table 3.1. This table lists the column cross-section types, the specified minimum yield stress, and the corresponding column category. Current (2007) fabrication practice uses predominantly steel with a yield stress of 50 ksi or higher. Only one shape type, namely a heavy welded built-up shape fabricated from universal mill plate, is in category 3 for the yield stress of 50 ksi. This method of using universal mill plates is no longer practiced because built-up shapes are made by welding flame-cut plates together. Thus current practice, as reflected by the 2005 AISC Specification, requires only the Category 2 column equation. The same is the case in the Canadian steel design standard.

The equations developed in the Bjorhovde thesis are also tabulated in the SSRC Guide (Galambos 1998). They are not reproduced here, but a simple accurate approximation that was developed for the Canadian steel design standard S16 (Galambos 1998, equation 3.15), is given as the following equation:

$$\frac{\sigma_{cr}}{\sigma_y} = \left(1 + \lambda^{2n}\right)^{\frac{-1}{n}} \tag{3.105}$$

where λ is the slenderness parameter defined by equation 3.85. The exponent $n = 2.24$ for column Curve 1, $n = 1.34$ for Column Curve 2, and $n = 1.00$ for Column Curve 3. The three column curves are shown in Figure 3.28.

Studies similar to those performed by Bjorhovde were also executed in Europe, and these researchers came up with similar curves and equations, except that the mesh of possibilities was refined to give five column categories. Details of these curves and the corresponding equations are also presented in the SSRC Guide. (Galambos 1998)

The modern column formulas used by the world's structural design standards are all based on essentially the same data and the same computational methodology, yet they employ remarkably different formulas, as seen in Table 3.2.

The diversity of formulas derives from the past history of the standard development in the different parts of the world. In practice, the designer uses tables or spreadsheet applications. Despite the different looking formulas, the resulting column strength curves are remarkably similar, as seen in Figure 3.29 that shows the variation of the design strength $\phi \frac{\sigma_{cr}}{\sigma_y} = \phi \frac{P_{cr}}{P_y}$ with the slenderness parameter λ, where ϕ is the code-specified resistance factor.

TABLE 3.1 Structural Stability Council Column Categories (Figure three.27, Galambos, 1998.)

Column Curve Selection Table

Fabrication Details		Axis	Specified Minimum Yield Stress of Steel (ksi)				
			$\leq$36	37 to 49	50 to 59	60 to 89	$\geq$90
Hot-rolled W-shapes	Light and medium W-shapes	Major	2	2	1	1	1
		Minor	2	2	2	1	1
	Heavy W-shapes (flange over 2 in.)	Major	3	2	2	2	2
		Minor	3	3	2	2	2
Welded Built-up H-shapes	Flame-cut plates	Major	2	2	2	1	1
		Minor	2	2	2	2	1
	Universal mill plates	Major	3	3	2	2	2
		Minor	3	3	3	2	2
Welded Box shapes	Flame-cut and universal mill plates	Major	2	2	2	1	1
		Minor	2	2	2	1	1
Square and rectangular tubes	Cold-formed	Major	N/A	2	2	2	2
		Minor	N/A	2	2	2	2
	Hot-formed and Cold-formed heat-treated	Major	1	1	1	1	1
		Minor	1	1	1	1	1
Circular tubes	Cold-formed	N/A	2	2	2	2	2
	Hot-formed	N/A	1	1	1	1	1
All stress-relieved shapes		Major and Minor	1	1	1	1	1

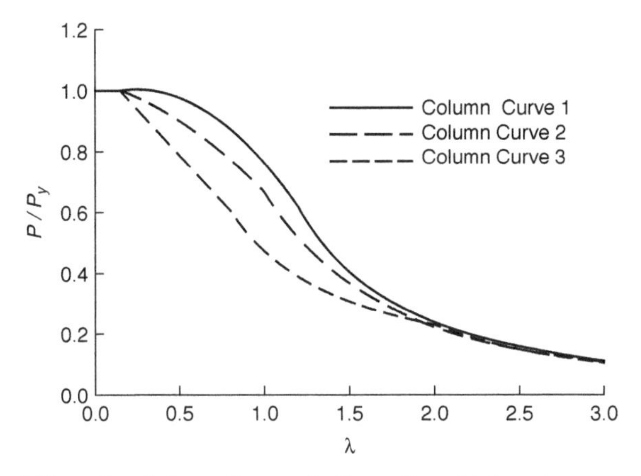

Fig. 3.28 The Column Research Council multiple column curves.

TABLE 3.2 Modern Column Formulas

<table>
<tr><th>Code Authority</th><th>Resistance Factor ϕ</th><th>Column formula $\frac{P_{cr}}{P_y}$</th><th>Comments</th></tr>
<tr><td>AISC, AISI, AASHTOUSA</td><td>0.9</td><td>0.658^{λ^2}</td><td>$\lambda \leq 1.5$</td></tr>
<tr><td></td><td></td><td>$\frac{0.877}{\lambda^2}$</td><td>$\lambda > 1.5$</td></tr>
<tr><td>CSA, CANADA, SOUTH AFRICA</td><td>0.9</td><td>$\left[1+\lambda^{2n}\right]^{\frac{-1}{n}}$</td><td><table><tr><th>SSRC Curve</th><th>n</th></tr><tr><td>I</td><td>2.24</td></tr><tr><td>II</td><td>1.34</td></tr><tr><td>II</td><td>1.00</td></tr></table></td></tr>
<tr><td>EC, EUROPE</td><td>0.909</td><td>$\frac{1}{Q+\sqrt{Q^2-\lambda^2}} \leq 1.0$</td><td>$Q = 0.5\left[1+\alpha(\lambda-0.2)+\lambda^2\right]$<table><tr><th>European Column Curve</th><th>α</th></tr><tr><td>a</td><td>0.21</td></tr><tr><td>b</td><td>0.34</td></tr><tr><td>c</td><td>0.49</td></tr><tr><td>d</td><td>0.76</td></tr></table></td></tr>
<tr><td>AS, AUSTRALIA</td><td>0.9</td><td>$\xi\left[1-\sqrt{1-\left(\frac{90}{\xi\bar{\lambda}}\right)^2}\right]$</td><td>$\bar{\lambda} = \pi\lambda\sqrt{800}$
$\eta = 0.00326(\bar{\lambda}-13.5) \geq 0$
$\xi = \frac{(\bar{\lambda}/90)^2+1+\eta}{2(\bar{\lambda}/90)^2}$</td></tr>
<tr><td>AIJ, JAPAN</td><td>0.9</td><td>1.0</td><td>$\lambda \leq 0.15$</td></tr>
<tr><td></td><td>$0.9-0.05\left[\frac{\lambda-0.15}{\frac{1}{\sqrt{0.6}}-0.15}\right]$</td><td>$1.0-0.5\left[\frac{\lambda-0.15}{\frac{1}{\sqrt{0.6}}-0.15}\right]$</td><td>$0.15 < \lambda \leq \frac{1}{\sqrt{0.6}}$</td></tr>
<tr><td></td><td>0.85</td><td>$\frac{1.0}{1.2\lambda^2}$</td><td>$\lambda \leq \frac{1}{\sqrt{0.6}}$</td></tr>
</table>

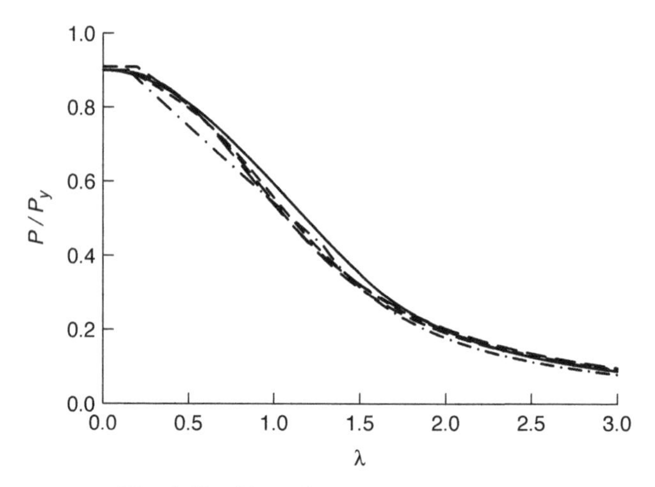

Fig. 3.29 The column curves compared.

The curves are for column category 2, or its equivalent in the various column standards.

The column design procedures just described pertain to columns of hot-rolled structural steel shapes. Residual stresses and initial curvatures were included in the development of the formulas. In contrast, U.S. design standards for aluminum columns and stainless steel columns assume ideally straight columns and use either the explicit (stainless steel) or the indirect application of the Ramberg–Osgood tangent modulus.

In concluding this chapter, it is interesting to note that despite the extensive knowledge of just about everything on the behavior of metal columns, design engineers use simple curve-fit equations that have a close resemblance to the empirical formulas used by designers 150 years ago.

3.8 SUMMARY

This chapter presents three models of column inelastic behavior: the tangent modulus, reduced modulus, and Shanley models. It discusses the derivation and limitations of each. In addition, the effects of inelastic behavior, which can derive from load conditions, residual stresses or both, and column out-of-straightness are taken into account when considering overall column strength. The development of the current U.S. column strength curve, as well as other code-based design curves, is presented to establish how the theoretical impacts on column strength are implemented in practice.

PROBLEMS

3.1. Plot the tangent modulus and reduced modulus column curves for a column of solid circular cross-section of radius R. Use the Ramberg–Osgood stress–strain relationship for which the tangent modulus ratio is expressed by the following equation:

$$\frac{E_t}{E} = \tau = \frac{1}{1 + 60\left(\frac{\sigma}{\sigma_y}\right)^{19}}$$

Formulas for cross-section properties:

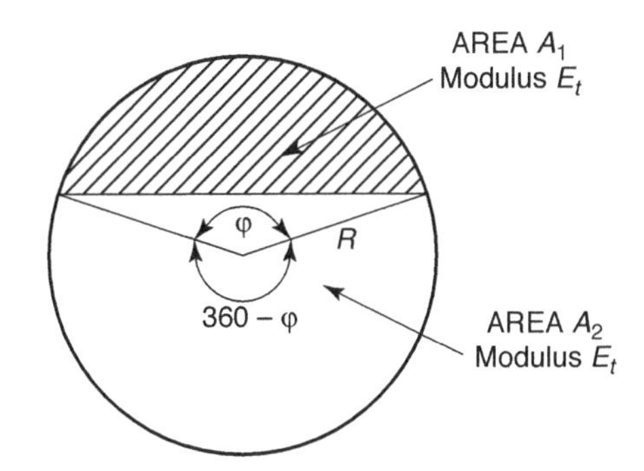

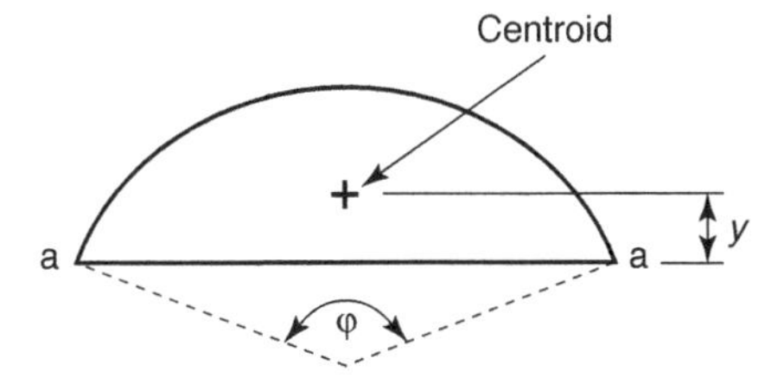

$$A = \frac{R^2}{2}(\varphi - sin\ \varphi)$$

$$y = \frac{2}{3}\left[\frac{R^3\ sin^3\ (\varphi/2)}{A}\right] - R\ cos\ (\varphi/2)$$

$$I_m = Ay^2 + \frac{R^4}{16}(2\varphi - sin\ 2\varphi) - R^4\left[\frac{(1 - cos\ \varphi)^3}{9(\varphi - sin\ \varphi}\right]$$

Fig. P3.1

3.2. Derive equations for the tangent modulus and the reduced modulus column strength for a pinned-end column of length L and rectangular cross section $b \times d$, where $b > d$. The material stress-strain curve follows a Ramberg–Osgood model according to the following equation:

$$\text{Strain: } \varepsilon = \frac{\sigma}{E} + \frac{3\sigma_y}{7E}\left(\frac{\sigma}{\sigma_y}\right)^{10}$$

Plot the column curve for the tangent modulus and the reduced modulus theory on the same plot. Use as the ordinate the nondimensional strength ratio $\frac{\sigma}{\sigma_y} = \frac{bd\sigma}{bd\sigma_y} = \frac{P}{P_y} = p$ and as the abscissa the nondimensional slenderness parameter $\lambda = \frac{L}{\pi r}\sqrt{\frac{\sigma_y}{E}}$. Also draw separately the stress-strain (p versus ε) and the tangent modulus (p versus τ) curves.

$$\varepsilon_y = \frac{\sigma_y}{E}$$

$$\text{Hint: } \frac{d\varepsilon}{d\sigma} = \frac{1}{E}\left[1 + \frac{30p^9}{7}\right]$$

$$\tau = \frac{d\sigma}{d\varepsilon} = \frac{1}{1 + \frac{30p^9}{7}}$$

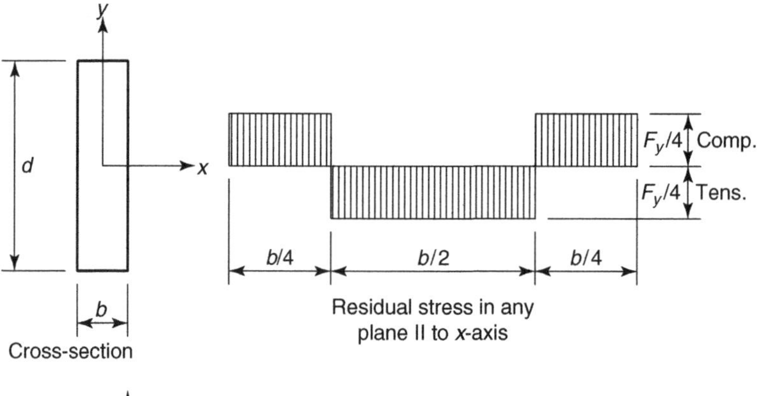

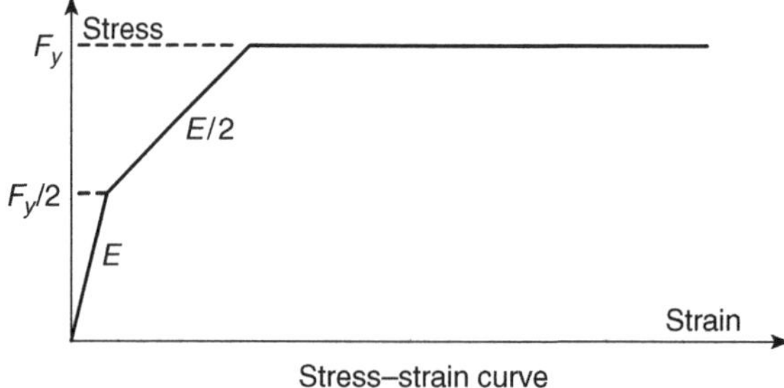

Fig. P3.3

3.3. For the cross-section, residual stress pattern and the stress–strain curve shown, determine the $\frac{P}{P_y} = \frac{P}{bdF_y}$ versus $\lambda = \frac{L}{r}\frac{1}{\pi}\sqrt{\frac{F_y}{E}}$ relationships and the ranges of application of these equations. Buckling is about the y-axis or the x-axis. The ends of the column of length L are pinned. In the inelastic range, calculate the tangent modulus loads only.

3.4.

a. Derive the equations for the reduced modulus theory for the problem in Figures 3.10 and 3.11.

b. Derive equations in Figure 3.17 and redraw Figure 3.20.

c. Derive the equations in Figure 3.23.

3.5. Calculate ϕP_{cr} for a 15 foot-long $W14X109(F_y = 50\,\text{ksi})$ pin-ended column buckling about its y-axis, using the five national column formulas in Table 3.2.

3.6. Determine the Tangent Modulus and Reduced Modulus column curve of a simply supported axially loaded column for buckling about both the x-axis and the y-axis

$$\left(\frac{L}{r}\text{vs.}\frac{P}{P_y}\right) \quad (P_y = A\sigma_y).$$

There are no residual stresses.

$$\text{Given:} \quad \frac{(d-2t)w}{bt} = 1.0 \quad \frac{t}{d} = 0.05$$

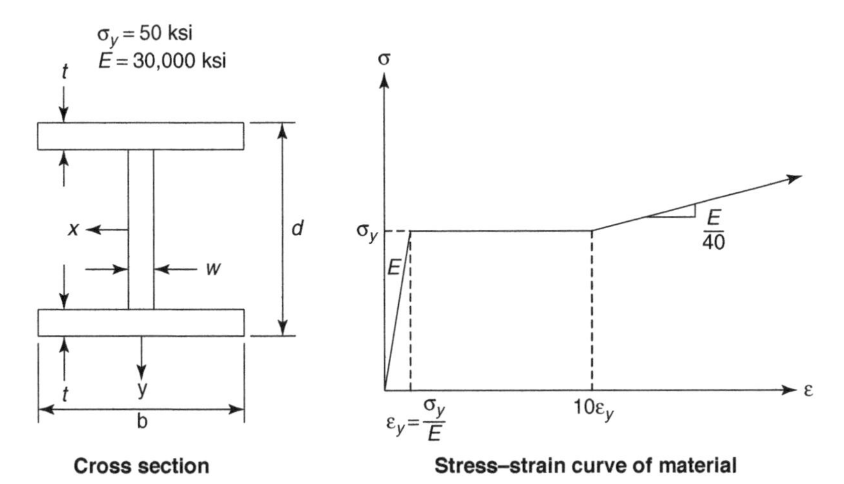

Fig. P3.6

CHAPTER FOUR

BEAM-COLUMNS

4.1 INTRODUCTION

The first three chapters discussed members to which only axial loads were applied. Although some moment may have been introduced by eccentric loads, it was presumed that the axial load was dominant and dictated the primary behavior of the member. Chapters 2 and 3 covered the stability behavior of individual columns, and, to a lesser extent, of frames and trusses that contained axially loaded compression elements. In Chapter 2 the columns were assumed to be elastic, while Chapter 3 expanded the treatment to inelastic members. This chapter considers members that are subjected to both axial load and bending moment referred to as beam-columns.

In a practical sense all members in a metal frame or truss are beam-columns. Axial force, bending moment, shear force and torque are always present, either by plan or accidentally. For the cases discussed in this chapter, it is assumed that neither the moment nor the axial load dominates to the extent that the other can be neglected (as is the case with simply supported beams or columns).

This chapter presents the theory of the behavior of elastic beam-columns, followed by the behavior of inelastic beam-columns. In the treatment of elastic behavior expressions for deformations and forces in beam-columns are developed. Interaction relationships between axial force, bending moment, and member slenderness are presented for the limit states of stability and first yield. Problems of more complex loading are demonstrated by using the principle of superposition. Statically indeterminate frames and trusses are also solved. Flexibility and stiffness approaches, systematic methods of analyzing structures known to most structural engineers, are derived and used to demonstrate efficient ways of stability analysis.

Metal beam-columns generally reach their ultimate strength after considerable yielding has taken place. Interaction relationships are developed for rectangular steel beam-columns to illustrate the principles of behavior. Interaction curves and equations are then be presented for practical shapes.

The last part of the chapter presents the approaches used in the world's structural design specifications. For the sake of ease in understanding, only in-plane behavior will be discussed in the chapter.

4.2 GENERAL DISCUSSION OF THE BEHAVIOR OF BEAM-COLUMNS

As previously stated, most frame or truss members are subjected to a number of types of forces; the extent to which those forces affect the overall member behavior dictates the classification of the member and how it is analyzed and designed. Here, we deal with idealizations where the forces are predominantly due to bending and compression. Beam-columns are important components of rigid frames (Figure 4.1) and of trusses (Figure 4.2). In laterally braced frames (as illustrated by the bottom story in the structure of Figure 4.1) the top of the column remains in the same position, that is, it does not move laterally with respect to its bottom. The top story of the frame in Figure 4.1 illustrates that the column-top displaces laterally when there is no bracing that prevents this motion. This distinction is important

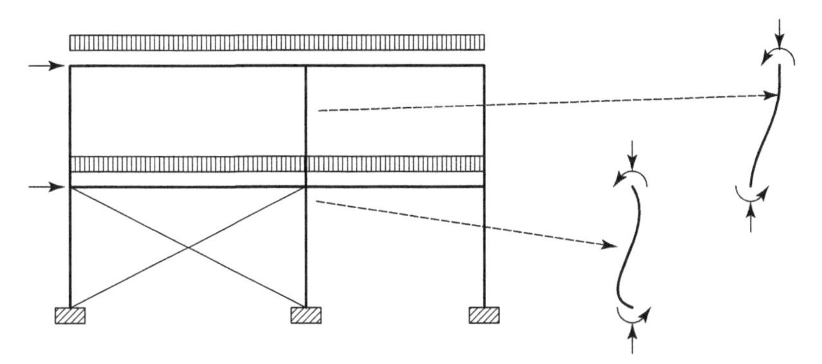

Fig. 4.1 Beam-columns in rigid frame.

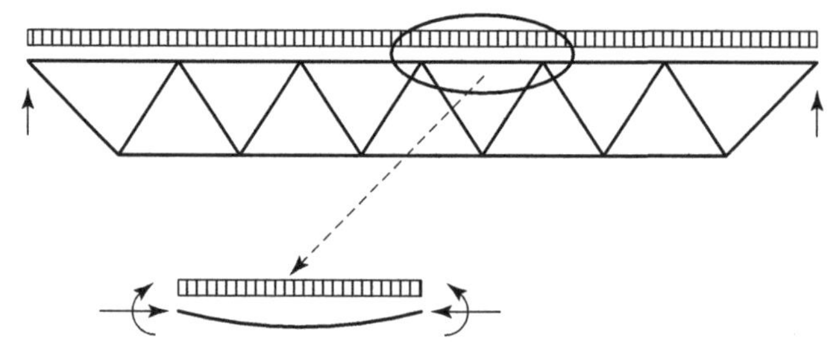

Fig. 4.2 Beam-column in a truss.

for the investigation of the stability of frames, as will be shown later in the chapter. In the case of beam-columns in trusses (Figure 4.2), it is usually assumed the ends of the member remain in the same position for the purposes of a stability investigation. There are many other structural situations where beam-columns exist, or where the complicated real condition is idealized as such a member.

The behavior of beam-columns is different from that of beams or columns. On the one hand, the axial load is smaller than the maximum force that can be carried by a column, and thus there is a reserve of capacity to carry some bending moment. On the other hand, the moment that can be supported is less than the plastic moment that could be supported if there were no axial force present.

The bending moment inside the span of the beam-column is composed of the *primary moment*, or often called the *first-order moment*, and the additional moment caused by the product of the axial force and the deflection. This is the *second-order moment*. It plays an increasingly important role as the deflection increases due to an increase in load. The relationship between the applied loads and the resulting deflections is nonlinear if both axial force and bending moment increase, even if the material remains elastic. This nonlinearity becomes even more pronounced once a portion of the beam-column yields. The bending moment demands a larger and larger proportion of the flexural capacity, until the internal stiffness cannot keep up with the demand of the external forces. Thus, a maximum moment is reached. Beyond the deformation at the maximum moment, the moment capacity is reduced. In case of pure gravity loading, such as under snow load, the attainment of the maximum capacity will result in immediate collapse of the member, and perhaps the whole structure. The moment-versus-end slope behavior of a planar beam-column is illustrated in Figure 4.3.

The solid-line curve in Figure 4.3 represents the in-plane behavior of a member that is subjected first to an axial force P, less than the critical load discussed in the previous two chapters. Subsequently, a bending moment M_o

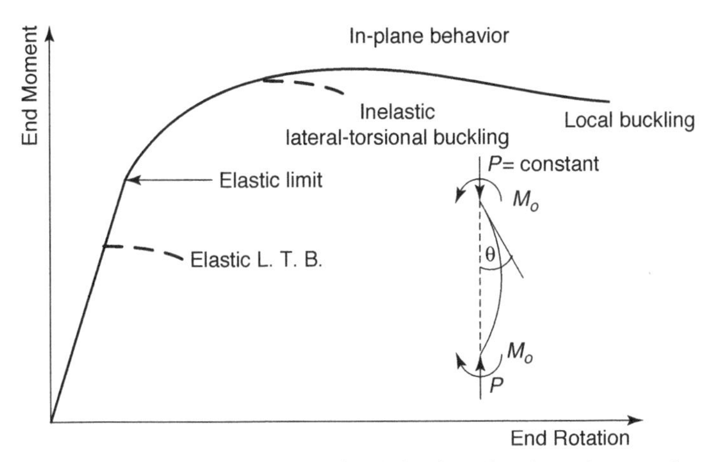

Fig. 4.3 Moment versus end-rotation behavior of a planar beam-column.

is applied to each end of the member, resulting in a monotonically increasing end slope θ. This order of application of forces (rather than an increase of the loads at the same time) is called *nonproportional loading*. The in-plane moment-rotation curve is linear (if the axial force remains constant) until the most stressed fiber in the member begins to exceed the yield strain of the material. The stiffness begins to degrade as the rotation is continued, first gradually, then more and more until the stiffness is zero as the peak moment capacity is attained. Further deformation can be sustained only as the moment is reduced. Eventually the most yielded part of the member will experience local buckling, resulting in an even steeper descent toward collapse. If the beam-column is not braced laterally, then there is a possibility of lateral-torsional buckling (discussed in Chapter 6), either in the elastic range of behavior or in the inelastic range, thus resulting in a reduced maximum capacity.

The behavior is further illustrated in Figure 4.4. The curve in this figure represents an experimentally obtained (Van Kuren and Galambos 1964) moment rotation curve. In the experiment a steel wide-flange beam-column was first subjected to a concentric axial force $P = 0.49P_y$. While P was kept constant, a moment M_o was applied about the major axis of the member at one end. This moment was applied through rotating the end of the beam-column by monotonically increasing the rotation θ_o. The measured relationship between M_o and θ_o is the curve in Figure 4.4. Lateral bracing was provided along the length of the member to curb deformation out of the plane of bending; therefore, the test illustrates in-plane bending only, with failure by lateral-torsional buckling prevented.

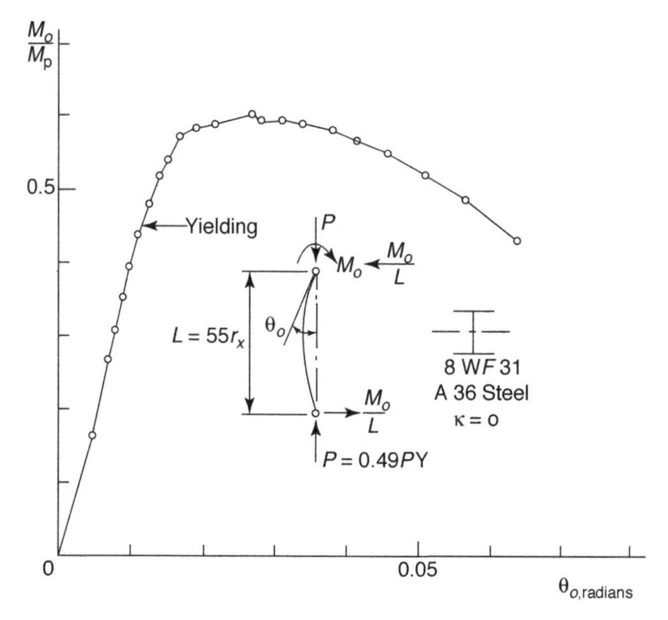

Fig. 4.4 Experimental moment-end rotation curve for a beam-column.

4.3 ELASTIC IN-PLANE BEHAVIOR OF BEAM-COLUMNS

The first case considered is a prismatic member of length L that is subjected at its ends by moments M_o and κM_o, where κ is the ratio of the two end moments. The range of values of κ is $-1 \leq \kappa \leq +1$, where the value $+1$ represents the case where the two end moments cause single curvature deflection under uniform moment, as shown in Figure 4.5.

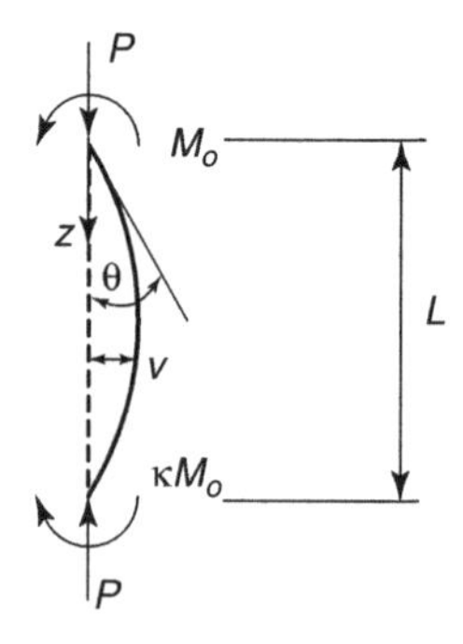

Fig. 4.5 Beam-column with end moments.

The differential equation of in-plane behavior is given by equation 2.9 with $\alpha = q = 0$, that is, there is no foundation modulus and no distributed load:

$$EIv^{iv} + Pv'' = 0$$

As in Chapter 2, this formula is rearranged by dividing each term by EI to get the following equation:

$$v^{iv} + k^2 v'' = 0$$

where

$$k^2 = \frac{P}{EI}$$

The solution of the differential equation is

$$v = A + Bz + C \sin kz + D \cos kz$$

and the constants of integration are defined by the following boundary conditions:

$$v(0) = v(L) = 0$$

$$v''(0) = -\frac{M_o}{EI}; \quad v''(L) = -\frac{\kappa M_o}{EI}$$

By solving for the constants, the following formulas are derived for the deflection and the moment, respectively:

$$v(z) = \frac{M_o}{P}\left\{\left[\frac{\kappa - \cos kL}{\sin kL}\right] \sin kz + \cos kz + \frac{z}{L}(1 - \kappa) - 1\right\} \tag{4.1}$$

$$M(z) = -EIv''(z) = M_o\left\{\left[\frac{\kappa - \cos kL}{\sin kL}\right] \sin kz + \cos kz\right\} \tag{4.2}$$

Two types of moment distribution along the length of the beam-column are possible from equation 4.2:

1. The moment is maximum inside the span.
2. The moment is maximum at the end.

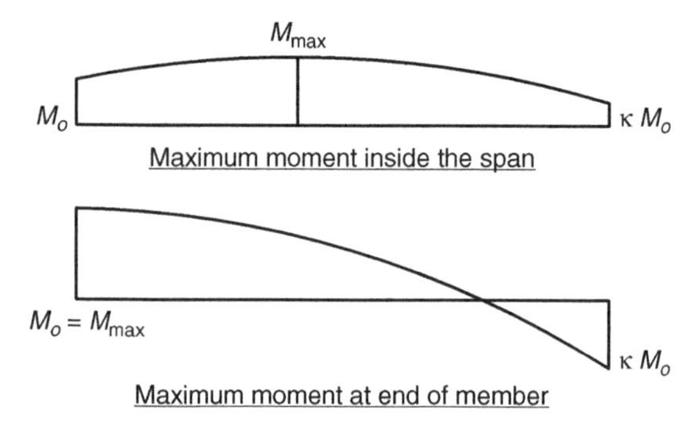

Fig. 4.6 Possible moment diagrams.

These two situations are illustrated in Figure 4.6.

The location of the maximum moment along the z-axis is determined by taking the derivative of the moment (equation 4.2) with respect to z, setting this expression equal to zero, and solving for the location $\bar{z}$ of the maximum moment.

$$\frac{dM}{dz} = 0 = M_o k \left\{ \left[\frac{\kappa - \cos kL}{\sin kL} \right] \cos k\bar{z} - \sin k\bar{z} \right\}$$

$$\tan k\bar{z} = \left[\frac{\kappa - \cos kL}{\sin kL} \right]$$

The sine and cosine functions are obtained from the triangle relations depicted in Figure 4.7.

The maximum moment is then obtained by substituting $z = \bar{z}$ into the moment equation. After some algebra, the resulting equation is equation 4.3.

$$M_{max} = M_o \left\{ \frac{\sqrt{1 + \kappa^2 - 2\kappa \cos kL}}{\sin kL} \right\} \tag{4.3}$$

When $\bar{z} = 0$ the value of $\cos kL = \kappa$. The maximum moment of the beam-column with end moments is, finally, equal to

$$\begin{aligned} M_{max} &= M_o \quad \text{if } \kappa \leq \cos kL \\ M_{max} &= \varphi M_o \quad \text{if } \kappa \geq \cos kL \end{aligned} \tag{4.4}$$

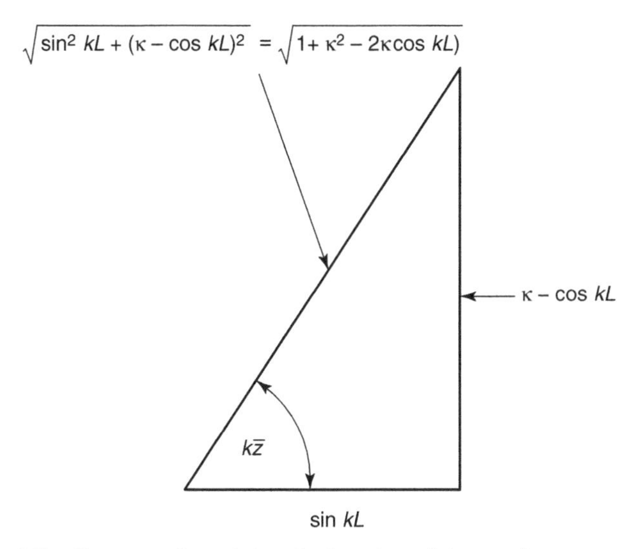

Fig. 4.7 Geometry for solving the location of the maximum moment.

where

$$\varphi = \frac{\sqrt{1+\kappa^2 - 2\kappa\cos kL}}{\sin kL} \tag{4.5}$$

As we mentioned before, the moment ratio is in the range $-1 \leq \kappa \leq 1$. Maximum moment multipliers φ are listed for three other loading conditions in Table 4.1. The first column depicts the structure, while the second column gives the formula for the maximum moment obtained from statics for the cases when the axial force is zero. The third column is the formula for the moment magnification factor. This factor is dependent on the axial load and the loading condition. The axial load effect is defined by

$$kL = \sqrt{\frac{PL^2}{EI}} = \sqrt{\frac{P}{P_E} \cdot \frac{\pi^2 EI}{L^2} \cdot \frac{L^2}{EI}} = \pi\sqrt{\frac{P}{P_E}} \tag{4.6}$$

Each of the moment amplification factors approaches infinity as the axial load approaches the Euler load. This signifies that it will not be possible to exceed the elastic buckling load of a pinned-end column unless additional restraints are added. Such a case was already discussed in Chapter 2 in connection with the effects of *initial imperfections* (see Figure 2.8). The curves of φ versus P/P_E for the four cases in Table 4.1 are plotted in Figure 4.8.

TABLE 4.1 Moment Amplification Factors

Loading case	Maximum First-order Moment	Moment Amplification Factor
M_o, κM_o	M_o	$\varphi = 1.0 \quad \text{if} \quad \kappa \le \cos kL$ $\varphi = \dfrac{\sqrt{1+\kappa^2-2\kappa\cos kL}}{\sin kL} \quad \text{if} \quad \kappa \ge \cos kL$
q	$M_o = \dfrac{qL^2}{8}$	$\varphi = \dfrac{8\left(1-\cos\dfrac{kL}{2}\right)}{(kL)^2\cos\dfrac{kL}{2}}$
q_o	$M_o = \dfrac{q_oL^2}{9\sqrt{3}}$	$\varphi = \dfrac{9\sqrt{3}}{(kL)^2}\left[\dfrac{\sqrt{(kL)^2-\sin^2 kL}}{kL\sin kL} - \dfrac{\arccos\dfrac{\sin kL}{kL}}{kL}\right]$
$L/2$, Q, $L/2$	$M_o = \dfrac{QL}{4}$	$\varphi = \dfrac{2\tan\dfrac{kL}{2}}{kL}$

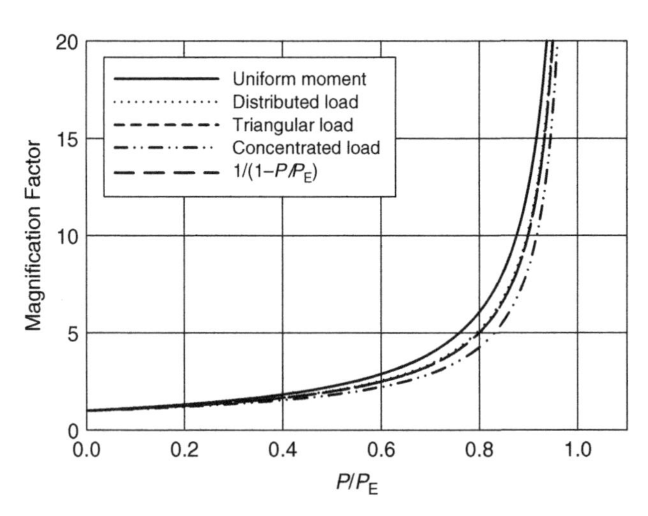

Fig. 4.8 Relationship between axial load and moment magnifiers for various loading conditions.

It is obvious that the four amplification factors do not differ from each other a great deal. For practical purposes, the four amplification factors are adequately represented by the much simpler formula

$$\varphi = \frac{1}{1 - P/P_E} \tag{4.7}$$

This formula appears to be intuitive, and it was probably so conceived in the first beam-column interaction equations used in design specifications, but it also has a mathematical significance, as discussed next.

Assume that the deflected shape of the beam-column is represented by the series of n sine shapes with amplitudes a_n, where $n = 1, 2, 3, \ldots$

$$v = \sum_1^n a_n \sin \frac{n\pi z}{L} \tag{4.8}$$

The first and second derivatives of v are equal to

$$v' = \sum_1^n \frac{n\pi a_n}{L} \cos \frac{n\pi z}{L}$$

$$-v'' = \sum_1^n \frac{n^2\pi^2 a_n}{L^2} \sin \frac{n\pi z}{L}$$

The strain energy of bending is equal to

$$U = \frac{1}{2}\int_0^L EI(v'')^2 dz = \frac{EI\pi^4}{2L^4} \cdot \frac{L}{2}\sum_1^n a_n^2 n^4$$

because

$$\int_0^L \sin^2 \frac{n\pi z}{L} dz = \frac{L}{2}$$

$$\int_0^L \sin \frac{n\pi z}{L} \cdot \sin \frac{m\pi z}{L} dz = 0, n \neq m$$

The strain energy is thus given by:

$$U = \frac{\pi^4 EI}{4L^3} \sum_1^n a_n^2 n^4$$

The potential energy of the end-moment is equal to the product of the applied end-moment and the end-rotation:

$$V_{P-M_o} = -M_o \cdot v'(0) + M_o \cdot v'(L)$$

$$v'(0) = \frac{\pi}{L} \sum_1^n a_n n; \quad v'(L) = \frac{\pi}{L} \sum_1^n a_n n \cdot (-1)^n$$

$$V_{P-M_o} = \frac{-\pi M_o}{L} \sum_1^n a_n n [1 - (-1)^n]$$

The potential energy of the axial force is the product of the axial force and the shortening of the member:

$$V_{P-P} = \frac{-P}{2} \int_0^L (v')^2 dz = \frac{-P}{2} \cdot \frac{\pi^2}{L^2} \cdot \frac{L}{2} \sum_1^n a_n^2 n^2 = -\frac{\pi^2 P}{4L} \sum_1^n a_n^2 n^2$$

because

$$\int_0^L \cos^2 \frac{n\pi z}{L} dz = \frac{L}{2}$$

$$\int_0^L \cos \frac{n\pi z}{L} \cdot \cos \frac{m\pi z}{L} dz = 0, \quad n \neq m$$

The total potential is the sum of the strain energy and the potential energy of the applied loads:

$$\begin{aligned} \Pi &= U + V_P \\ &= \frac{\pi^4 EI}{4L^3} \sum_1^n a_n^2 n^4 - \frac{\pi M_o}{L} \sum_1^n a_n n \cdot [1 - (-1)^n] - \frac{\pi^2 P}{4L} \sum_1^n a_n^2 n^2 \end{aligned} \tag{4.9}$$

According to the *principle of virtual work*, the derivative of the total potential with respect to any coefficient a_i defines equilibrium for that shape component:

$$\frac{\partial \Pi}{\partial a_i} = 0 = \frac{\pi^4 EI}{4L^3}(2a_i i^4) - \frac{\pi M_o}{L}\{i[1-(-1)^i]\} - \frac{\pi^2 P}{4L}(2a_i i^2)$$

Solving for a_i

$$a_i = \frac{2M_o L^2[1-(-1)^i]}{\pi^3 EIi\left(i^2 - \frac{PL^2}{\pi^2 EI}\right)} = \frac{2M_o L^2[1-(-1)^i]}{\pi^3 EIi\left(i^2 - \frac{P}{P_E}\right)}$$

The deflection is, therefore

$$v = \frac{2M_o L^2}{\pi^3 EI}\sum_1^n \frac{1-(-1)^n}{n\left(n^2 - \frac{P}{P_E}\right)} \cdot \sin\frac{n\pi z}{L} \tag{4.10}$$

The maximum deflection occurs at $z = L/2$:

$$v_{\max} = \frac{2M_o L^2}{\pi^3 EI}\sum_1^n \frac{1-(-1)^n}{n\left(n^2 - \frac{P}{P_E}\right)} \tag{4.11}$$

The maximum moment also occurs at $z = L/2$ (see Figure 4.9), and it is equal to the following:

$$M_{\max} = M_o + Pv_{\max}$$

$$M_{\max} = M_o\left[1 + \frac{2PL^2}{\pi^3 EI}\sum_1^n \frac{(1-(-1)^n)\sin\frac{n\pi}{2}}{n\left(n^2 - \frac{P}{P_E}\right)}\right]$$

$$= M_o\left[1 + \frac{2P}{\pi P_E}\sum_1^n \frac{[1-(-1)^n]\sin\frac{n\pi}{2}}{n\left(n^2 - \frac{P}{P_E}\right)}\right]$$

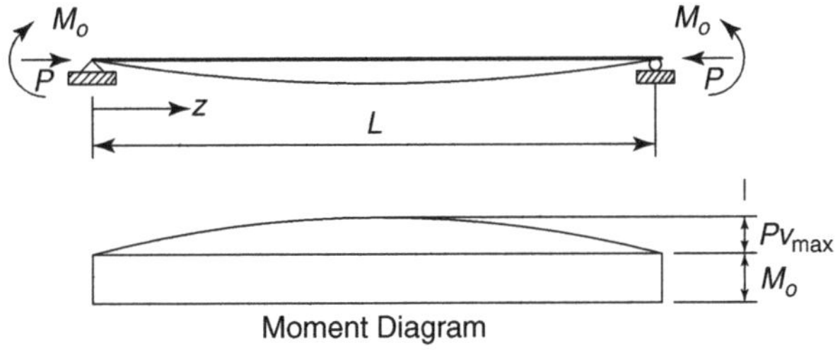

Fig. 4.9 Beam-column under uniform end-moment.

$$M_{\max} = M_o \varphi_{\text{approx}}$$

$$\varphi_{\text{approx}} = 1 + \frac{2P}{\pi P_{\text{E}}} \sum_{1}^{n} \frac{[1-(-1)^n]\sin\frac{n\pi}{2}}{n\left(n^2 - \frac{P}{P_{\text{E}}}\right)}$$

The approximate amplification factor is equal to zero if n is even. Thus

$$\varphi_{\text{approx}} = 1 + \frac{4}{\pi} \cdot \frac{P}{P_{\text{E}}} \left[\frac{1}{1 - \frac{P}{P_{\text{E}}}} - \frac{1}{3\left(9 - \frac{P}{P_{\text{E}}}\right)} + \frac{1}{5\left(25 - \frac{P}{P_{\text{E}}}\right)} - \cdots \right] \tag{4.12}$$

The analytically exact value of the moment amplification factor is given by equation 4.5. For the case of uniform bending, this becomes equal to

$$\varphi = \frac{\sqrt{2(1 - \cos kL)}}{\sin kL} \tag{4.13}$$

Table 4.2 compares the amplification factors. The values for $n = 5$ are very close to the exact ones. The usual approximation of $1/(1 - P/P_{\text{E}})$ is adequate, however, for most design applications.

It is interesting to note that in the worst case, where $P/P_{\text{E}} = 0.8$, the error with the approximation is over 17 percent. However, at this point, even a small moment will be amplified to the extent that the interaction between axial load and moment will cause failure of the beam-column. Application of a safety factor to this instance in practical cases makes this level of axial load in a beam-column unobtainable.

TABLE 4.2 Magnification Factors Compared

P/P_{E}	Exact	$1/(1 - P/P_{\text{E}})$	$n = 1$	$n = 3$	$n = 5$
0.2	1.310	1.250	1.318	1.309	1.311
0.4	1.832	1.667	1.849	1.849	1.833
0.6	2.884	2.500	2.910	2.880	2.886
0.8	6.058	5.000	6.093	6.052	6.060

Summary of Important Points from Sections 4.1–4.3

- Beam columns are members that resist both axial load and moment. They can resist neither the critical load nor the plastic moment that the member could resist if only axial load or moment were applied.
- The application of the axial load on the deflected member causes a second order moment. The maximum moment can be obtained by multiplying the first order (applied) moment by a moment amplifier. In double curvature bending, the maximum moment is sometimes equal to the applied end moment. In transversely loaded members without applied end moments, the maximum moment will occur within the span of the beam.

4.4 ELASTIC LIMIT INTERACTION RELATIONSHIPS

In the previous section we discussed the amplification of elastic moments and deflections in prismatic beam-columns due to the presence of a compressive axial force. Here, we determine the interaction between axial and flexural loads at the limit when the sum of the elastic stresses due to axial force and bending moment equals the yield stress, as expressed in the following equation:

$$\sigma = \frac{P}{A} + \frac{M_{\max}}{S} = F_y \tag{4.14}$$

In equation 4.14 σ is the normal stress, P is the axial force, and A is the cross-sectional area. $M_{\max}$ is the maximum moment on the member, S is the elastic section modulus, and F_y is the yield stress. (This notation will be used here rather than the symbol σ_y.) If the equation is divided by F_y, and introducing the cross-section limit forces $P_y = AF_y$ (the *squash load*) and $M_y = SF_y$ (the *yield moment*), also noting that the maximum moment equals $M_{\max} = \varphi M_o$, then the following interaction equation is obtained:

$$\frac{P}{P_y} + \varphi \frac{M_o}{M_y} = 1.0 \tag{4.15}$$

In the strict sense this equation applies to the case of beam-columns with end moments. However, it is also generally a good approximation as long as the amplification factor considers the loading, such as the values listed in Table 4.1 for three other cases. The interaction equation is conventionally

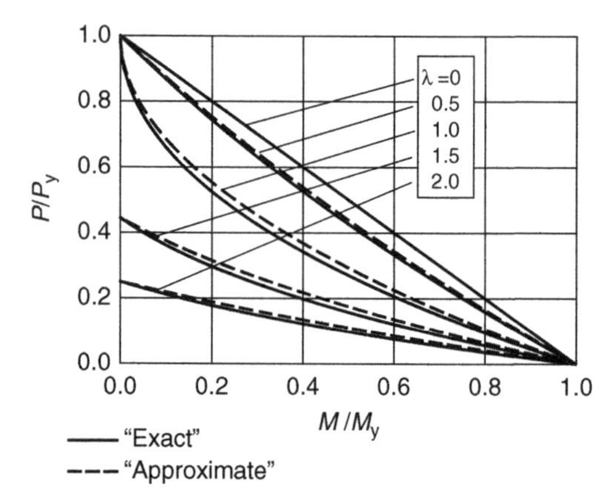

Fig. 4.10a Exact and approximate interaction equations compared for beam-columns under uniform moment ($\kappa = 1.0$).

plotted with the moment ratio as the abscissa and the axial ratio as the ordinate, as illustrated in Figure 4.10.

The curves in Figure 4.10a are for the case of a beam-column under equal end-moments. The *exact* amplification factor is equation 4.13, and the *approximate* amplification factor is equation 4.7. The comparison of the solid and the dashed lines illustrates that the difference between the two is not significant, so the simpler equation can be confidently used in everyday design practice. The symbol λ is a length parameter. The larger the value of λ, the less capacity the beam-column has. The formula for λ is nondimensional, and it is also universal for different material strengths and stiffnesses.

$$\lambda = \frac{1}{\pi} \cdot \frac{L}{r} \cdot \sqrt{\frac{F_y}{E}} \tag{4.16}$$

For example, if $\lambda = 1.0$, $F_y = 50\,\text{ksi}$ and $E = 29{,}000\,\text{ksi}$, then the slenderness ratio $L/r = 76$.

When the axial load is zero, the beam-column is actually a beam and it can support the full yield moment. When the bending moment is zero, the axial capacity will either equal the squash load P_y or the Euler buckling load P_E, whichever is smaller. The nondimensional buckling load is

$$\frac{P_E}{P_y} = \frac{\pi^2 EI}{L^2} \cdot \frac{1}{AF_y} = \frac{\pi^2 E}{F_y\left(\frac{L}{r}\right)^2} = \frac{1}{\lambda^2} \tag{4.17}$$

The curves in Figure 4.10b show the effect of moment gradient on the yield limit of beam-columns having a slenderness ratio $L/r = 100$ and end-moment

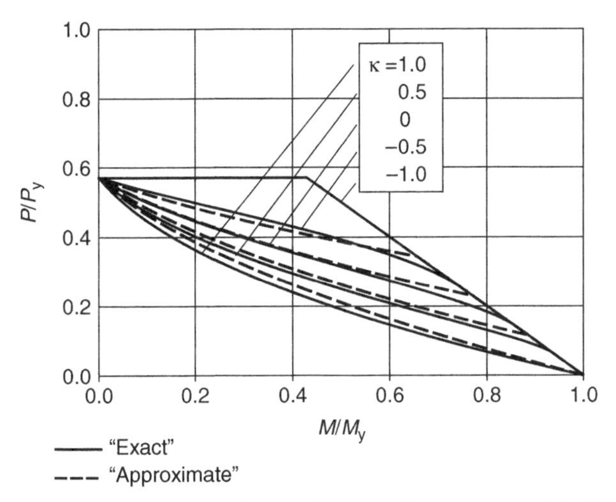

Fig. 4.10b Exact and approximate interaction equations compared for beam-columns of slenderness $\lambda = 1.322$ ($L/r = 100$ for $F_y = 50$ ksi).

ratios varying from +1 (equal end-moments causing single curvature deflection) to −1 (equal end-moments causing an S-shaped deflection curve). The trend here is that the member becomes stronger as the moment ratio moves from single-curvature to double-curvature bending. The approximate curves, shown as dashed lines, are in good agreement with the *exact* curves. These latter curves were computed using the amplification factor of equation 4.3. The approximate amplification factor was determined empirically by Austin (1961). The formula is the following:

$$\varphi_{\text{approx}} = \frac{C_m}{1 - \frac{P}{P_E}} \tag{4.18}$$

$$C_m = 0.6 + 0.4\kappa \geq 0.4$$

4.5 EXAMPLE PROBLEMS OF BEAM-COLUMN STRENGTH

Three problems are presented that further illustrate the use of the differential equation approach to solving different loading cases:

1. Example 1 is the solution of a pinned-end beam-column subjected to a linearly varying distributed load.

2. Example 2 presents a statically indeterminate member, namely, a propped cantilever with a uniformly distributed load.
3. Example 3 illustrates the principle of superposition for a beam-column with multiple loads.

These examples are presented prior to the following sections, where systematic methods for structural analysis are discussed.

4.5.1 Example 4.1: Amplification factor for Linearly Varying Load

In this example, we derive the formula for the moment amplification factor for a pinned-end beam-column with triangular loading shown in Figure 4.11.

The distributed load: $q(z) = \frac{q_o z}{L}$

The differential equation:

$$EIv^{iv} + Pv'' = q(z) = \frac{q_o z}{L}$$

With $k^2 = \frac{P}{EI}$ the differential equation is

$$v^{iv} + k^2 v'' = \frac{q_o k^2 z}{PL}$$

The homogeneous solution is

$$v_H = A + Bz + C \sin kz + D \cos kz$$

The particular solution is

$$v_P = C_1 + C_2 z + C_3 z^2 + C_4 z^3$$
$$v_P'' = 2C_3 + 6C_4 z$$

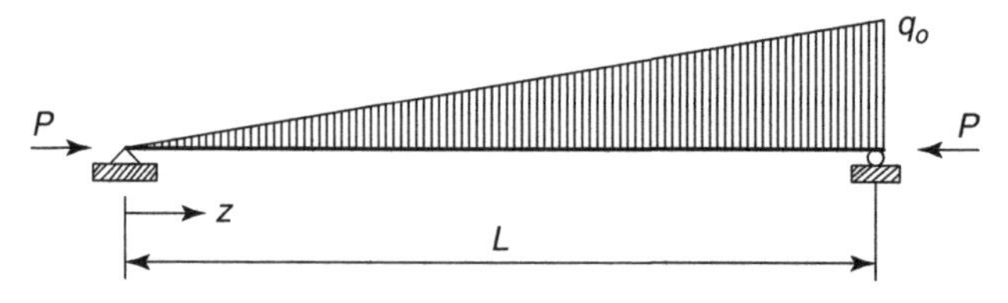

Fig. 4.11 Pinned-end beam-column with triangular load.

$$v_P^{iv} = 0$$

$$v_P^{iv} + k^2 v_P'' = 0 + k^2(2C_3 + 6C_4 z) = \frac{q_o k^2 z}{PL}$$

$$C_3 = 0; \quad C_4 = \frac{q_o}{6PL}$$

$$v_P = \frac{q_o z^3}{6PL}$$

The deflection therefore equals

$$v = A + Bz + C \cdot \sin kz + D \cdot \cos kz + \frac{q_o z^3}{6PL}$$

The boundary conditions are

$$v(0) = v(L) = v''(0) = v''(L) = 0$$

Substitution of the boundary conditions leads to the following formulas for the deflection and the bending moment:

$$v(z) = -\frac{q_o L}{P}\left(\frac{1}{6} + \frac{1}{(kL)^2}\right) z + \frac{q_o \sin kz}{Pk^2 \sin kL} + \frac{q_o z^3}{6PL}$$

$$M(z) = -EIv'' = \frac{q_o}{k^2}\left(\frac{\sin kz}{\sin kL} - \frac{z}{L}\right)$$

The maximum moment occurs at the location $z = \bar{z}$, where $\frac{dM}{dz} = 0$. After differentiation and rearrangement, the following relationships result:

$$\cos k\bar{z} = \frac{\sin kL}{kL}$$

$$k\bar{z} = \arccos\left(\frac{\sin kL}{kL}\right)$$

$$\sin k\bar{z} = \frac{\sqrt{(kL)^2 - \sin^2 kL}}{kL}$$

$$M_{\max} = M(\bar{z}) = \frac{q_o}{k^2}\left[\frac{\sqrt{(kL)^2 - \sin^2 kL}}{kL \sin kL} - \frac{1}{kL}\arccos\left(\frac{\sin kL}{kL}\right)\right]$$

The first-order maximum moment—that is, the maximum moment when $P = 0$ can be looked up in the AISC Manual, and it is equal to $\frac{q_o L^2}{9\sqrt{3}}$. The second-order maximum moment, that is, $P \neq 0$, is equal to

$$M_{\max} = \frac{q_o L^2}{9\sqrt{3}} \varphi$$

where the moment magnification factor is given by

$$\varphi = \frac{9\sqrt{3}}{(kL)^2} \left[\frac{\sqrt{(kL)^2 - \sin^2 kL}}{kL \sin kL} - \frac{1}{kL} \arccos\left(\frac{\sin kL}{kL} \right) \right]$$

An approximation is derived in the following manner from the first-order maximum moment and the maximum deflection:

$$M_{\max} = \frac{q_o L^2}{9\sqrt{3}} + \frac{P v_{\max}^{(1)}}{1 - \frac{P}{P_E}}$$

$$v_{\max}^{(1)} = 0.00652 \frac{q_o L^4}{EI}$$

$$M_{\max} = \frac{q_o L^2}{9\sqrt{3}} \left[1 + \frac{9\sqrt{3}}{q_o L^2} \cdot \frac{P}{1 - \frac{P}{P_E}} \cdot 0.00652 \frac{q_o L^4}{EI} \cdot \frac{\pi^2 EI}{L^2} \cdot \frac{1}{P_E} \right]$$

$$= \frac{1 + 0.003 \frac{P}{P_E}}{1 - \frac{P}{P_E}} \cong \frac{1}{1 - \frac{P}{P_E}}$$

$$\varphi_{\text{approximate}} = \frac{1}{1 - \frac{P}{P_E}}$$

From this problem we learn that the approximate moment amplification factor is appropriate for the beam-column with a linearly varying load. This was already demonstrated in connection with Figure 4.8. Thus, the interaction curves of Figure 4.10 also apply for the triangular load pattern, as well as for the other loadings in Table 4.1, as long as the first-order maximum moments listed in that table are used in the elastic limit interaction equation (equation 4.15).

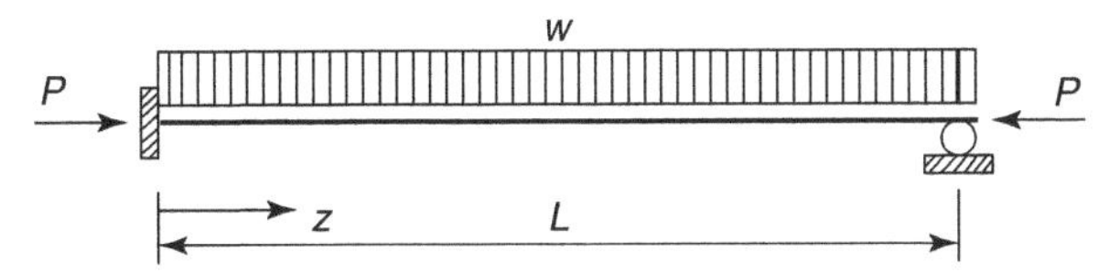

Fig. 4.12 Uniformly loaded propped cantilever.

4.5.2 Example 4.2: Propped Cantilever

Example 4.2 describes the derivation of expressions for the moment diagram of a propped cantilever subjected to a uniformly distributed load. Figure 4.12 illustrates the example problem; Figure 4.13 and Figure 4.14 break the problem into the simply supported case and end-moment case, respectively.

Step 1: Uniformly Distributed Load Case. The deflection equals

$$v(z) = \frac{wL^4}{EI(kL)^4}\left\{\left(\frac{1-\cos kL}{\sin kL}\right)\sin kz + \cos kz - 1 + \frac{(kL)^2}{2}\left[\left(\frac{z}{L}\right)^2 - \frac{z}{L}\right]\right\}$$

$$(kL)^2 = \frac{PL^2}{EI} = \pi^2 \cdot \frac{P}{P_E}$$

Let $b = \dfrac{1-\cos kL}{\sin kL}$

$$v'(z) = \frac{wL^4}{EI(kL)^4}\left\{b \cdot k \cdot \cos kz - k \cdot \sin kz + \frac{(kL)^2}{2}\left[\frac{2z}{L^2} - \frac{1}{L}\right]\right\}$$

The slope at the left support is

$$v'(0) = \frac{wL^3}{EI \cdot (kL)^3}\left[b - \frac{kL}{2}\right]$$

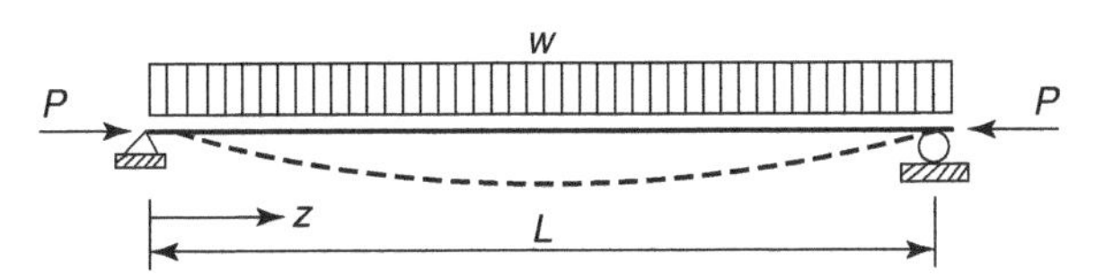

Fig. 4.13 Uniformly distributed load case.

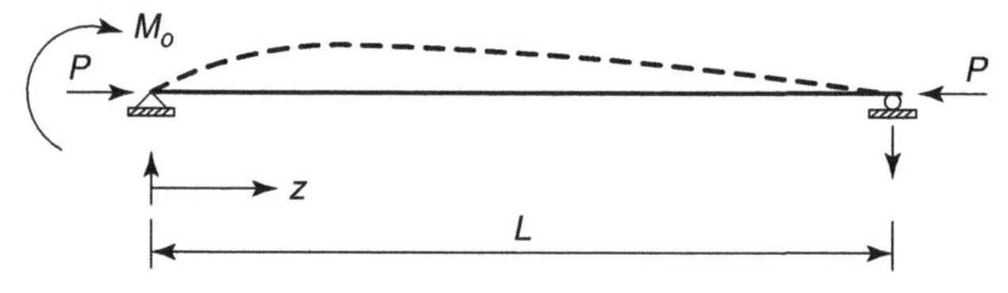

Fig. 4.14 End-moment case.

Step 2: End-Moment Case. The deflection equals

$$v(z) = \frac{M_o}{P}\left(\cos kz - \frac{\sin kz}{\tan kL} + \frac{z}{L} - 1\right)$$

$$v'(z) = \frac{M_o}{P}\left(-k \cdot \sin kz - \frac{k \cdot \cos kz}{\tan kL} + \frac{1}{L}\right)$$

The slope at the left support is

$$v'(0) = \frac{M_o}{PL}\left(1 - \frac{kL}{\tan kL}\right)$$

For a fixed-end condition, the sum of the two slopes is zero.

$$\frac{wL^3}{EI \cdot (kL)^3}\left[b - \frac{kL}{2}\right] = \frac{M_o}{PL}\left(1 - \frac{kL}{\tan kL}\right)$$

Solving for M_o

$$\frac{M_o}{wL} = \frac{-1 + \cos kL + \frac{kL}{2}\sin kL}{kL(\sin kL - kL \cdot \cos kL)} = a$$

Moment at any location z is computed as follows:

From the Distributed Load:

$$\frac{M_w(z)}{wL^2} = -\frac{EI}{wL^2} \cdot v''(z) = \frac{1 - b \cdot \sin kz - \cos kz}{(kL)^2}$$

From the Fixed-End-Moment:

$$\frac{M(z)}{wL^2} = a\left(\cos kz - \frac{\sin kz}{\tan kL}\right)$$

The total moment is then

$$\frac{M(z)}{wL^2} = -a\left(\cos kz - \frac{\sin kz}{\tan kL}\right) + \frac{-1 + b \cdot \sin kz + \cos kz}{(kL)^2}$$

This is the analytically exact solution for the second-order moment diagram. The fixed-end-moment approximation given for this example in the Commentary to the AISC Specification (2005) is

$$M_o^{\mathrm{F}} = \frac{wL^2}{8}\left(\frac{1 - 0.4\frac{P}{P_{\mathrm{E}}}}{1 - \frac{P}{P_{\mathrm{E}}}}\right)$$

The total approximate second-order moment diagram is therefore equal to

$$\left[\frac{M(z)}{wL^2}\right]_{\mathrm{Approx}} = \left\{\frac{1 - 0.4\frac{P}{P_{\mathrm{E}}}}{1 - \frac{P}{P_{\mathrm{E}}}}\right\}\left[\frac{1}{8}\left(\frac{z}{L} - 1\right) + \frac{1}{2}\left(\frac{z}{L} - \frac{z^2}{L^2}\right)\right]$$

The moment diagram, that is, the moment at any location z along the member, is shown in Figure 4.15a for the exact solution. The nine curves represent the effect of the axial load as it varies from 0 to $0.9P_{\mathrm{E}}$. The comparison between the exact (solid line) and the approximate (dashed line) moment diagrams for $P = 0.2P_{\mathrm{E}}$, $0.4P_{\mathrm{E}}$ and $0.6P_{\mathrm{E}}$ is given in Figure 4.15b. The approximation is seen to give larger values of the moment. Thus, the approximation is conservative. The simpler approximate method is acceptably accurate for the design office application.

4.5.3 Example 4.3: Superposition of Several Load Effects

In Example 4.3, superposition is used for a beam with multiple transverse loads shown in Figure 4.16. A beam-column is subjected to an axial force P at its ends. In addition, there are two concentrated loads and a uniformly distributed load acting transversely on the member. There is also a moment applied at the left end of the member. The geometry and loading are shown later in this chapter. Superposition requires that the total moments are the

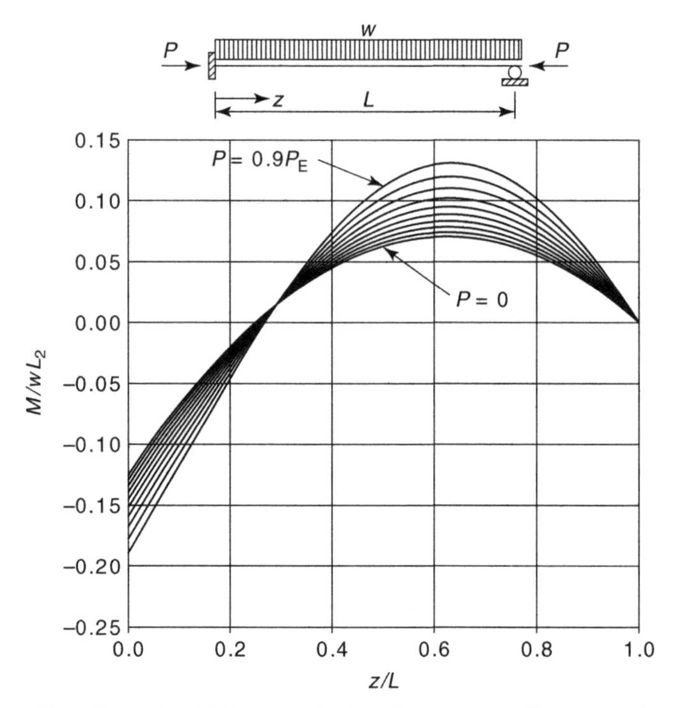

Fig. 4.15a The effect of axial force on the bending moment diagram, using the "exact" method.

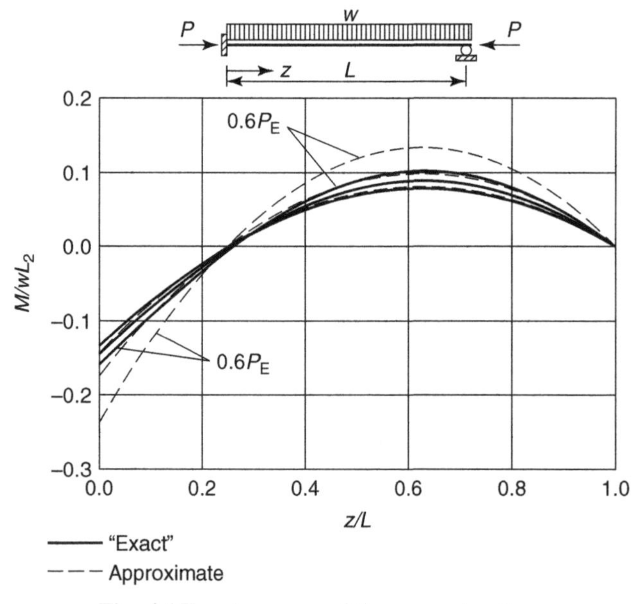

Fig. 4.15b Accuracy of the approximation.

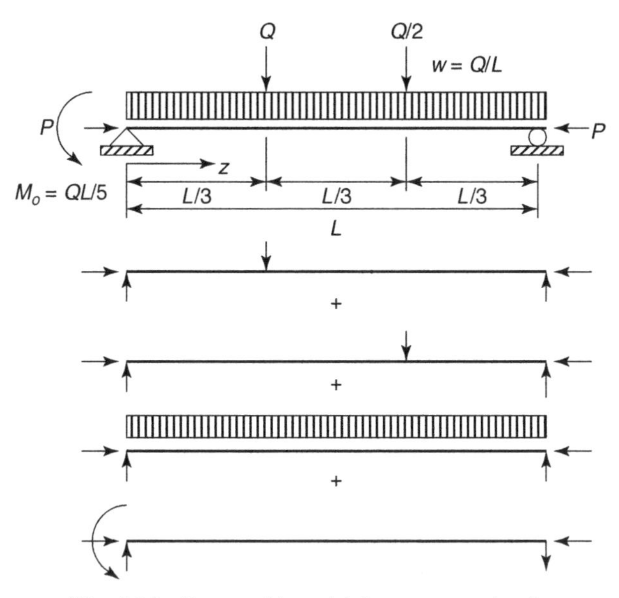

Fig. 4.16 Beam with multiple transverse loads.

sum of the moments from the four separate loading cases shown in the sketches below the picture of the beam.

A W14 × 64 section is bent about its x-axis. We determine the value Q such that the maximum normal stress is equal to the yield stress.

$$F_y = 50\,\text{ksi} \qquad r_x = 5.98\,\text{in.} \qquad i = 0 \ldots 120$$

$$E = 29{,}000\,\text{ksi} \qquad A = 17.9\,\text{in.}^2 \qquad L = 360\,\text{in.}$$

$$S_x = 92.2\,\text{in.}^2 \qquad I_x = 640\,\text{in.}^2 \qquad P = 700\,\text{kip}$$

$$z_i = \frac{i}{120}$$

$$kL = \sqrt{\frac{PL^2}{EI_x}}$$

$$kL = 2.211$$

$$Q = 1\,\text{kip} \qquad w = \frac{Q}{L} \qquad M_o = \frac{QL}{5}$$

Q Acting to the Left of Center

$$M_{1_i} = \begin{vmatrix} \dfrac{QL}{kL}\left(\dfrac{\sin\left(kL - \frac{kL}{3}\right)}{\sin(kL)}\sin(kL\,z_i)\right) & \text{if } i \leq 40 \\ \dfrac{QL}{kL}\left[\dfrac{\sin\left(\frac{kL}{3}\right)}{\sin(kL)}\sin[kL(1 - z_i)]\right] & \text{otherwise} \end{vmatrix}$$

Q/2 Acting to the Right of Center

$$M_{2_i} = \begin{vmatrix} \dfrac{QL}{2kL}\left(\dfrac{\sin\left(kL - \frac{2kL}{3}\right)}{\sin(kL)}\sin(kL\,z_i)\right) & \text{if } i \leq 80 \\ \dfrac{QL}{2kL}\left[\dfrac{\sin\left(\frac{2kL}{3}\right)}{\sin(kL)}\sin[kL(1 - z_i)]\right] & \text{otherwise} \end{vmatrix}$$

Distributed Load

$$M_{3_i} = \frac{wL^2}{kL^2}\left(\frac{1 - \cos(kL)}{\sin(kL)}\sin(kL\,z_i) + \cos(kL\,z_i) - 1\right)$$

Moment at Left End

$$M_{4_i} = M_o\left(\cos(kL\,z_i) - \frac{\cos(kL)\sin(kL\,z_i)}{\sin(kL)}\right)$$

Total Moment

$$M_i = M_{1_i} + M_{2_i} + M_{3_i} - M_{4_i}$$

$$M_{\max} = \max(M) \qquad M_{\max} = 193.317\,\text{kip-in}$$

$$Q_T = \left(F_y - \frac{P}{A}\right)\frac{S_x}{M_{\max}}$$

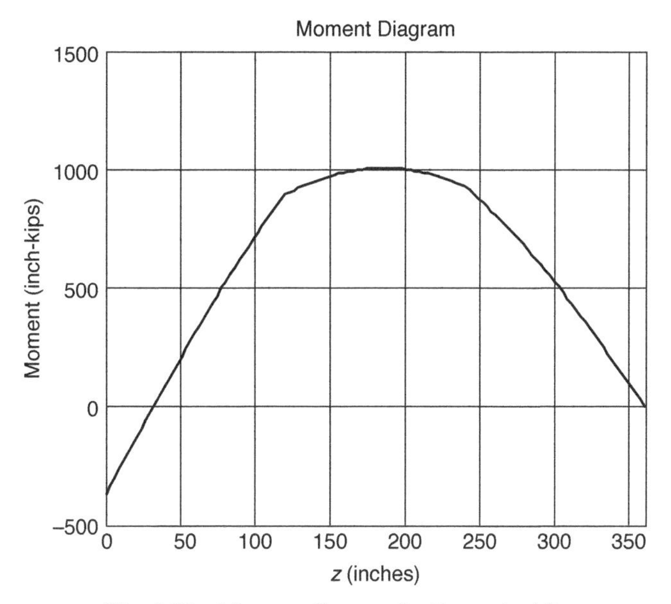

Fig. 4.17 Moment diagram for Example 4.3.

$$Q_T = 5.196\,\text{kip}$$

The value of Q to cause a stress equal to the yield stress is 5.196 kips. The maximum moment equals $193.317 \times 5.196 = 1{,}004$ kip-inches. The moment diagram is shown in Figure 4.17.

4.6 SYSTEMATIC METHODS OF ANALYSIS: FLEXIBILITY METHOD

Results from the theory presented in the previous parts of this chapter are now used to develop systematic methods of structural analysis. These will be familiar to anyone who has mastered the flexibility and the stiffness methods of linear structural analysis. To practicing engineers the two methods are known as the *three-moment equations* and the *slope-deflection methods*. We first develop the three-moment equation method. This type of analysis uses the compatibility of slopes at common joints in the frame to develop simultaneous equations for the calculation of the indeterminate moments.

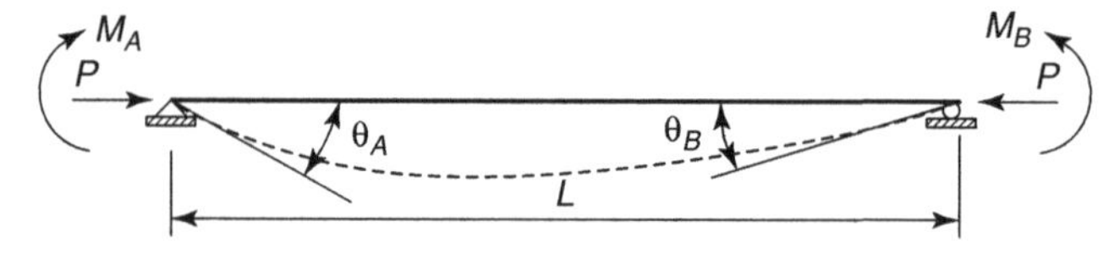

Fig. 4.18 Three-moment equation method sign conventions.

The beam-column in Figure 4.18 shows the notations and sign-conventions. The equations for the end-slopes are equal to

$$\theta_A = \frac{M_A L}{3EI} \cdot \frac{3}{(kL)^2}\left[1 - \frac{kL \cdot \cos kL}{\sin kL}\right] + \frac{M_B L}{6EI} \cdot \frac{6}{(kL)^2}\left[-1 + \frac{kL}{\sin kL}\right]$$

$$\theta_B = \frac{M_A L}{6EI} \cdot \frac{6}{(kL)^2}\left[-1 + \frac{kL}{\sin kL}\right] + \frac{M_B L}{3EI} \cdot \frac{3}{(kL)^2}\left[1 - \frac{kL \cdot \cos kL}{\sin kL}\right]$$

Introducing the functions c and s to simplify the equations, where

$$c = \frac{1}{(kL)^2}\left[1 - \frac{kL}{\tan kL}\right] \tag{4.19}$$

$$s = \frac{1}{(kL)^2}\left[\frac{kL}{\sin kL} - 1\right] \tag{4.20}$$

The end-slopes due to the end-moments are, therefore, equal to

$$\theta_A = \frac{L}{EI}(c \cdot M_A + s \cdot M_B) \tag{4.21}$$

$$\theta_B = \frac{L}{EI}(s \cdot M_A + c \cdot M_B) \tag{4.22}$$

When the axial force is zero,

$$kL \rightarrow 0, \quad P \rightarrow 0, \quad c = \frac{1}{3}, \quad s = \frac{1}{6}$$

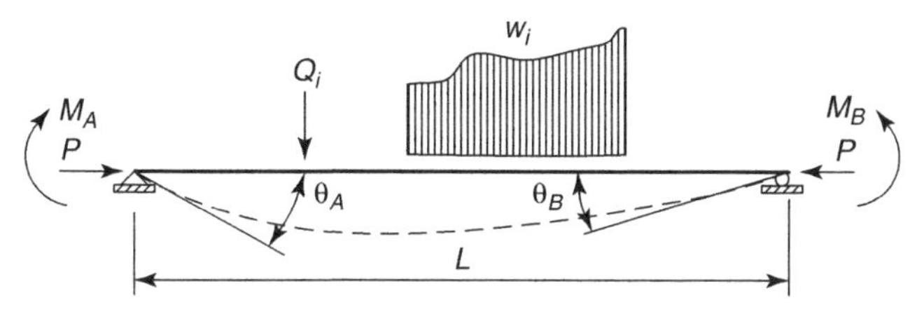

Fig. 4.19 General loading on the beam-column.

resulting in the familiar expressions of linear structural analysis. It should be pointed out that it takes some mathematical manipulations using L'Hospital's theorem to arrive from equations 4.19 and 4.20 to the previous values. Numerically, it is easy to show this, however, by solving for c and s using $kL = 0.01$.

If there are transverse loads on the span (Figure 4.19), the equations for the end-slopes are

$$\theta_A = \frac{L}{EI}(c \cdot M_A + s \cdot M_B) + \theta_{oA} \tag{4.23}$$

$$\theta_B = \frac{L}{EI}(s \cdot M_A + c \cdot M_B) + \theta_{oB} \tag{4.24}$$

The slopes θ_{oA} and θ_{oB} are due to the loads within the span. Table 4.3 gives a list of some of the more frequently occurring load types. The slopes shown are positive. Two consecutive spans are shown in Figure 4.20.

Slope compatibility at the common joint requires that $\theta_{n-1} = \theta_n$, where

$$\theta_{n-1} = \theta_{o(n-1)}^{\text{right}} + \frac{L_{n-1}}{EI_{n-1}}(s_{n-1}M_{n-1} + c_{n-1}M_n)$$

$$\theta_n = \theta_{o(n)}^{\text{left}} + \frac{L_n}{EI_n}(c_nM_n + s_nM_{n+1})$$

The three-moment equation for nonmoving supports (Figure 4.20, top sketch) is, therefore, defined by the following equation:

$$M_{n-1}\left[\frac{L_{n-1}s_{n-1}}{EI_{n-1}}\right] + M_n\left[\frac{L_{n-1}c_{n-1}}{EI_{n-1}} + \frac{L_nc_n}{EI_n}\right] + M_{n+1}\left[\frac{L_ns_n}{EI_n}\right]$$
$$= -\theta_{o(n-1)}^{\text{right}} - \theta_{o(n)}^{\text{left}} \tag{4.25}$$

TABLE 4.3 End Slopes for Transverse Loads

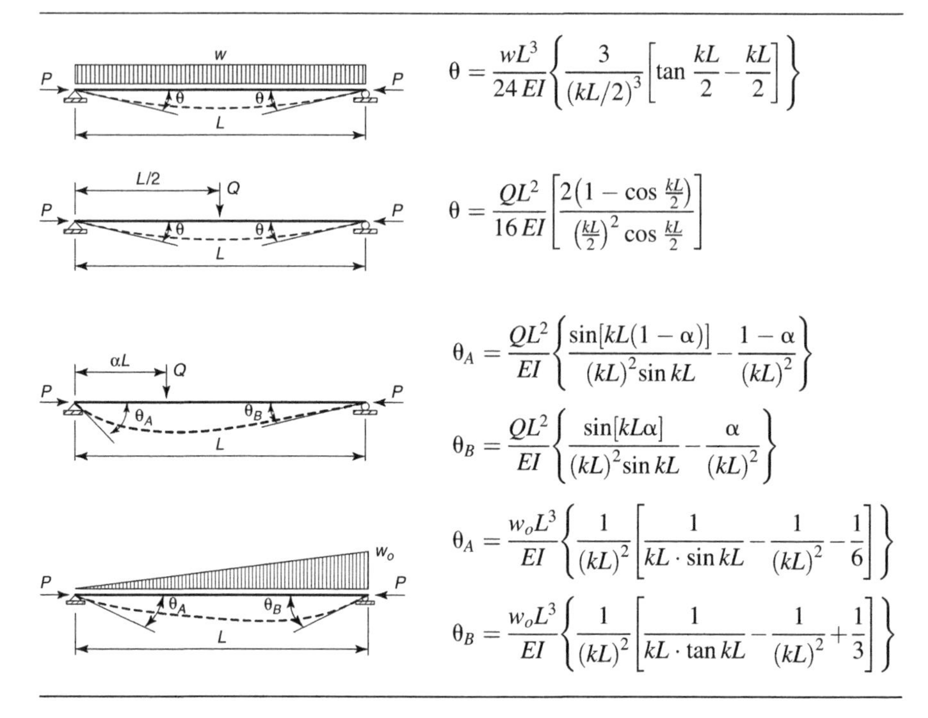

$$\theta = \frac{wL^3}{24\,EI}\left\{\frac{3}{(kL/2)^3}\left[\tan\frac{kL}{2} - \frac{kL}{2}\right]\right\}$$

$$\theta = \frac{QL^2}{16\,EI}\left[\frac{2\left(1-\cos\frac{kL}{2}\right)}{\left(\frac{kL}{2}\right)^2\cos\frac{kL}{2}}\right]$$

$$\theta_A = \frac{QL^2}{EI}\left\{\frac{\sin[kL(1-\alpha)]}{(kL)^2\sin kL} - \frac{1-\alpha}{(kL)^2}\right\}$$

$$\theta_B = \frac{QL^2}{EI}\left\{\frac{\sin[kL\alpha]}{(kL)^2\sin kL} - \frac{\alpha}{(kL)^2}\right\}$$

$$\theta_A = \frac{w_oL^3}{EI}\left\{\frac{1}{(kL)^2}\left[\frac{1}{kL\cdot\sin kL} - \frac{1}{(kL)^2} - \frac{1}{6}\right]\right\}$$

$$\theta_B = \frac{w_oL^3}{EI}\left\{\frac{1}{(kL)^2}\left[\frac{1}{kL\cdot\tan kL} - \frac{1}{(kL)^2} + \frac{1}{3}\right]\right\}$$

For the case of moving supports (Figure 4.21), replace the right side of equation 4.25 by

$$= -\theta_{o(n-1)}^{\text{right}} - \theta_{o(n)}^{\text{left}} + \rho_{n-1} - \rho_n \tag{4.26}$$

Next we illustrate the method of the three-moment equation with two examples.

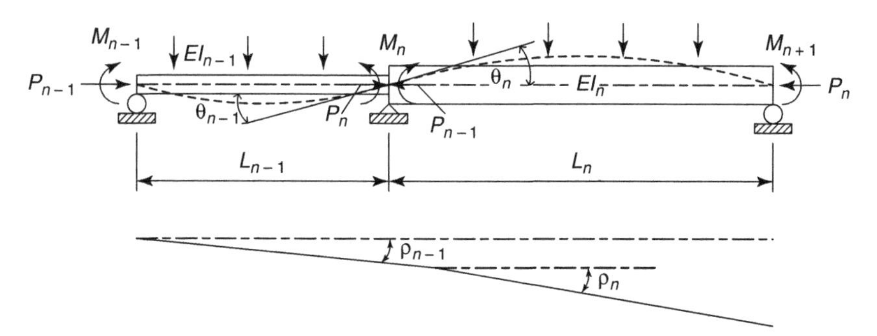

Fig. 4.20 Two adjacent spans for developing the three-moment equation.

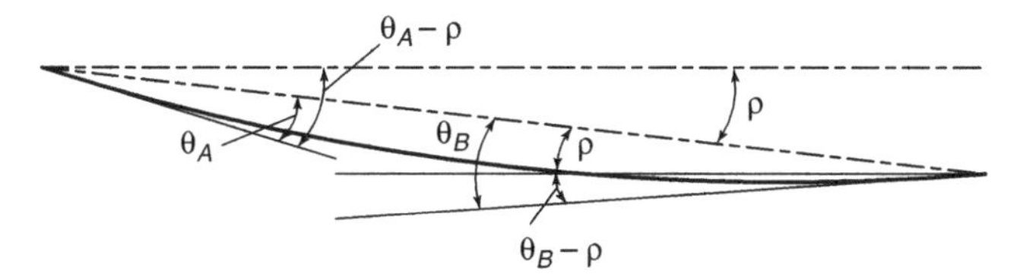

Fig. 4.21 Geometry of moving supports.

4.6.1 Example 4.4: Stability of a Compression Chord in a Truss

A truss example is provided here to demonstrate the application of the three-moment equation. In this example, we determine the elastic buckling load of the top chord of the truss shown in Figure 4.22.

The truss is symmetric about its center. The dimensions of the centers of the cross-sections are shown in the drawing: The depth of the truss is d and the lengths of the top chord panels are $3d$, $2d$, $2d$, $2d$, $3d$, respectively. All the panels of the top chord are in compression. The top chord is continuous and has the same cross-section throughout its length. For the purpose of axial force analysis, it is appropriate to assume that the panel points act as pins and that they do not translate. By a statical analysis, the axial forces in the panels of the top chords can be shown to be equal to the values shown in the lower part of Figure 4.22. The buckled shape of the top chord is also shown in this figure. Because of the symmetry of the axial forces about the center, the buckled shape is symmetrical, and therefore the panel point moments are also symmetrical. That is, $M_1 = M_6 = 0,\ M_2 = M_5$, and $M_3 = M_4$.

For the purpose of setting up the three-moment equations for panels 1-2-3 and 2-3-4, it is necessary to calculate the stability functions c and s. The following equations apply for the solution:

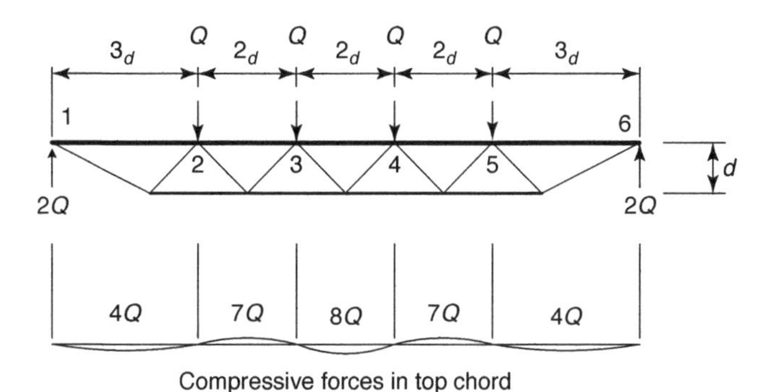

Fig. 4.22 Example truss.

$$kL = \sqrt{\frac{PL^2}{EI}}; \quad c = \frac{1}{(kL)^2}\left[1 - \frac{kL}{\tan kL}\right]; \quad s = \frac{1}{(kL)^2}\left[\frac{kL}{\sin kL} - 1\right]$$

The specific values for the truss top chord are listed in the table. Note that $Z = \sqrt{Qd^2/EI}$.

Panel	P	kL	c	s
1–2	$4Q$	$3d\sqrt{\frac{4Q}{EI}} = 6Z$	$\frac{1}{(6 \cdot Z)^2}\left[1 - \frac{6 \cdot Z}{\tan(6 \cdot Z)}\right]$	$\frac{1}{(6 \cdot Z)^2}\left[\frac{6 \cdot Z}{\sin(6 \cdot Z)} - 1\right]$
2–3	$7Q$	$2d\sqrt{\frac{7Q}{EI}} = 2\sqrt{7} \cdot Z$	$\frac{1}{(2\sqrt{7} \cdot Z)^2}\left[1 - \frac{2\sqrt{7} \cdot Z}{\tan(2\sqrt{7} \cdot Z)}\right]$	$\frac{1}{(2\sqrt{7} \cdot Z)^2}\left[\frac{2\sqrt{7} \cdot Z}{\sin(2\sqrt{7} \cdot Z)} - 1\right]$
3–4	$8Q$	$2d\sqrt{\frac{8Q}{EI}} = 4\sqrt{2} \cdot Z$	$\frac{1}{(4\sqrt{2} \cdot Z)^2}\left[1 - \frac{4\sqrt{2} \cdot Z}{\tan(4\sqrt{2} \cdot Z)}\right]$	$\frac{1}{(4\sqrt{2} \cdot Z)^2}\left[\frac{4\sqrt{2} \cdot Z}{\sin(4\sqrt{2} \cdot Z)} - 1\right]$

The three-moment equations for panels 1-2-3 and 2-3-4, respectively, give the following two simultaneous equations. Since there are no applied loads between the panel points, the right sides of the equations are zero.

$$\textit{Span}\ 1\text{-}2\text{-}3: \quad M_2\left[\frac{3d \cdot c_{1-2}}{EI} + \frac{2d \cdot c_{2-3}}{EI}\right] + M_3\left[\frac{2d \cdot s_{2-3}}{EI}\right] = 0$$

$$\textit{Span}\ 2\text{-}3\text{-}4: \quad M_2\left[\frac{2d \cdot s_{2-3}}{EI}\right] + M_3\left[\frac{2d \cdot c_{2-3}}{EI} + \frac{2d \cdot c_{3-4}}{EI}\right] + M_3\left[\frac{2d \cdot s_{3-4}}{EI}\right] = 0$$

These equations are homogeneous simultaneous equations in terms of the unknown buckling load Q for the truss. The determinant of the coefficients of the two moments M_2 and M_3 are set equal to zero in order to arrive at the value of the unknown.

$$\begin{vmatrix} (3c_{1-2} + 2c_{2-3}) & (2s_{2-3}) \\ (2s_{2-3}) & 2(c_{2-3} + c_{3-4} + s_{3-4}) \end{vmatrix} = 0$$

The solution was obtained numerically using MATHCAD, as follows:

$$Z = \sqrt{\frac{Qd^2}{EI}} = 0.721$$

$$\frac{Qd^2}{EI} = 0.520$$

The most compressed panel is the center panel, 3-4, with an axial load of

$$P_{cr} = 8Q = 4.158\frac{EI}{d^2} = \frac{\pi^2 EI}{(K_{eff} \cdot 2d)^2}$$

$$K_{eff} = \sqrt{\frac{\pi^2}{4 \cdot 4.158}} = 0.77$$

The effective length factor to be used in the design of the top chord is thus 0.77.

4.6.2 Example 4.5: Stability of a Rigid Frame

This example illustrates the application of the three-moment equation to a rigid frame that is subject to a lateral load H and to two equal forces P applied downward at the tops of the columns, as shown in Figure 4.23. In this example we do two things:

1. Determine the elastic buckling load.
2. Calculate the joint moments as a function of the axial loads P.

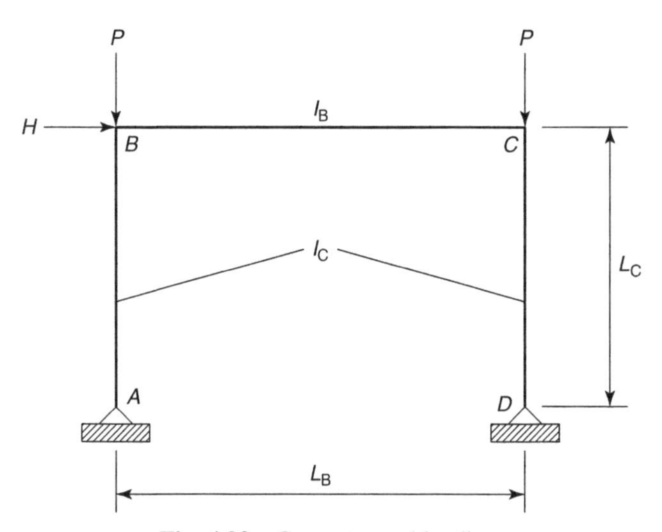

Fig. 4.23 Geometry and loading.

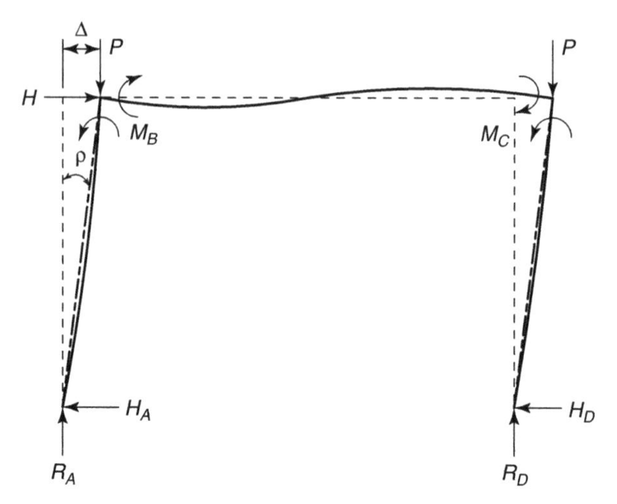

Fig. 4.24 Deformed shape of the frame.

The equilibrium of the external and internal forces is determined for the deflected shape, shown in Figure 4.24. In order to simplify the calculations it will be assumed that the lateral load H is much smaller than the load P. Therefore, the axial force in the horizontal member will be small enough so that, for all practical purposes, the stability functions c and s will be equal to their first-order values: $c_{BC} = 1/3$ and $s_{BC} = 1/6$. We assume that the axial shortening of the members is negligible compared to the flexural deflections, and thus the deflections Δ at the two column tops are equal, and the rotation ρ is the same for both columns.

In order to solve the problem, it is necessary to apply static equilibrium. This leads to the following equations:

$$\text{Summation of vertical forces:} \quad R_A + R_D = 2P$$

$$\text{Summation of horizontal forces:} \; H_A + H_D = H$$

$$\text{Summation of moments about } A: \; HL_C + P\Delta + P(L_B + \Delta) - R_D L_B = 0$$

From these equations we find the following values for the axial loads in the two columns:

$$R_A = P - \frac{HL_C}{L_B} - \frac{2P\Delta}{L_B}$$

$$R_D = P + \frac{HL_C}{L_B} + \frac{2P\Delta}{L_B}$$

Two further steps are necessary:

$$\text{Sum of moments about the top of the left column: } R_A\Delta + H_A L_C - M_B = 0$$

$$\text{Sum of moments about the top of the right column: } R_D\Delta + H_D L_C - M_C = 0$$

Adding these two equations, and noting that $R_A + R_B = 2P,\ H_A + H_B = H$, and $\rho = \frac{\Delta}{L_C}$, the following relationship is obtained for the rotation of the columns:

$$\rho = \frac{M_B + M_C}{2PL_C} - \frac{H}{2P}$$

The three-moment equation can now be written for the two spans *A-B-C* and *B-C-D*. For span *A-B-C*: $M_{n-1} = 0;\ M_n = M_B;\ M_{n+1} = -M_C$, and therefore,

$$M_B\left[\frac{L_C c_{AB}}{EI_C} + \frac{L_B c_{BC}}{EI_B}\right] - M_C\left[\frac{L_B s_{BC}}{EI_B}\right] = \rho = \frac{M_B + M_C}{2PL_C} - \frac{H}{2P}$$

For span $B\text{-}C\text{-}D: M_{n-1} = M_B;\ M_n = -M_C;\ M_{n+1} = 0$, and so

$$M_B\left[\frac{L_B s_{BC}}{EI_B}\right] - M_C\left[\frac{L_B c_{BC}}{EI_B} + \frac{L_C c_{CD}}{EI_C}\right] = -\rho = -\frac{M_B + M_C}{2PL_C} + \frac{H}{2P}$$

These two equations can now be rearranged, with the introduction of these parameters:

$$\gamma = \frac{L_B I_C}{L_C I_B} \quad \text{and} \quad kL_C = \sqrt{\frac{PL_C^2}{EI_C}}$$

Also noting that $c_{BC} = 1/3$ and $s_{BC} = 1/6$, we get the following two simultaneous equations:

$$\begin{bmatrix} \left(c_{AB} + \frac{\gamma}{3} - \frac{1}{2(kL_C)^2}\right) & -\left(\frac{\gamma}{6} + \frac{1}{2(kL_C)^2}\right) \\ \left(\frac{\gamma}{6} + \frac{1}{2(kL_C)^2}\right) & \left(-c_{CD} - \frac{\gamma}{3} + \frac{1}{2(kL_C)^2}\right) \end{bmatrix} \begin{bmatrix} M_B \\ M_C \end{bmatrix} = \begin{bmatrix} -\frac{HL_C}{2(kL_C)^2} \\ \frac{HL_C}{2(kL_C)^2} \end{bmatrix}$$

The first use of these equations is to determine the elastic buckling load. For this case, $R_A = R_D = P$; $c_{AB} = c_{CD} = c$; $H = 0$. The buckling condition requires that the determinant of the coefficients must equal zero:

$$\begin{vmatrix} \left(c + \frac{\gamma}{3} - \frac{1}{2(kL_C)^2}\right) & -\left(\frac{\gamma}{6} + \frac{1}{2(kL_C)^2}\right) \\ \left(\frac{\gamma}{6} + \frac{1}{2(kL_C)^2}\right) & \left(-c - \frac{\gamma}{3} + \frac{1}{2(kL_C)^2}\right) \end{vmatrix} = 0$$

This determinant is decomposed, and, after some algebra, we obtain the following buckling condition:

$$c + \frac{\gamma}{6} - \frac{1}{(kL_C)^2} = 0$$

From this equation the following final equation is equal to

$$kL_C \cdot \tan kL_C - \frac{\gamma}{6} = 0$$

For the following numerical values $I_B = 2I_C$; $L_B = 3L_C$, we get

$$\gamma = 1.5$$

and the buckling equation becomes equal to

$$kL_C \cdot \tan kL_C - 4 = 0$$

The lowest value of the eigenvalue from this equation is

$$(kL_C)^2 = 1.599$$

$$P_{cr} = \frac{1.599EI_C}{L_C^2} = \frac{\pi^2 EI_C}{(K_{eff}L_C)^2}$$

$$K_{eff} = \frac{\pi}{\sqrt{1.599}} = 2.48$$

This same effective length factor is obtained from the AISC sway permitted nomographs with $G_{Top} = 1.5$ and G_{Bottom} of infinity.

The next operation is the calculation of the bending moments at the joints B and C for the case where

$$H = P/10$$

and

$$P = \frac{1.0EI_{\mathrm{C}}}{L_{\mathrm{C}}^2}$$

The stability functions c and s are determined for the actual axial force in each column. Previously, it was shown that

$$R_A = P - \frac{HL_{\mathrm{C}}}{L_{\mathrm{B}}} - \frac{2P\Delta}{L_{\mathrm{B}}}$$

$$R_D = P + \frac{HL_{\mathrm{C}}}{L_{\mathrm{B}}} + \frac{2P\Delta}{L_{\mathrm{B}}}$$

Thus

$$\frac{R_A L_{\mathrm{C}}^2}{EI_{\mathrm{C}}} = \frac{PL_{\mathrm{C}}^2}{EI_{\mathrm{C}}} - \frac{HL_{\mathrm{C}}^3}{L_{\mathrm{B}}EI_{\mathrm{C}}} - \frac{2PL_{\mathrm{C}}^2\Delta}{L_{\mathrm{B}}EI_{\mathrm{C}}} = \frac{PL_{\mathrm{C}}^2}{EI_{\mathrm{C}}}\left[1 - \frac{H}{P}\cdot\frac{L_{\mathrm{C}}}{L_{\mathrm{B}}} - \frac{2\Delta}{L_{\mathrm{B}}}\right]$$

Since $\frac{2\Delta}{L_{\mathrm{B}}} \ll 1.0$ and $H = P/10$,

$$\frac{R_A L_{\mathrm{C}}^2}{EI_{\mathrm{C}}} = \frac{PL_{\mathrm{C}}^2}{EI_{\mathrm{C}}}\left[1 - \frac{H}{P}\cdot\frac{L_{\mathrm{C}}}{L_{\mathrm{B}}} - \frac{2\Delta}{L_{\mathrm{B}}}\right] = 1.0\left[1 - \frac{L_{\mathrm{C}}}{10\cdot 3L_{\mathrm{C}}}\right] = 0.967$$

$$\frac{R_D L_{\mathrm{C}}^2}{EI_{\mathrm{C}}} = 1.033$$

The corresponding values of the applicable stability functions are then $c_{AB} = 0.357$ and $c_{CD} = 0.359$. We now solve the simultaneous equations for the joint moments, as follows:

$$\begin{bmatrix} \left(0.357 + \frac{1.5}{3} - \frac{1}{2}\right) & -\left(\frac{1.5}{6} + \frac{1}{2}\right) \\ \left(\frac{1.5}{6} + \frac{1}{2}\right) & \left(-0.359 - \frac{1.5}{3} + \frac{1}{2}\right) \end{bmatrix} \begin{bmatrix} M_B \\ M_C \end{bmatrix} = \begin{bmatrix} -\frac{HL_{\mathrm{C}}}{2} \\ \frac{HL_{\mathrm{C}}}{2} \end{bmatrix}$$

The moments are then equal to

$$M_B = 1.276HL_{\mathrm{C}} = 0.128PL_{\mathrm{C}}$$
$$M_C = 1.274HL_{\mathrm{C}} = 0.127PL_{\mathrm{C}}$$

The corresponding first-order moments are equal to $M_B^{(1)} = M_C^{(1)} = 0.5HL_C$. The moment amplification due to the presence of the vertical loads is thus very significant.

4.7 SYSTEMATIC METHODS OF ANALYSIS: THE STIFFNESS METHOD

The advantages and disadvantages of the two methods of linear structural analysis have been discussed many times in the literature. With the widespread use of computerized structural calculations, the *stiffness method* has been the predominantly used method, and it is the basis of the modern structural analysis software. This chapter introduces the stiffness method by way of *slope-deflection equations*. These are used to solve stability problems; the commonly applied generalization of the method into the *matrix methods of structural analysis* lies outside the scope of this discussion.

The basic structural component used in this discussion is the beam-column subjected to end-moments and to in-plane forces shown in Figure 4.25. The member is subjected to end-moments that result in end rotations, θ. The end-moments and end-rotations are positive as shown in the figure: They are positive if they act and rotate in a clockwise sense.

The relationship between the end-moments and rotations has been developed in prior parts of this chapter, and they are reproduced next, taking into account the sign convention already defined:

$$\begin{aligned} \theta_A &= \frac{L}{EI}[cM_A - sM_B] + \rho \\ \theta_B &= \frac{L}{EI}[-sM_A + cM_B] + \rho \end{aligned} \tag{4.27}$$

The additional slope due to a clockwise bar-rotation ρ is also included in equation 4.27.

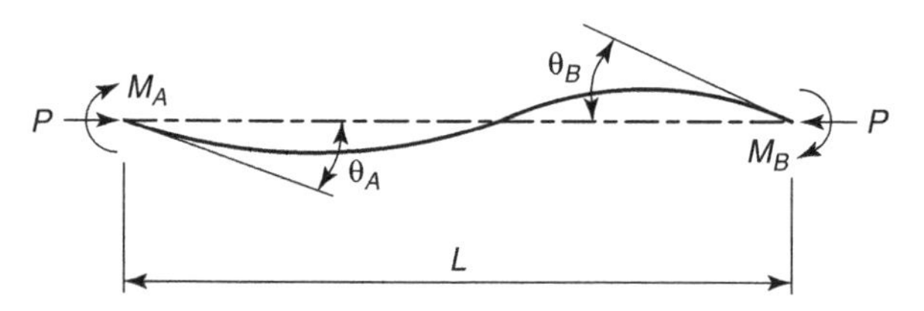

Fig. 4.25 Slope-deflection sign convention.

The functions c and s have been derived previously (equations 4.19 and 4.20), and they are reproduced below:

$$c = \frac{1}{(kL)^2}\left[1 - \frac{kL}{\tan kL}\right]$$
$$s = \frac{1}{(kL)^2}\left[\frac{kL}{\sin kL} - 1\right] \tag{4.28}$$

where $kL = \sqrt{PL^2/EI}$. Equation 4.27 will be rearranged to define the end-moments in terms of the bar and joint rotations, and the following *slope-deflection equations* result:

$$M_A = \frac{EI}{L}\{C\theta_A + S\theta_B - [C+S]\rho\} - M_A^F$$
$$M_B = \frac{EI}{L}\{S\theta_A + C\theta_B - [C+S]\rho\} + M_B^F \tag{4.29}$$

The *fixed-end moments* M_A^F and M_B^F have been added to account for the presence of the in-plane loads between the beam-column ends. The quantities C and S are *stability functions* that depend on the axial parameter kL, and they are given by the following equations:

$$C = \frac{c}{c^2 - s^2}$$
$$S = \frac{s}{c^2 - s^2} \tag{4.30}$$

It can be shown that in the absence of axial load (when $P = 0$),

$$C = 4 \quad \text{and} \quad S = 2$$

Numerically, this can be demonstrated by setting $kL = 0.01$ and solving for C and S. Analytically, it will be necessary to invoke L'Hopital's theorem a few times. It may sometimes occur that in a frame or in a truss, a member is under *axial tension*. In such a case, equations 4.30 still apply, but equations 4.28 are changed into the following expressions:

$$c = c_{\text{tension}} = \frac{-1}{(kL)^2}\left[1 - \frac{kL}{\tanh(kL)}\right]$$
$$s = s_{\text{tension}} = \frac{-1}{(kL)^2}\left[\frac{kL}{\sinh(kL)} - 1\right] \tag{4.31}$$

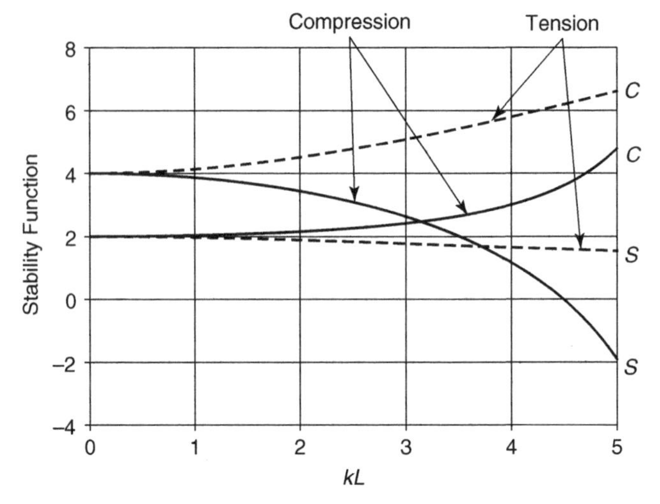

Fig. 4.26 Stability functions S and C.

The trigonometric functions are replaced by the corresponding hyperbolic functions, and the signs are changed. The variation of C and S for the cases of axial compression and axial tension are shown in Figure 4.26. It can be observed that for kL less than about 1, or P/P_E less than about 0.1, there is little difference in S and C from $S = 2$ and $C = 4$, the values when there is no axial force present.

The fixed-end moments are derived by superposition of the end slopes from the intrapanel loads and the moments. For example, the fixed-end moments for the uniformly distributed load is illustrated in Figure 4.27.

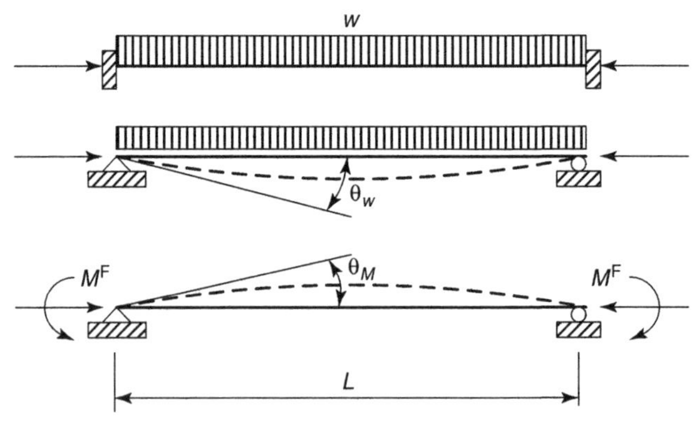

Fig. 4.27 Superposition for fixed-end moment.

The end-slope for the uniformly distributed simply-supported beam-column is obtained from Table 4.3.

$$\theta_w = \frac{wL^3}{24EI}\left\{\frac{3}{\left(\frac{kL}{2}\right)^3}\left[\tan\frac{kL}{2} - \frac{kL}{2}\right]\right\}$$

The end-slope for the end moment equals (equations 4.23 and using the definitions of c and s from equations 4.19 and 4.20)

$$\theta_M = \frac{-L}{EI}\left[c \cdot M^{\mathrm{F}} + s \cdot M^{\mathrm{F}}\right] = \frac{-LM^{\mathrm{F}}}{EI}\left[\frac{\frac{1}{\sin kL} - \frac{1}{\tan kL}}{kL}\right]$$

Noting that $\theta_w + \theta_M = 0$, we obtain, after some algebraic and trigonometric manipulations, the following expression for the fixed-end moment for the case of the uniformly loaded beam-column:

$$M^{\mathrm{F}} = \pm\frac{wL^2}{12}\left[\frac{12}{(kL)^2}\right]\left[1 - \frac{kL}{2 \cdot \tan\frac{kL}{2}}\right] \tag{4.32}$$

In this equation, $\frac{wL^2}{12}$ is the fixed-end moment when the axial load is zero. The rest of the equation is thus a moment-amplification factor. It can be shown that by setting kL equal to a small number, say 0.01, the amplification factor is indeed equal to 1.0. It approaches infinity when $kL = 2\pi$, that is, when $P = 4P_{\mathrm{E}}$, the buckling load of a fixed-end column $(P_{\mathrm{E}} = \frac{\pi^2 EI}{L^2})$. While the bracketed parts of equation 4.32 are not particularly complicated, an even simpler, and very accurate, approximation is given in the Commentary of the AISC Specification (2005):

$$M^{\mathrm{F}} = \left\{\frac{1 - 0.4\left(\frac{P}{4P_{\mathrm{E}}}\right)}{1 - \frac{P}{4P_{\mathrm{E}}}}\right\}\frac{wL^2}{12} \tag{4.33}$$

The curves of the two amplifiers of the first-order fixed-end moment are essentially on top of each other, so it is not necessary to use the more complicated equation 4.32. The dashed curve in Figure 4.28 shows the amplification factor for the case of the uniformly loaded fixed-end beam-column.

The fixed-end moment for the beam-column with one concentrated load Q located a distance αL from the left support can be derived to give the following equation:

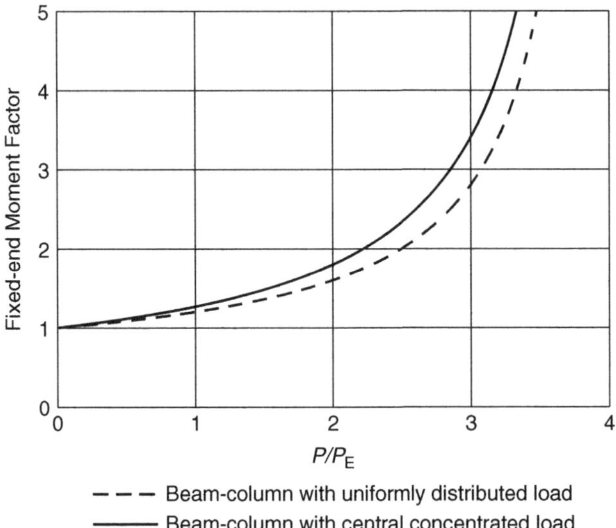

Fig. 4.28 Fixed-end moment amplification factors.

$$M_A^F = -QL\left(\frac{c \cdot u_1 - s \cdot u_2}{c^2 - s^2}\right)$$
$$M_B^F = -QL\left(\frac{c \cdot u_2 - s \cdot u_1}{c^2 - s^2}\right) \quad (4.34)$$

In these equations the subscript A refers to the fixed-end moment at the left support, and the subscript B to the right support. The terms c and s are defined by equations 4.28, and (Timoshenko and Gere 1961):

$$u_1 = \frac{\sin[kL(1-\alpha)]}{(kL)^2 \sin kL} - \frac{1-\alpha}{(kL)^2}$$
$$u_2 = \frac{\sin(kL \cdot \alpha)}{(kL)^2 \sin kL} - \frac{\alpha}{(kL)^2} \quad (4.35)$$

The first-order fixed-end moments are tabulated in the AISC Manual (2005), and they are equal to the following expressions:

$$M_{A(1)}^F = -QL[\alpha(1-\alpha)^2]$$
$$M_{B(1)}^F = -QL[\alpha^2(1-\alpha)] \quad (4.36)$$

If the concentrated load is in the center (i.e., $\alpha = 0.5$), then the fixed-end moments are represented very accurately by

$$M_F = \frac{QL}{8}\left[\frac{1 - 0.2\frac{P}{4P_E}}{1 - \frac{P}{4P_E}}\right] \quad (4.37)$$

The bracket in equation 4.37 is the amplification factor given in the Commentary of the AISC Specification 2005. The first order fixed-end moment is $QL/8$. The curve of this amplification factor is shown in Figure 4.28 as the solid curve. When the concentrated load is not in the center, then it is necessary to use the more complicated equations 4.34. The error is not significant, however, if Q is within 10 percent of the center of the span on either side. For other types of loading, as, for example a triangular distributed load, an expression would need to be derived along the lines already illustrated.

Following are two examples that will illustrate the application of the slope-deflection equations. There are extensive further applications of these equations in the chapter on frame stability (Chapter 5).

4.7.1 Example 4.6 Column with a Central Spring

The case studied in Example 4.6 is a column that is simply supported at its ends, and it has a lateral spring at its center. The geometry, loading, and the deformed shape are shown in Figure 4.29. In this example, we derive expressions for the stiffness and strength of the lateral brace required to force a node of the buckled shape of a column at its center, so the buckling load will equal the Euler load.

The slope-deflection equations (equations 4.29) are equal to

$$M_{AB} = \frac{EI}{L}\left\{C \cdot \theta_A + S \cdot \theta_B - [C + S] \cdot \frac{\Delta}{L}\right\} = 0$$

$$M_{BA} = \frac{EI}{L}\left\{S \cdot \theta_A + C \cdot \theta_B - [C + S] \cdot \frac{\Delta}{L}\right\}$$

$$M_{BC} = \frac{EI}{L}\left\{C \cdot \theta_B + S \cdot \theta_C + [C + S] \cdot \frac{\Delta}{L}\right\}$$

$$M_{CB} = \frac{EI}{L}\left\{S \cdot \theta_B + C \cdot \theta_C + [C + S] \cdot \frac{\Delta}{L}\right\} = 0$$

From the first and last of these equations, we solve for the end-slopes:

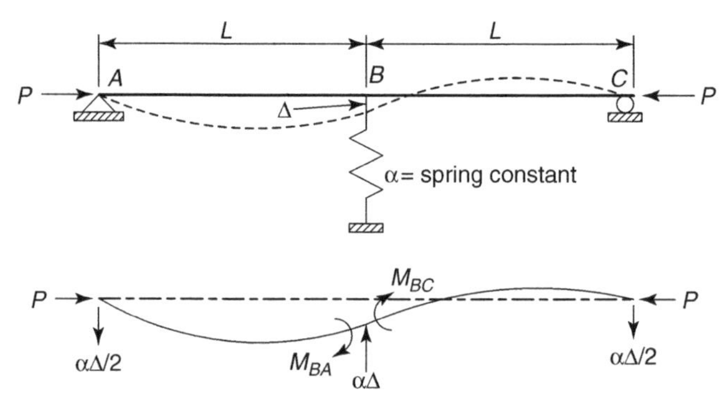

Fig. 4.29 Column with spring restraint at the center.

$$\theta_A = -\frac{S}{C}\theta_B + \frac{C+S}{C}\cdot\frac{\Delta}{L}$$

$$\theta_C = -\frac{S}{C}\theta_B - \frac{C+S}{C}\cdot\frac{\Delta}{L}$$

Substituting these equations into the second and third equation, respectively, we get

$$M_{BA} = \frac{EI}{L}\left[\frac{C^2 - S^2}{C}\right]\left[\theta_B - \frac{\Delta}{L}\right]$$

$$M_{BC} = \frac{EI}{L}\left[\frac{C^2 - S^2}{C}\right]\left[\theta_B + \frac{\Delta}{L}\right]$$

From equation 4.23, it can be demonstrated that

$$\frac{C^2 - S^2}{C} = \frac{1}{c} = \frac{(kL)^2}{1 - \frac{kL}{\tan kL}} \tag{4.38}$$

and, therefore

$$M_{BA} = \frac{1}{c}\cdot\frac{EI}{L}\left(\theta_B - \frac{\Delta}{L}\right)$$
$$M_{BC} = \frac{1}{c}\cdot\frac{EI}{L}\left(\theta_B + \frac{\Delta}{L}\right) \tag{4.39}$$

The next step is to consider equilibrium of the forces acting on the deformed member, as shown in Figure 4.29. Taking moments about a point just left of the center spring location, and also taking moments about a point just to the right of the center, we arrive at the following two equilibrium equations:

$$P \cdot \Delta - \frac{\alpha \cdot \Delta \cdot L}{2} + M_{BA} = 0$$

$$P \cdot \Delta - \frac{\alpha \cdot \Delta \cdot L}{2} - M_{BA} = 0$$

Substitution of the formulas for the two moments results in the following two homogeneous simultaneous equations:

$$\begin{bmatrix} \frac{EI}{cL} & PL - \frac{\alpha L^2}{2} - \frac{EI}{cL} \\ -\frac{EI}{cL} & PL - \frac{\alpha L^2}{2} - \frac{EI}{cL} \end{bmatrix} \begin{bmatrix} \theta_B \\ \frac{\Delta}{L} \end{bmatrix} = 0$$

A displacement is only possible if the determinant of their coefficients is equal to zero:

$$\begin{vmatrix} \frac{EI}{cL} & PL - \frac{\alpha L^2}{2} - \frac{EI}{cL} \\ -\frac{EI}{cL} & PL - \frac{\alpha L^2}{2} - \frac{EI}{cL} \end{vmatrix} = 0$$

Decomposition of the determinant, rearrangement, and substitution of c leads to the following equation:

$$\frac{1}{c}\left[\frac{PL^2}{2EI} - \frac{\alpha L^3}{4EI} - \frac{1}{2c}\right] = 0$$

This equation gives two solutions:

$$\frac{1}{c} = \frac{(kL)^2}{1 - \frac{kL}{\tan kL}} = \frac{(kL)^2 \tan kL}{\tan kL - kL} = 0 \rightarrow \tan kL = \frac{\sin kL}{\cos kL} = 0$$

$$\sin kL = 0$$

$$kL = \pi$$

$$k^2L^2 = \frac{PL^2}{EI} = \pi^2 \rightarrow P_{cr} = \frac{\pi^2 EI}{L^2} = P_E$$

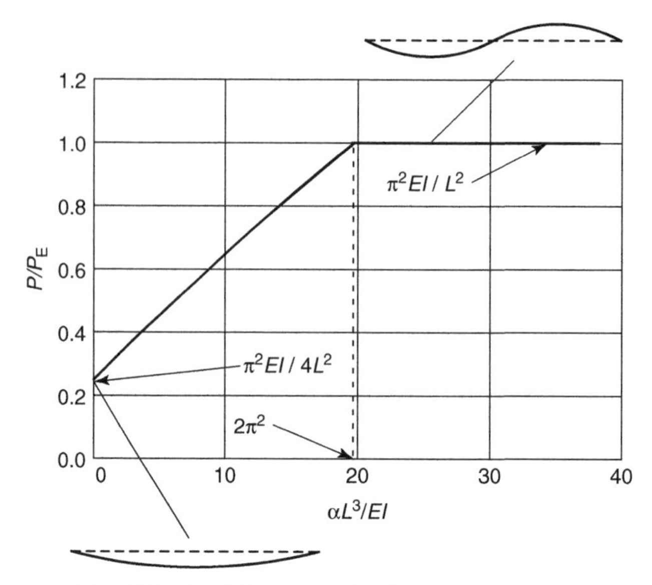

Fig. 4.30 Buckling strength of restrained column.

One of the critical loads is thus the Euler load. The buckled shape consists of a complete sine wave, with a node in the center of the column at the location of the restraining spring. (see Figure 4.29). The other equation furnishes another critical load:

$$\frac{PL^2}{2EI} - \frac{\alpha L^3}{4EI} - \frac{1}{2c} = 0 \rightarrow (kL)^2 - \frac{\alpha L^3}{2EI} - \frac{(kL)^2}{1 - \frac{kL}{\tan kL}} = 0$$

Finally, this buckling condition is equal to

$$\frac{\alpha L^3}{2EI}[kL - \tan kL] - (kL)^3 = 0$$

Since $kL = \pi\sqrt{\frac{P}{P_E}}$, the relationship between the nondimensional critical load and the nondimensional spring constant can be plotted, as shown in Figure 4.30. When $\alpha = 0$, the buckling length is equal to $2L$. As α increases, the critical load increases almost linearly until the buckling length is L when $P = P_E$. There is no advantage of increasing the spring constant beyond $2\pi^2$, because the column will buckle regardless at $P = P_E$. The maximum necessary spring constant to attain P_E is therefore

$$\alpha_{ideal} = \alpha_{id} = \frac{2\pi^2 EI}{L^3} = \frac{2P_E}{L} \qquad (4.40)$$

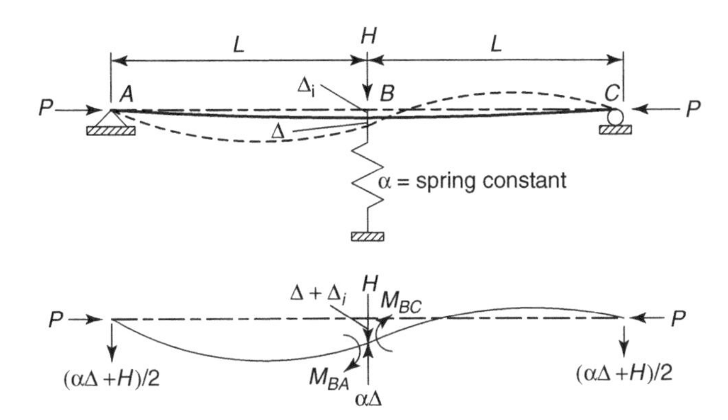

Fig. 4.31 Column with initial deflection.

Since the relationship between $\frac{P}{P_E}$ and $\frac{\alpha L^3}{EI}$ is almost linear. $\frac{\alpha L^3}{2EI}[kL - \tan kL] \; -(kL)^3 = 0$ can be replaced by $\frac{P}{P_E} = 0.55 + 0.75\frac{\alpha}{\alpha_{id}}$. In summary, then, $\frac{P}{P_E} = 0.55 + 0.75\frac{\alpha}{\alpha_{id}}$ if $\alpha \leq \alpha_{id}$ and $\frac{P}{P_E} = 1$ if $\alpha > \alpha_{id}$, where $\alpha_{id} = \frac{2P_E}{L}$ and $P_E = \frac{\pi^2 EI}{L^2}$. Winter (1958) arrived at the same result in a much simpler, but more intuitive, manner. The subject of bracing will be elaborated in much more detail in Chapter 7.

The previous discussion found the ideal spring constant necessary to force the braced column into a buckled mode with a node in the center. The next question to consider is what the strength of the bracing needs to be in order to assure that each half of the member will support the Euler load. The model shown in Figure 4.31 is examined to arrive at the solution. The column is subjected to a transverse load H at the location of the restraint, and it also has at that location an initial deflection Δ_i.

Taking moments about both sides of the point of support in the center, we obtain the following equilibrium equations:

$$P(\Delta + \Delta_i) + \left[\frac{H - \alpha\Delta}{2}\right]L + M_{BA} = 0$$

$$P(\Delta + \Delta_i) + \left[\frac{H - \alpha\Delta}{2}\right]L - M_{BC} = 0$$

The following equations are developed by repeating the same operations as before for the buckling analysis:

$$\begin{bmatrix} \frac{1}{c} & (kL)^2 - \frac{1}{c} - \frac{\alpha L^3}{2EI} \\ \frac{-1}{c} & (kL)^2 - \frac{1}{c} - \frac{\alpha L^3}{2EI} \end{bmatrix} \begin{bmatrix} \theta_B \\ \frac{\Delta}{L} \end{bmatrix} = \begin{bmatrix} -\frac{HL^2}{2EI} + (kL)^2 \frac{\Delta_i}{L} \\ -\frac{HL^2}{2EI} + (kL)^2 \frac{\Delta_i}{L} \end{bmatrix}$$

We are interested in the deflection Δ at the center of the member when the critical load is the Euler load (i. e., when $kL = \pi$ and when $H = 0$). From the previous equations, we show that

$$\frac{\Delta}{L} = \frac{-\pi^2 \cdot \frac{\Delta_i}{L}}{\pi^2 - \frac{\alpha L^3}{2EI}} = \frac{\Delta_i}{L} \left[\frac{1}{1 - \frac{\alpha L^3}{2\pi^2 EI}} \right]$$

Since $\alpha_{\text{id}} = \frac{2\pi^2 EI}{L^3} = \frac{2P_{\text{E}}}{L}$ from equation 4.40, the equation above can be written as

$$\frac{\Delta}{L} = \frac{\Delta_i}{L} \left[\frac{1}{\frac{\alpha}{\alpha_{\text{id}}} - 1} \right]$$

The spring constant α is the actual spring constant, and it is larger than the ideal value. The force to be resisted by the spring is

$$F_{\text{req}} = \alpha \cdot \Delta = \Delta_i \left[\frac{\alpha_{\text{id}}}{1 - \frac{\alpha_{\text{id}}}{\alpha}} \right] \tag{4.41}$$

This same equation was derived by Winter (1958) by employing a simpler intuitive model. The models of the determination of the ideal required bracing stiffness and of the required bracing strength have been used to derive design criteria in the AISC Specification. There it is assumed that the actual value of the spring constant is twice the ideal value, $\alpha = 2\alpha_{\text{id}}$. It is also assumed that the initial deflection is the mill tolerance, $\Delta_i = \frac{L}{1000}$. Thus

$$\begin{aligned} F_{\text{req}} &= \alpha_{\text{id}} \left[\frac{L}{1000(1 - 1/2)} \right] = 0.002 \cdot L\alpha_{\text{id}} \\ &= 0.002L \cdot \frac{2P_{\text{E}}}{L} = 0.004P_{\text{E}} \end{aligned}$$

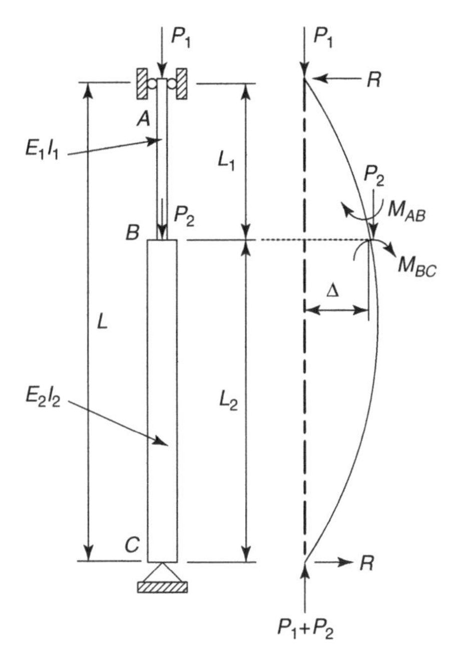

Fig. 4.32 Details of a stepped column.

4.7.2 Example 4.7 Stability of a Stepped Column

Example 4.7 illustrates the determination of the buckling load of a stepped column. Such members are found often in industrial halls. Design specifications have little or no guidance on how to handle such a problem. (A similar problem was solved in Chapter 2 as Example 2.4 by the differential equation method.) The example illustrates how to solve this problem with the slope-deflection method. The geometry, loading and deflection of the stepped column is shown in Figure 4.32. The ends of the column are pinned, and there is one step, or change in column size, within the column height. An axial load is applied at the top and another one at the point of the step.

Applying equilibrium, taking moments about point C defines the horizontal reaction R.

$$\sum M_C = 0 = R \cdot L - P_2 \cdot \Delta \rightarrow R = P_2 \cdot \frac{\Delta}{L}$$

Next, we take moments about points just above and just below B, respectively:

$$P_1\Delta + RL_1 - M_{BA} = 0$$

$$(P_1 + P_2)\Delta - RL_2 + M_{BC} = 0$$

It was shown in Example 4.6 that the moments at the end of a member segment where the other end is pinned are equal to

$$M_{BA} = \frac{E_1 I_1}{L_1} \cdot \frac{1}{c_{AB}} \left[\theta_B + \frac{\Delta}{L_1} \right]$$

$$M_{BC} = \frac{E_2 I_2}{L_2} \cdot \frac{1}{c_{BC}} \left[\theta_B - \frac{\Delta}{L_2} \right]$$

After introducing the following definitions

$$\rho = \frac{\Delta}{L_1}$$

$$\beta = \frac{L_1}{L}$$

$$\phi_{AB}^2 = \frac{P_1 L_1^2}{E_1 I_1}$$

$$\phi_{BC}^2 = \frac{(P_1 + P_2) L_2^2}{E_2 I_2}$$

$$\varphi_1 = \frac{P_2 \beta L_1^2}{E_1 I_1}$$

$$\varphi_2 = \frac{P_2 \beta L_1^2}{E_2 I_2}$$

We find the following buckling determinant:

$$\begin{vmatrix} \left[\dfrac{1}{c_{AB}} \right] & \left[\phi_{AB}^2 + \varphi_1 - \dfrac{1}{c_{AB}} \right] \\ \left[\dfrac{-1}{c_{BC}} \right] & \left[\left(\dfrac{\beta}{1-\beta} \right) \left(\phi_{BC}^2 - \dfrac{1}{c_{BC}} \right) - \varphi_2 \right] \end{vmatrix} = 0$$

The buckling load is finally obtained from solving this equation numerically. The procedure is demonstrated in the following numerical example

Stepped Column Buckling Analysis (see Figure 4.32)
For both sections of the column:

$$F_y = 50\,\text{ksi}$$
$$E = 29{,}000\,\text{ksi}$$
$$P_1 = P \text{ and } P_2 = 2P$$

$TOP : W8 \times 40$	$BOTTOM : W14 \times 109$
$A_1 = 11.7\,\text{in.}^2$	$A_1 = 32.0\,\text{in.}^2$
$I_1 = 146\,\text{in.}^4$	$I_1 = 1,240\,\text{in.}^4$
$r_{x1} = 3.53\,\text{in.}$	$r_{x2} = 6.22\,\text{in.}$
$r_{y1} = 2.04\,\text{in.}$	$r_{y2} = 3.73\,\text{in.}$
$L_1 = 10\,\text{ft.}$	$L_2 = 20\,\text{ft.}$

Lateral bracing is provided in the out-of-plane direction at the juncture between the two-column section. The columns are oriented for strong axis bending in-plane.

Minor Axis Buckling. Because of the lateral-bracing at the joint between the two sections in the out-of-plane direction, each column segment has an effective length of L_1 and L_2, respectively. The strength is determined by the 2005 AISC Specification.

$$F_{e1} = \frac{\pi^2 E}{\left(\frac{L_1}{r_{y1}}\right)^2} \qquad F_{cr1y} = \left| \begin{array}{ll} 0.658^{\frac{F_y}{F_{e1}}} & F_y \quad \text{if } F_{e1} \geq 0.44F_y \\ 0.877\,F_{e1} & \text{otherwise} \end{array} \right.$$

$$P_{1\,cry} = A_1\,F_{cr1y} \qquad \boxed{P_{1\,cry} = 454.233\,\text{kip}}$$

$$F_{e2} = \frac{\pi^2 E}{\left(\frac{L_2}{r_{y2}}\right)^2} \qquad F_{cr2y} = \left| \begin{array}{ll} 0.658^{\frac{F_y}{F_{e2}}} & F_y \quad \text{if } F_{e2} \geq 0.44F_y \\ 0.877\,F_{e2} & \text{otherwise} \end{array} \right.$$

$$P_{2\,cry} = A_2\,F_{cr2\,y} \qquad \boxed{P_{2\,cry} = 1.182 \times 10^3\text{kip}}$$

Axial strength of top column, governed by y-axis buckling: $P = 454\,\text{kip}$

Axial strength of bottom column, governed by y-axis buckling: $P = 1182/3 = 394\,\text{kip}$

Major Axis Buckling.

1. Elastic buckling

$$\beta = \frac{L_1}{L_1 + L_2}$$

Estimate of P for seed value for the iterative analysis: $P = 400\,\text{kips}$

$$P_1(P) = P \qquad P_2(P) = 2P$$

$$\phi_{AB}(P) = \sqrt{\frac{P_1(P)\,L_1^2}{E\,I_1}} \qquad \phi_{BC}(P) = \sqrt{\frac{(P_1(P) + P_2(P))L_2^2}{E\,I_2}}$$

$$\varphi_1(P) = \frac{P_2(P)\,\beta\,L_1^2}{E\,I_1} \qquad \varphi_2(P) = \frac{P_2(P)\,\beta\,L_2^2}{E\,I_2}$$

$$c_{AB}(P) = \frac{1}{\phi_{AB}(P)^2}\left(1 - \frac{\phi_{AB}(P)}{\tan(\phi_{AB}(P))}\right) \qquad c_{BC}(P) = \frac{1}{\phi_{BC}(P)^2}\left(1 - \frac{\phi_{BC}(P)}{\tan(\phi_{BC}(P))}\right)$$

$$\text{Answer}(P) = \text{root}\left[\left|\begin{bmatrix} \dfrac{1}{c_{AB}(P)} & \phi_{AB}(P)^2 - \dfrac{1}{c_{AB}(P)} + \varphi_1(P) \\ \dfrac{-1}{c_{BC}(P)} & \dfrac{\beta}{1-\beta}\left(\dfrac{-1}{c_{BC}(P)} + \phi_{BC}(P)^2\right) - \varphi_2(P) \end{bmatrix}\right|, P\right]$$

$$\boxed{\text{Answer}(P) = 550.06\,\text{kip}}$$

2. Inelastic buckling

$$A_1E_y = 585\,\text{kip} \qquad A_2F_y = 1.6 \times 10^3\text{kip}$$

$$P_1(P) = \frac{P_1(P)}{A_1\,F_y} \qquad P_2(P) = \frac{P_1(P) + P_2(P)}{A_2\,F_y}$$

The following model of the tangent modulus is introduced at the end of Chapter 3 (equations 3.103 and 3.104):

$$\tau = \frac{E_t}{E} = 1.0 \qquad \text{for} \quad P \le 0.5P_y$$

$$\tau = 4\left(\frac{P}{P_y}\right)\left(1 - \frac{P}{P_y}\right) \qquad \text{for} \quad 0.5\,P_y < P \le P_y$$

$$\tau_1(P) = \begin{vmatrix} 4p_1(P)(1 - p_1(P)) & \text{if } p_1(P) \geq 0.5 \\ 1.0 & \text{otherwise} \end{vmatrix}$$

$$\tau_2(P) = \begin{vmatrix} 4p_2(P)(1 - p_2(P)) & \text{if } p_2(P) \geq 0.5 \\ 1.0 & \text{otherwise} \end{vmatrix}$$

$$P_1(P) = P \qquad P_2(P) = 2P$$

$$\phi_{AB}(P) = \sqrt{\frac{P_1(P)\,L_1^2}{EI_1\tau_1(P)}} \qquad \phi_{BC}(P) = \sqrt{\frac{(P_1(P) + P_2(P))L_2^2}{EI_2\tau_2(P)}}$$

$$\Phi_1(P) = \frac{P_2(P)\,\beta\,L_1^2}{EI_1\tau_1(P)} \qquad \Phi_2(P) = \frac{P_2(P)\,\beta\,L_2^2}{EI_2\tau_2(P)}$$

$$c_{AB}(P) = \frac{1}{\phi_{AB}(P)^2}\left(1 - \frac{\phi_{AB}(P)}{\tan(\phi_{AB}(P))}\right) \qquad c_{BC}(P) = \frac{1}{\phi_{BC}(P)^2}\left(1 - \frac{\phi_{BC}(P)}{\tan(\phi_{BC}(P))}\right)$$

$$\text{Answer}(P) = \text{root}\left[\left|\begin{bmatrix} \frac{1}{c_{AB}(P)} & \phi_{AB}(P)^2 - \frac{1}{c_{AB}(P)} + \Phi_1(P) \\ \frac{-1}{c_{BC}(P)} & \frac{\beta}{1-\beta}\left(\frac{-1}{c_{BC}(P)} + \phi_{BC}(P)^2\right) - \Phi_2(P) \end{bmatrix}\right| P\right]$$

$$\boxed{\text{Answer}(P) = 421.91\ \text{kip}}$$

Elastic buckling load in top part of column $= 550\,\text{kip} < 585\,\text{kip}$, the yield load.

Elastic buckling load in bottom part of column $= 3 \times 550 = 1{,}650\,\text{kip} > 1{,}600\,\text{kip}$, the yield load.

Inelastic buckling load in top part of column $= 422\,\text{kip} < 585\,\text{kip}$, the yield load.

Inelastic buckling load in bottom part of column $= 3 \times 422 = 1{,}266\,\text{kip} < 1{,}600\,\text{kip}$

The critical buckling load of the column is $P = 394\,\text{kip}$, y-axis buckling of bottom column

4.7.3 Summary of Elastic Beam-column Behavior

Beam-columns are members whose behavior is dictated by neither axial load nor moment, but rather by the interaction between the two. Because an axial load acts on the deflected shape caused by flexure, second-order moments are developed that must be considered in the analysis and design of beam columns. The amplification of the moment may be determined through the use of a moment amplifier. Although exact formulas for moment amplifiers vary, depending on the type of loading on the beam, a simplified formula (equation 4.7) is derived that provides a good approximation for any general load case. The moment amplification is dependent on the ratio of the applied load to the Euler buckling load of the column (P_E)

Unlike the elastic columns discussed in Chapter 2, beam-columns do not have a bifurcation point that delineates when failure occurs. The interaction of axial load and flexure is addressed through the use of interaction equations. The application of interaction equations in establishing the strength of beam columns is discussed in greater detail in Chapter 8.

Traditional methods of structural analysis can be employed to determine the capacity of beam-columns. Two methods have been discussed: the flexibility approach, or three-moment equation method, and the stiffness-based slope-deflection approach. The latter is the most commonly used based on its application in matrix-based software algorithms. Both methods utilize stability functions that are a function of the axial parameter kL. Through application of equilibrium and these traditional approaches, the buckling load of beam-columns and simple frames can be determined.

4.8 INELASTIC STRENGTH OF BEAM-COLUMNS

In the previous sections of this chapter, it was assumed that the material is linearly elastic. The onset of yielding or elastic buckling was considered to be the limit of strength. It was already discussed in the beginning of the chapter (see Figure 4.3) that ductile materials, such as steel or aluminum, can carry forces beyond the yield point. In the case of steel members the presence of residual stresses can substantially reduce the range of elastic behavior. For aluminum alloys the stress-strain curve is nonlinear. The schematic curve in Figure 4.33 illustrates the end moment-versus-end-slope relationship of a beam-column, of length L and having a given moment ratio κ, that is subjected to a constant axial force P. The member has a doubly symmetric cross-section, and it is bent about one of its principal axes. Lateral bracing is assumed to prevent lateral-torsional and out-of-plane column buckling.

The $M_o - \theta_o$ relationship is linear until the maximum elastic stress in the member reaches a value equal to the yield stress minus the compressive residual stress. The stiffness of the beam-column decreases as the end-slope is further increased, resulting in a nonlinear deformation behavior. The stiffness continues to lessen until it reaches zero, and equilibrium can only be

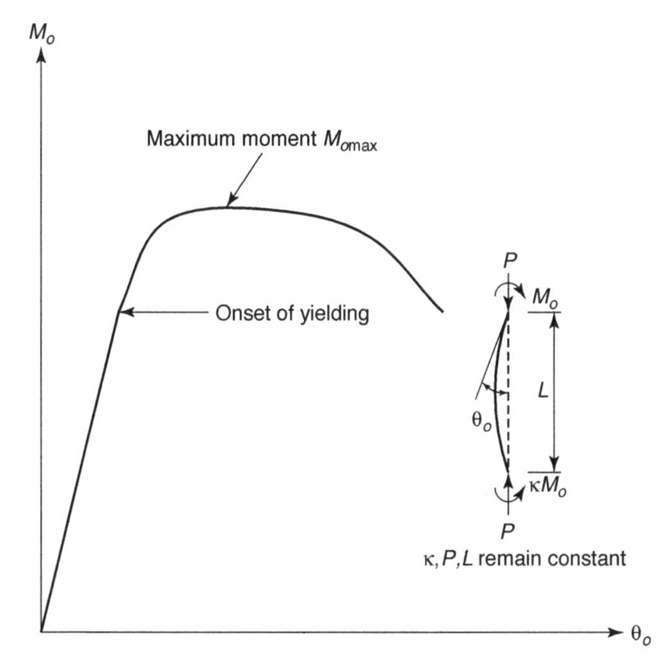

Fig. 4.33 Behavior of a beam-column.

maintained if the end moment is reduced. The maximum strength is the end-moment at the peak of the curve. Since both the material and the geometry are nonlinear, the determination of the maximum strength is a complicated process. In the following we approach the solution by first considering the plastic strength of cross-sections made from a material with an ideal elastic-plastic stress-strain curve (Figure 3.18). Next we determine the approximate strength of a beam-column of rectangular cross section. Finally, we introduce the design interaction equations.

The limits of the strength of a wide-flange cross-section are the yield moment and the plastic moment, as seen in Figure 4.34. The cross-section is subjected to a bending moment *M* and an axial force *P*. The latter is applied at the geometrical centroid of the section. It is important to keep this latter convention in mind, since moving the point of application of *P* away from the centroid will introduce an additional moment.

For the wide-flange section shown, the maximum elastic stress is in the extreme fiber of the top flange. Elastic behavior is terminated when $\frac{P}{A} + \frac{M}{S_x} = F_y - F_r, A$ is the area, S_x is the elastic section modulus about the major axis, F_y is the yield stress and F_r is the compressive residual stress at the tip of the flange. The plastic moment M_{pc} is reached when the whole cross-section is yielded in compression and tension.

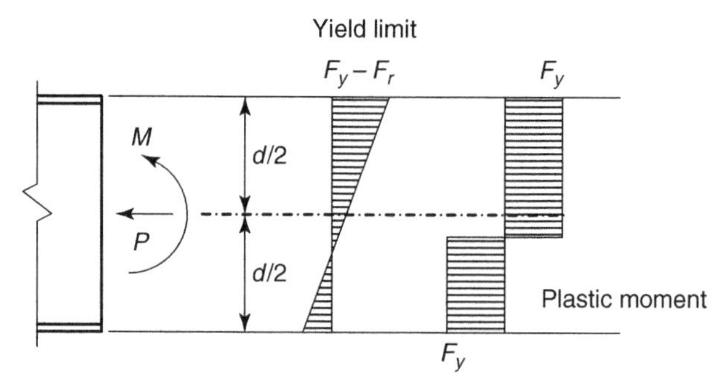

Fig. 4.34 Strength limits of the cross-section.

For a rectangular cross-section, the plastic moment has been derived already in Chapter 3 as the limiting case where the curvature approaches very large values (see Figure 3.18). It will be rederived here again based on equilibrium of the stress blocks. The location of P and M_{pc}, the cross-section and the stress distribution for the plastic moment condition are given in Figure 4.35.

The total stress distribution can be separated into the contribution due to the axial force and due to the bending moment.

$$\text{Yield load:} \quad P_y = bdF_y \tag{4.42}$$

$$\text{Plastic moment when } P = 0: \quad M_p = \frac{bd^2 F_y}{4} \tag{4.43}$$

$$\text{Axial stress equilibrium:} \quad P = (2y_p - d)bF_y$$

$$\frac{P}{bdF_y} = \frac{P}{P_y} = \frac{2y_p}{d} - 1$$

$$\frac{y_p}{d} = \frac{1}{2}\left(1 + \frac{P}{P_y}\right)$$

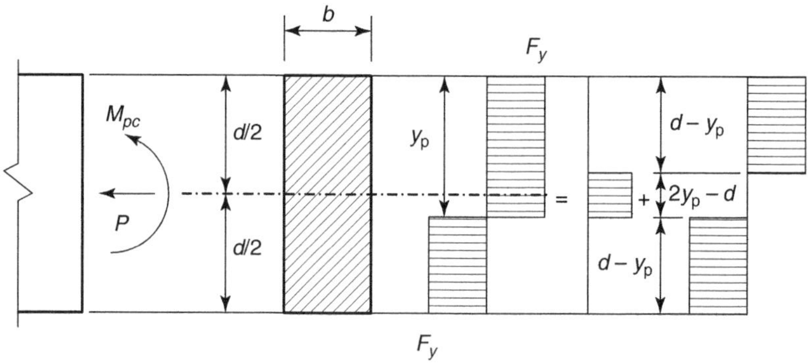

Fig. 4.35 Plastic moment of a rectangular cross-section.

$$\text{Moment equilibrium}: \quad M_{pc} = (d - y_p)bF_y \times \left(d - \frac{2(d - y_p)}{2}\right)$$

$$M_{pc} = (d - y_p)y_p \times bF_y = \frac{bd^2F_y}{4}\left(1 - \frac{P}{P_y}\right)\left(1 + \frac{P}{P_y}\right)$$

$$\frac{M_{pc}}{M_p} = 1 - \left(\frac{P}{P_y}\right)^2 \tag{4.44}$$

Equation 4.44 is the plastic moment in the presence of axial force for a rectangular cross-section. The formulas for M_{pc} for other cross-sections are derived in a similar manner, but they are more complicated, of course, depending on the cross-sectional geometry. The equations for the plastic moments for x-axis and y-axis bending of wide-flange shapes are in Table 4.4. Simpler approximate formulas are also given in the table (*Plastic Design in Steel* ASCE 1971). These approximate equations are compared with the analytically exact formulas in Figure 4.36, where the axial ratio is the ordinate and the bending ratio is the abscissa. The exact formulas were calculated for the geometry of a W14 × 99 rolled wide-flange shape. The curves for the approximate formulas are seen to be close enough for practical purposes.

The approximate formula for a solid circular cross-section of radius R is given as equation 4.45. This is a very good approximation of a complicated exact equation.

$$\frac{M_{pc}}{M_p} = 1 + 0.08\frac{P}{P_y} - 1.08\left(\frac{P}{P_y}\right)^2; \quad P_y = \pi R^2 F_y; \quad M_p = \frac{4R^3F_y}{3} \tag{4.45}$$

The curves from the analytically exact equations are shown in Figure 4.37, starting with the top dashed line, of a W14 × 99 wide-flange section bent about the minor axis, a solid circular section, a rectangular section, and a W14 × 99 wide-flange section bent about the major axis, respectively. The solid line represents the lower-bound interaction equation that is the basis of the AISC Specification for the design of beam-columns. As can be seen, the AISC equation closely replicates the x-axis interaction strength of the wide-flange shape. Since the AISC equation is a lower bound to the most frequent practical situation, it was adopted for use in the design standard. It should be realized, however, that for other shapes it can be very conservative. Equation 4.46 is the AISC basic interaction equation.

TABLE 4.4 Plastic Moments for Wide-flange Shapes

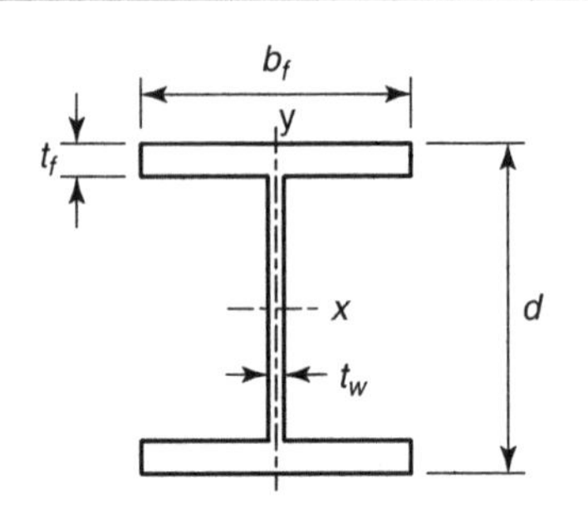

$M_{px} = Z_x F_y$

$M_{py} = Z_y F_y$

Bending about x-axis

$$\text{for } 0 \le \frac{P}{P_y} \le \frac{t_w(d - 2t_f)}{A}$$

$$\frac{M_{pcx}}{M_{px}} = 1 - \frac{A^2 \left(\frac{P}{P_y}\right)^2}{4t_w Z_x}$$

$$\text{for } \frac{t_w(d - 2t_f)}{A} \le \frac{P}{P_y} \le 1$$

$$\frac{M_{pcx}}{M_{py}} = \frac{A\left(1 - \frac{P}{P_y}\right)}{2Z_x}\left[d - \frac{A\left(1 - \frac{P}{P_y}\right)}{2b_f}\right]$$

Approximation:

$$\frac{M_{pcx}}{M_{px}} = 1.18\left(1 - \frac{P}{P_y}\right) \le 1.0$$

Bending about y-axis

$$\text{for } 0 \le \frac{P}{P_y} \le \frac{t_w d}{A}$$

$$\frac{M_{pc}}{M_p} = 1 - \frac{A^2 \left(\frac{P}{P_y}\right)^2}{4dZ_y}$$

$$\text{for } \frac{t_w d}{A} \le \frac{P}{P_y} \le 1$$

$$\frac{M_{pc}}{M_p} = \frac{A^2\left(1 - \frac{P}{P_y}\right)}{8t_f Z_y}\left[\frac{4b_f t_f}{A} - \left(1 - \frac{P}{P_y}\right)\right]$$

Approximation:

$$\frac{M_{pcy}}{M_{py}} = 1.19\left[1 - \left(\frac{P}{P_y}\right)^2\right] \le 1.0$$

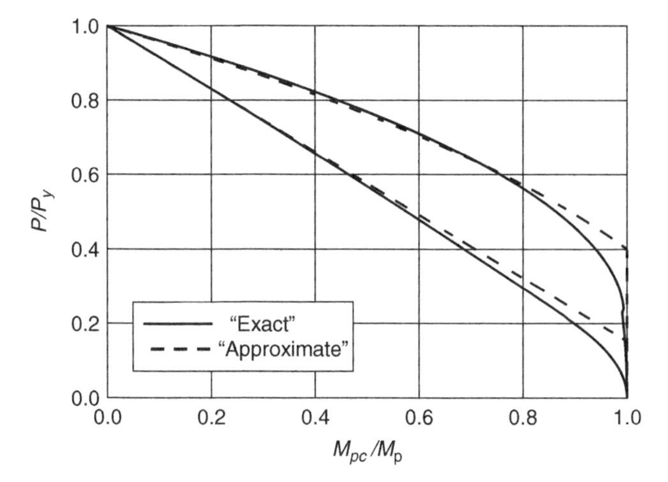

Fig. 4.36 Exact and approximate *M-P* relations for *x*-axis bending (lower curves) and *y*-axis bending (upper curves) of a W14 × 99 wide-flange section.

$$
\begin{aligned}
&\frac{1}{2}\frac{P}{P_y}+\frac{M_{pc}}{M_p}=1 \quad \text{if } \frac{P}{P_y}\leq 0.2 \\
&\frac{P}{P_y}+\frac{8}{9}\frac{M_{pc}}{M_p}=1 \quad \text{if } \frac{P}{P_y}\geq 0.2
\end{aligned}
\tag{4.46}
$$

The previous discussion above considered the plastic strength of a cross-section that is subjected to an axial force at its geometric centroid and to a

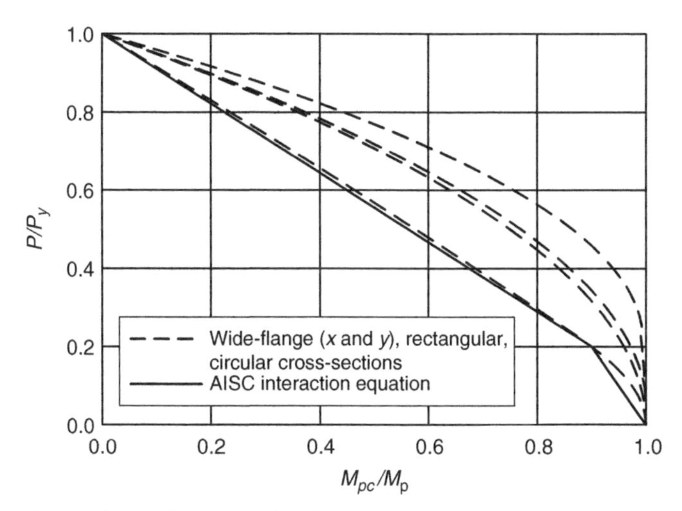

Fig. 4.37 Comparison of cross-section interaction curves with AISC interaction equation.

bending moment. Actual beam-columns have a length and the moment includes a component from the product of the axial force times the deflection. When some of the material in the beam-column is inelastic, the determination of the moment-deformation relationship, such as that shown in Figure 4.27, becomes very laborious, and exact closed form solutions are not available. Numerical integration methods, approximate simplified models, or empirical interaction equations are then employed to obtain solutions. (See Chapter 8, Galambos 1998.)

Numerical integration methods will not be considered here. We first present a simplified model for beam-columns with rectangular sections made of a perfectly elastic-plastic material in order to illustrate the types of solutions that can be achieved. Following that, empirical interaction equations are presented as a practical design method.

A subset of the rectangular beam-column problem has been solved in Chapter 3 (section 3.6), where inelastic columns with initial imperfections and load eccentricities were discussed. The start of the solution involves the moment-curvature-thrust relationships. The curves representing these relationships are shown in Figures 3.19 and 3.20. The formulas relating the non-dimensional axial load, P/P_y, bending moment M/M_y, and curvature ϕ/ϕ_y, are given in the form of a flowchart in Figure 3.17. These will be used in the ensuing derivations and they will therefore be repeated again here.

Range I: the cross-section is elastic (Figure 3.14)

$$\frac{M}{M_y} = \frac{\phi}{\phi_y}; \quad \text{for } 0 \leq \frac{\phi}{\phi_y} \leq 1 - \frac{P}{P_y} \tag{4.47}$$

Range II: the cross-section is yielded in compression (Figure 3.15)

$$\frac{M}{M_y} = 3\left(1 - \frac{P}{P_y}\right) - \frac{2\left(1 - \frac{P}{P_y}\right)^{\frac{3}{2}}}{\sqrt{\frac{\phi}{\phi_y}}}; \quad \text{for } 1 - \frac{P}{P_y} \leq \frac{\phi}{\phi_y} \leq \frac{1}{1 - \frac{P}{P_y}} \tag{4.48}$$

Range III: the cross-section is yielded in compression and tension (Figure 3.16)

$$\frac{M}{M_y} = \frac{3}{2}\left[1 - \left(\frac{P}{P_y}\right)^2\right] - \frac{1}{2\left(\frac{\phi}{\phi_y}\right)^2}; \quad \text{for } \frac{\phi}{\phi_y} \geq \frac{1}{1 - \frac{P}{P_y}} \tag{4.49}$$

In these equations $P_y = bdF_y$; $M_y = \frac{bd^2F_y}{6}$; $\phi_y = \frac{2F_y}{dE}$ and b is the width and d the depth of the rectangular section.

The beam-column to be analyzed is shown in the left sketch of Figure 4.38. The member is of length L and it is subjected to an axial load P and to

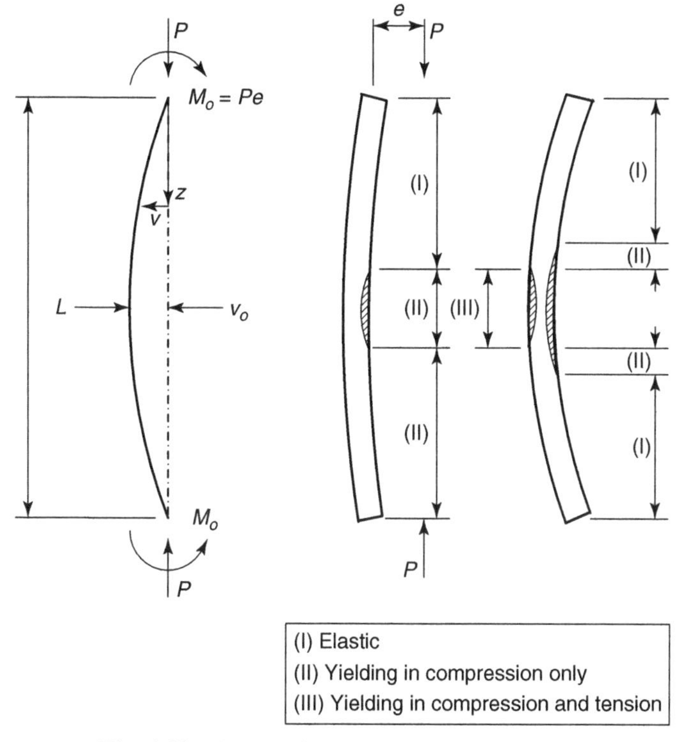

Fig. 4.38 Stages of yielding in a beam-column.

equal end-moments $M_o = Pe$ at each end, causing single-curvature bending. The central deflection is v_o. The other two sketches in this figure show the two possible stress distributions at the maximum moment: Case 1, yielding in compression only, and Case 2, yielding in compression and tension. The Roman numerals (I), (II), and (III) refer to the ranges of the moment formulas.

We assume that the beam-column will deflect in a sine wave: $v = v_o \sin \frac{\pi z}{L}$.

The curvature is equal to $\phi = -v'' = v_o \left(\frac{\pi}{L}\right)^2 \sin \frac{\pi z}{L}$. The maximum curvature occurs at $z = L/2$ and it is $\phi_o = \left(\frac{\pi}{L}\right)^2 v_o$. Dividing by $\phi_y = \frac{2F_y}{dE}$, noting that the radius of gyration of a rectangular section is $r = \frac{d}{\sqrt{12}}$ and introducing the slenderness parameter

$$\lambda = \frac{1}{\pi} \frac{L}{r} \sqrt{\frac{F_y}{E}} \tag{4.50}$$

It can be shown that

$$\frac{\phi_o}{\phi_y} = \frac{6}{\lambda^2}\frac{v_o}{d} \tag{4.51}$$

The object of this derivation is to calculate the maximum moment that the beam-column can support: The end-moment corresponding to the peak of the curve in Figure 4.33. In order to reach the limiting strength, it is necessary that the beam-column should yield. The two possibilities are depicted in the two sketches in Figure 4.38. At the top of the moment-deformation curve (Figure 4.33) the required stiffness equals the available stiffness: The change in the external moment equals the change in the internal moment.

The stability criterion is therefore

$$\frac{\delta M_{\text{exterior}}}{\delta v_o} = \frac{\delta M_{\text{interior}}}{\delta v_o} \rightarrow \frac{\delta\left(\frac{M_{\text{exterior}} - M_{\text{interior}}}{M_y}\right)}{\delta\left(\frac{v_o}{d}\right)} = 0$$

In addition, equilibrium must also be maintained:

$$\frac{M_{\text{exterior}}}{M_y} - \frac{M_{\text{interior}}}{M_y} = 0$$

1. *Yielding in compression only.*

$$M_{\text{exterior}} = M_o + Pv_o$$

$$\frac{M_{\text{exterior}}}{M_y} = \frac{M_o}{M_y} + \frac{Pv_o}{bd^2F_y/6} = \frac{M_o}{M_y} + 6 \cdot \frac{P}{P_y} \cdot \frac{v_o}{d}$$

The interior moment is given by equation 4.48. With equation 4.51 and the definitions

$$m_o = \frac{M_o}{M_y}; \quad p = \frac{P}{P_y}; \quad \eta = \frac{v_o}{d}$$

$$\frac{M_{\text{exterior}}}{M_y} - \frac{M_{\text{interior}}}{M_y} = 6p\eta + m_o - 3(1-p) + \frac{2(1-p)^{\frac{3}{2}}\lambda}{\sqrt{6\eta}}$$

Setting the derivative with respect to η equal to zero and solving for η:

$$\eta = \frac{(1-p)\lambda^{\frac{2}{3}}}{6p^{\frac{2}{3}}} \tag{4.52}$$

Equating the external and the internal moment, and substituting η (equation 4.52) leads to the following interaction equation between the end-moment, the axial force and the slenderness parameter:

$$m_o = 3(1 - p)[1 - p^{\frac{1}{3}}\lambda^{\frac{2}{3}}] \tag{4.53}$$

The upper limit of the applicability of equation 4.53 is $\frac{\phi}{\phi_y} = \frac{6\eta}{\lambda^2} \leq \frac{1}{1-p}$ (equations 4.48 and 4.51) and the lower limit is $\frac{\phi}{\phi_y} = \frac{6\eta}{\lambda^2} \geq 1 - p$. Substitution of η from equation 4.52 leads to the following range of application of equation 4.53:

$$\frac{(1-p)^3}{p} \leq \lambda^2 \leq \frac{1}{p}$$

2. *Yielding in compression and tension.*

The interior moment is given as equation 4.49.

$$\frac{M_{\text{exterior}}}{M_y} = 6p\eta + m_o$$

$$\frac{M_{\text{interior}}}{M_y} = \frac{3}{2}(1 - p^2) - \frac{1}{2\left(\frac{6\eta}{\lambda^2}\right)^2}$$

$$\frac{M_{\text{exterior}}}{M_y} - \frac{M_{\text{interior}}}{M_y} = 6p\eta + m_o - \frac{3}{2}(1 - p^2) + \frac{\lambda^4}{72\eta^2}$$

Setting the derivative with respect to η equal to zero leads to

$$\eta = \frac{\lambda^{\frac{4}{3}}}{6p^{\frac{1}{3}}} \tag{4.54}$$

Equating the external and the internal moment, and substituting η from equation 4.54 leads to the following interaction equation between the end-moment, the axial force and the slenderness parameter:

$$m_o = \frac{3}{2}(1 - p^2 - p^{\frac{2}{3}}\lambda^{\frac{4}{3}}) \tag{4.55}$$

The upper limit of applicability of equation 4.55 is when $\frac{\phi}{\phi_y} \rightarrow \infty$ resulting in $\lambda = 0$, and the lower limit is when $\frac{\phi}{\phi_y} = \frac{6\eta}{\lambda^2} \leq \frac{1}{1-p}$. Thus

$$0 \leq \lambda^2 \leq \frac{(1-p)^3}{p}$$

TABLE 4.5 Summary of Equations

Definitions	Strengths	Ranges
$m_o = \frac{M_o}{M_y}$	$m_o = 3(1-p)\left[1-(p\lambda^2)^{\frac{1}{3}}\right]$	$\frac{(1-p)^3}{p} \leq \lambda^2 \leq \frac{1}{p}$
$M_y = \frac{bd^2F_y}{6}$	$m_o = \frac{3}{2}\left(1-p^2-(p\lambda^2)^{\frac{2}{3}}\right)$	$0 \leq \lambda^2 \leq \frac{(1-p)^3}{p}$
$p = \frac{P}{P_y}$		
$P_y = bdF_y$		
$\lambda = \frac{L}{\pi r}\sqrt{\frac{F_y}{E}}$		

In summary, the interaction relationships between the applied end-moment, the axial force, and the length is shown in Table 4.5, and the resulting interaction curves are presented in Figure 4.39. Also shown in this figure are the results from numerical integration. The approximate curves are close to the numerically obtained values.

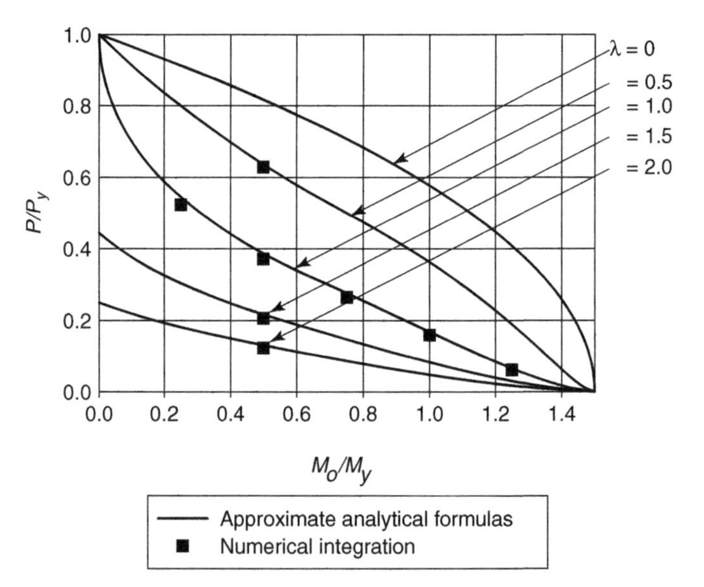

Fig. 4.39 Interaction curves for rectangular beam-columns.

4.9 DESIGN OF BEAM-COLUMNS

The prediction of the behavior and strength of beam-columns is not trivial. Even as simple a problem as the determination of the strength of a member of rectangular section subjected to a concentric axial force and to equal end-moments involved a complicated set of mathematical manipulations. Because of this complexity, the study of the beam-columns has attracted the talents of structural engineering researchers for well over a hundred years. Literally hundreds of test-reports, theses, journal articles, and books have resulted in very diverse, and often extremely ingenious recommendations for use by design engineers who have to design beam-columns rapidly and economically in their everyday work. This text does not attempt to even scratch the surface of the history of beam-columns. The readers are referred to the literature in case of interest. The best place to start is in Chapter 8 "Beam-columns" of the Structural Stability Research Council (SSRC), "Guide to Stability Design Criteria for Metal Structures" (Galambos 1998). Further study could make use of the many classical textbooks on structural stability theory that were listed at end of Chapter 2. If the interest is in the description and comparison of design methods in different parts of the world, then the SSRC "World View" (SSRC 1991) is recommended as a starting point.

In the previous section of this book it was pointed out that metal beam-columns of practical dimensions always reach their maximum strength after some parts of the member have undergone some yielding. Except for the simplest cases, there is no convenient closed-form solution for such nonlinear problems. Many hundreds of variations of numerical or semi-analytical methods have been advanced, but in today's world of fast computers there is no obstacle to using numerical integration to obtain the desired answers. Perhaps the best summary of such methods is provided in the two-volume text on beam-columns by Chen and Atsuta (1976). A flow-diagram illustrating the steps in the numerical scheme is shown in Figure 4.40.

The first numerical integration is performed with the data defining the material, the geometry and the residual stresses in the cross-section to produce the cross-section force-versus-strain relationship. The second integration is performed over the length of the beam-column with the just computed constitutive information with the forces on the member, its geometry, and its initial deformation. This process results in the force-versus-deformation relations of the whole beam-column. From this information, the engineer can extract whatever is needed to define a limiting strength or limiting deformation. Many such programs have been developed and used by researchers to compare test results to prediction and to verify less rigorous computations or to test the accuracy of empirical equations. Even

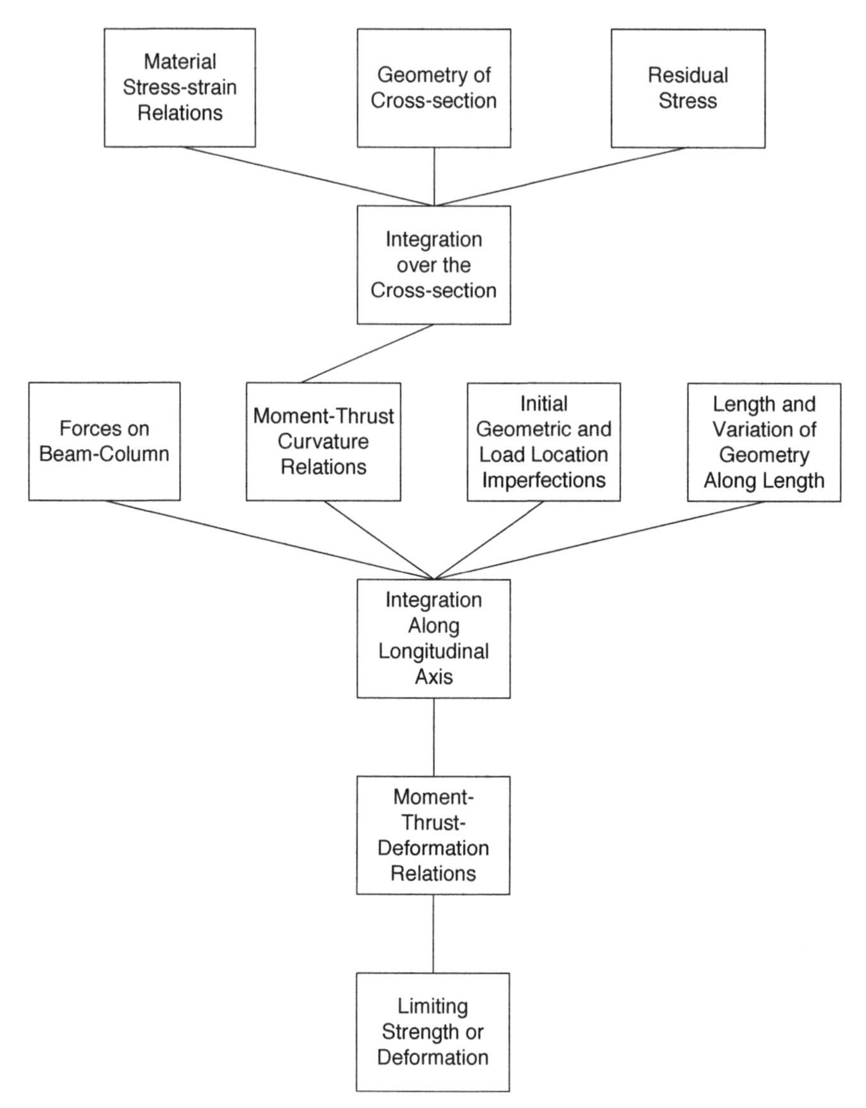

Fig. 4.40 Flowchart of numerical procedures and data for beam-column analysis.

though the results are usually achieved in the blink of an eye, these programs are not suitable for general design office use, though they are convenient for studies where "exact" answers are required. In the design office environment it is as yet more desirable to employ simple formulas, both from the point of view of efficient design and of control and supervision of the work by other agencies such as building authorities.

Many such formulas have been recommended and adopted in design codes over the years. Some of them were based on the attainment of an allowable stress based on elastic beam-column theory that contained artificial initial deflections so that the results would reproduce test strengths. Some specifications used tables or charts that were based on numerical or experimental work, while others adopted curve-fit formulas. Mostly, however, design standards all over the world use interaction equations that relate the axial force and the flexural forces such that the sum of ratios must be less than or equal to unity. There are a great many variations of the interaction equations over time and around the world. Although a study of this history would be very interesting, it will not be pursued here. Instead, the scheme that has been part of the AISC *LRFD Specification* since 1986, and that is now used in the 2005 AISC *Specification for Steel Buildings* is presented in Chapter 8.

PROBLEMS

4.1. Calculate the tangent-modulus buckling load of the column shown. Use the slope-deflection method to formulate the buckling determinant. The lower segment of the column has a moment of inertia twice that of the upper segment of the column. Use a spread-sheet to obtain numerical results. Calculate also the buckling load assuming elastic behavior of the material.

Data: Solid circular aluminum column

Diameter of upper segment $= 4''$

Length of column, $L = 100''$

$F_y = 35\,\text{ksi}$

$E = 10,000\,\text{ksi}$

Ramberg-Osgood parameter, $n = 20$

Fig. p4.1

$$\lambda = \frac{4}{3} \cdot \frac{L}{d_1} \sqrt{\frac{F_y}{E}} \quad p = \frac{P}{A_1 F_y}$$

$$\tau_1(p) = \frac{1}{1 + 0.002\,n \cdot \frac{E}{F_y} \cdot p^{n-1}} \qquad \tau_2(p) := \frac{1}{1 + 0.002\,n \cdot \frac{E}{F_y} \cdot (\sqrt{2} \cdot p)^{n-1}}$$

4.2. Calculate the tangent modulus strength of a stepped aluminum column with a square cross-section and specified Ramberg-Osgood tangent modulus values.

Data:

$$E = 10,000\,\text{ksi}$$

$$F_y = 35\,\text{ksi}$$

$$L = 10\,\text{ft.}$$

$$l_1 = 0.3L$$

Section 1:

$$d_1 = 2.8''$$
$$A_1 = d_1^2$$

Section 2:

$$d_2 = 4''$$
$$A_2 = d_2^2$$
$$l_2 = 0.2L$$

$$\tau = \frac{1}{1 + 12(p^{19})},\ p = \frac{P}{A_1 F_y}$$

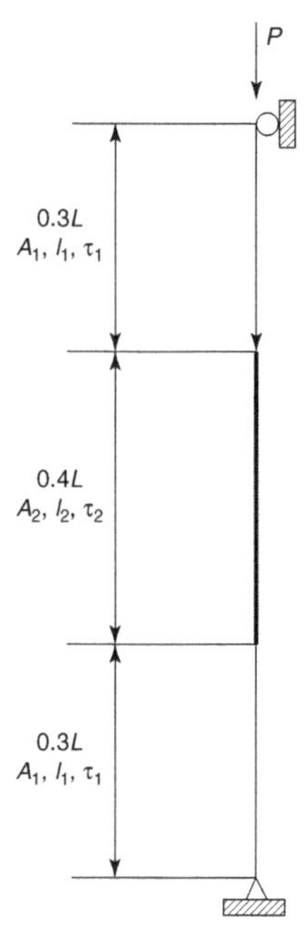

Fig. p4.2

Buckling Condition (Timeshenko and Gere 1961)

$$\tan(k_1 l_1)\tan(k_2 l_2) = \frac{k_1}{k_2}$$

$$k_1 l_1 = l_1\sqrt{\frac{P}{E\tau_1 I_1}} = 0.3L\sqrt{\frac{P}{E\tau_1 I_1}} = K_1$$

$$k_2 l_2 = l_1\sqrt{\frac{P}{E\tau_2 I_2}} = 0.2L\sqrt{\frac{P}{E\tau_2 I_2}} = K_2$$

4.3. Develop a spreadsheet program that will calculate the inelastic strength of a steel stepped column. The dimensions and properties are general. Use the Column Research Council (CRC) model for the tangent modulus (equations 3.103 and 3.104).

Data:

$$L = 15\,\text{ft.}$$
$$l_1 = 0.2\text{L}$$
$$l_2 = 0.3\text{L}$$

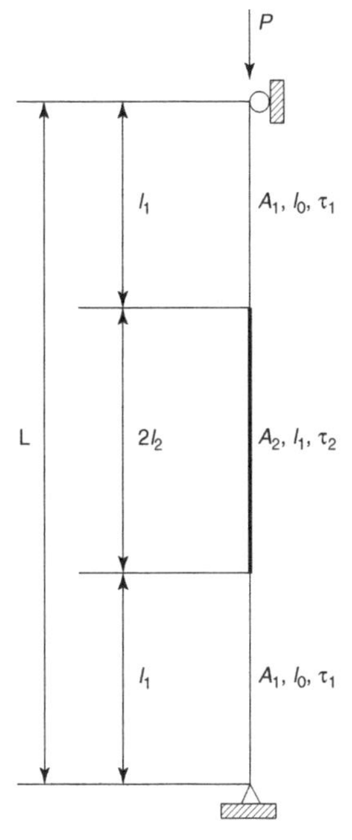

Fig. p4.3

$$F_y = 50\,\text{ksi}$$
$$E = 29{,}000\,\text{ksi}$$
$$\text{Section 1: } W8 \times 31$$
$$\text{Section 2: } W8 \times 67$$

4.4. **a.** Derive the equations in Table 4.1 for the case of uniformly distributed load.

b. Derive a formula for exact amplification factor for the case shown

c. Derive an expression for the fixed-end moment for case shown

d. Derive the formulas in Table 4.3.

e. Derive the exact expressions for the plastic moments M_{pcx} and M_{pcy} in Table 4.4.

f. Derive the equations in Table 4.5.

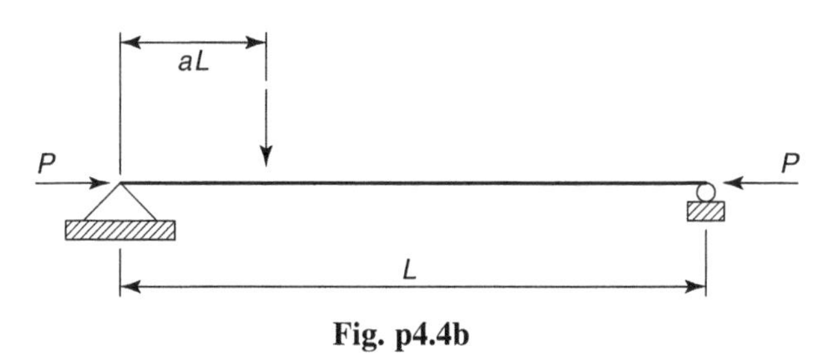

Fig. p4.4b

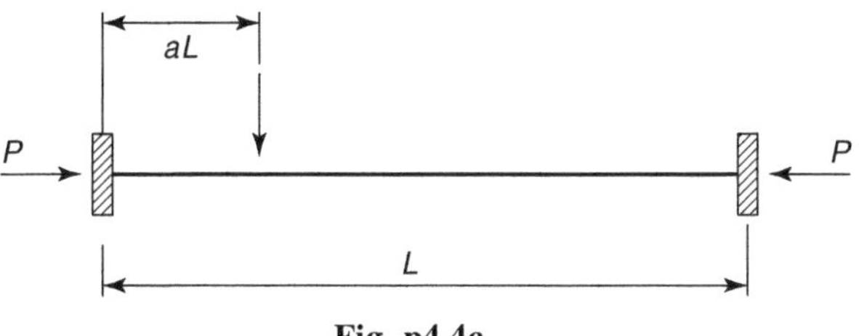

Fig. p4.4c

4.5. Draw the first-order and second-order moment diagrams and discuss the differences.

Given:

W14 × 145 section

$L_1 = 180\,\text{in.}$
$L_2 = 90\,\text{in.}$
$L_3 = 160\,\text{in.}$
$P = 5000\,\text{kip}$
$Q = 60\,\text{kip}$
$w_1 = 0.6\,\text{kip/in.}$
$w_2 = 1.2\,\text{kip/in.}$
$E = 29,000\,\text{ksi}$

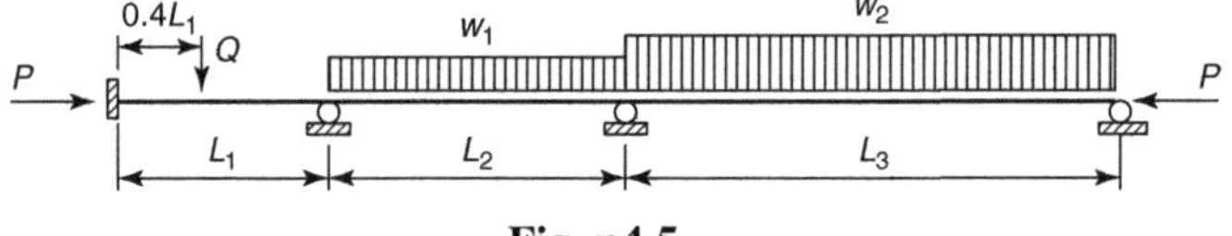

Fig. p4.5

4.6. Determine the equation for elastic buckling of a pinned-end column subject to concentric axial loads of P applied at the top of the column and at the midpoint. The member has constant stiffness EI throughout its length.

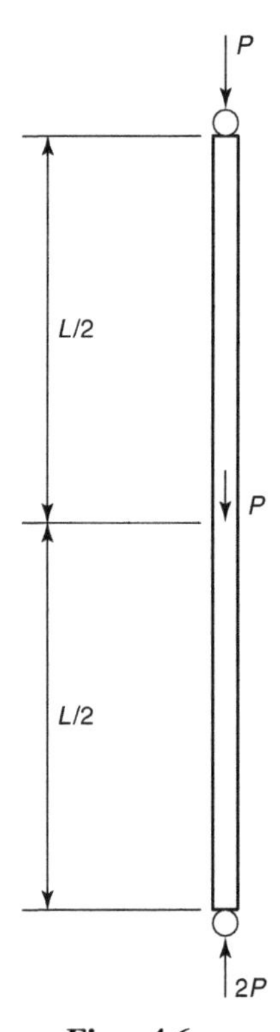

Fig. p4.6

CHAPTER FIVE

FRAME STABILITY

5.1 INTRODUCTION

In earlier chapters, the focus was on single members that failed primarily due to axial load. Unlike isolated axially loaded members, the primary means by which framed structures resist applied loads is flexure. The behavior is still affected by the presence of axial load and shear force, however. In elastic design, the maximum flexural strength (or alternatively, the ultimate strength) is defined as the condition when the plastic moment of a member is reached at the point of maximum moment. In plastic design, the maximum strength of the frame is determined when a plastic mechanism forms that leads to instability of the system.

In this chapter, the focus is on elastic behavior and the limit state of elastic frame buckling. When frame buckling occurs, it is often during erection when the structure is most vulnerable because bracing has not been installed yet; frame buckling also must be considered as one of the possible limit states in the design of the final structure.

In the following example, in-plane behavior is illustrated. That is, the structure will not buckle out of the plane of the frame. Two cases are considered. In the *symmetric* case, the structure deflects under load in one pattern of deflection until a critical load is reached (the bifurcation point, as described in Chapter 1), and if the structure is not braced, it suddenly assumes

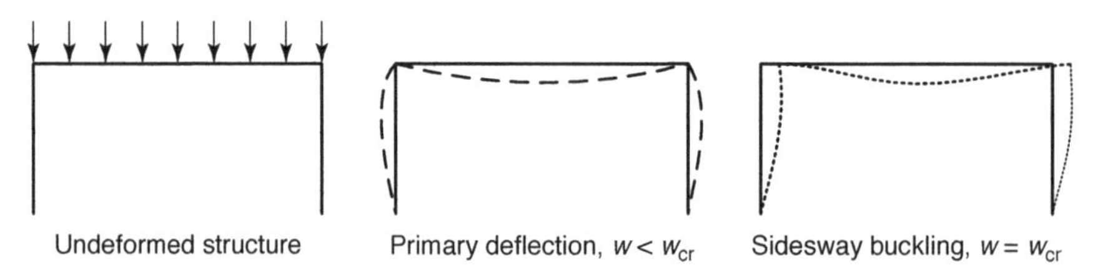

Fig. 5.1 *Symmetric* primary deflection and *unsymmetric* sidesway buckling.

another deflection pattern as it fails by sidesway. This case is shown in Figure 5.1.

In the *unsymmetric* case, shown in Figure 5.2, the normal pattern of deflection is greatly amplified as the critical load is approached. Unlike single columns or the symmetric case, the unsymmetric case does not have a bifurcation point.

The curves in Figure 5.3 illustrate the load-deflection relationships for the two types of behavior: In the *symmetric* buckling case, there is no lateral deflection Δ until the critical load w_{cr} is reached. Bifurcation takes place at this load. That is, the frame may be either undeformed, or it may take a slightly deflected position. The *unsymmetric* case, by contrast, deflects laterally from the start of loading.

Two types of curves are presented for the unsymmetric case. The linear load-deflection graph represents the *first-order response*. In first-order analysis, the load-deflection response is based on the original undeformed configuration where equilibrium is formulated on the deflected structure; if the system is elastic, the response is linear. The nonlinear curve represents *second-order response*; with each increment of load the incremental deflection is a little more than in the previous load increment. If the material remains elastic, the load-deflection curve will asymptotically approach the critical buckling load, a shown in Figure 5.3.

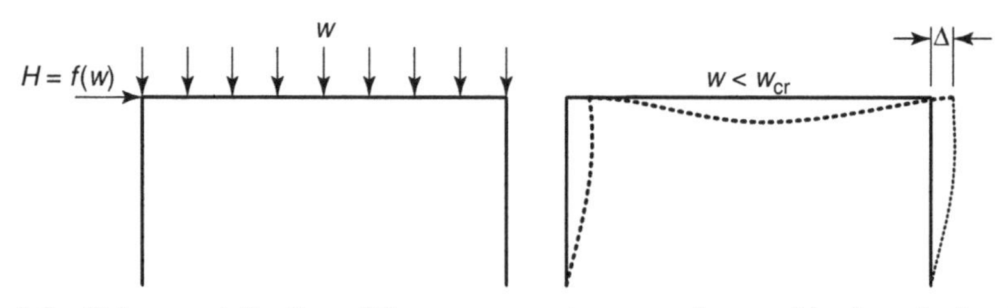

Fig. 5.2 Sidesway deflection of the *unsymmetric* case under combined vertical and lateral load.

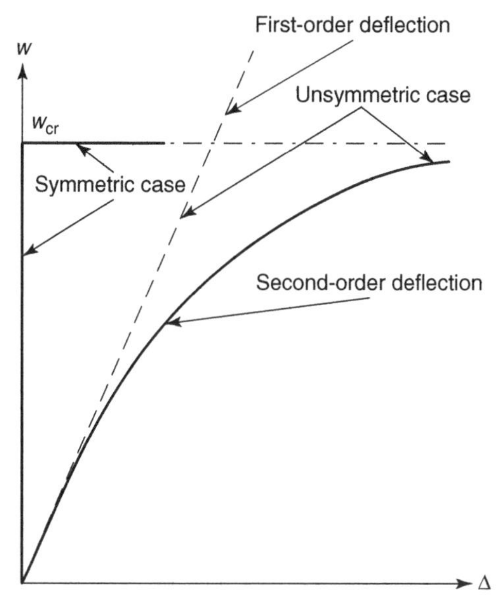

Fig. 5.3 Load-versus-lateral deflection curves.

This chapter presents various methods whereby buckling behavior is considered in the analysis of frames. Figure 5.4 illustrates in broad outline the principles involved.

- In *Step 1* the axial forces and the bending moments are determined by either first- or second-order analysis. For our purposes it is acceptable to ignore the shear forces and the effects of axial shortening. These are generally small and can, in most cases, be neglected. We are also excluding the out-of-plane effects.
- *Step 2* involves the determination of the buckling load of the frame with discrete axial loads applied to the top of the column. This idealization results in a more tractable analytical exertion as compared to finding the buckling load of the frame with a distributed load, and it is only slightly unconservative. The inaccuracy is within the accuracy most other structural calculations.
- In *Step 3,* an interaction equation of the form presented in Figure 5.4 is checked for compliance once $P_{cr} = w_{cr}L$ has been found. The moment M^* in this equation is the second-order moment found either through a second-order analysis or by amplifying the first-order moment to approximate the second order moment. M_p is the plastic moment of the member being investigated. This interaction gives the pertinent parameters

Step 1: Conduct first- or second-order force analysis, P and M

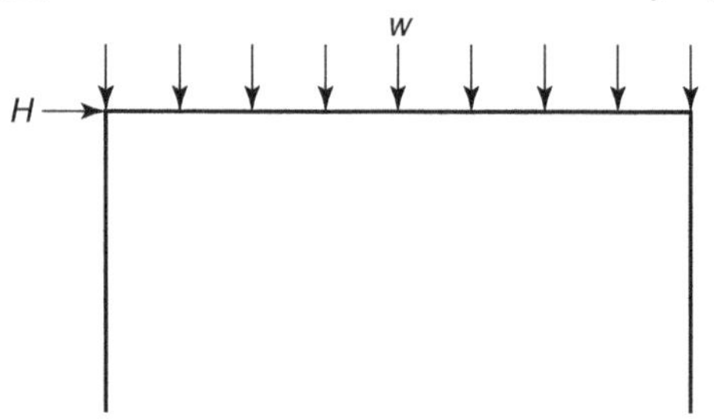

Step 2: Calculate P_{cr} by frame stability analysis.

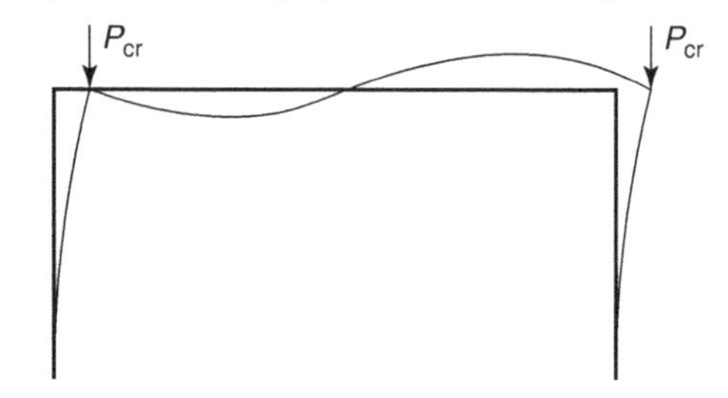

Step 3: Do interaction equation check.

$$\frac{P}{P_{cr}} + \frac{M^*}{M_p} \leq 1.0$$

Fig. 5.4 Steps for frame stability analysis.

for illustration. It is the simplest form of the equation types discussed in Chapter 4.

In the following sections of this chapter we consider rigorous and approximate ways of determining the buckling loads of frames loaded by axial forces applied on the top of the columns.

5.2 TWO-BAY FRAME EXAMPLES

As an illustration of the stability of a rigid frame structure, the two-bay single-story frame shown in Figure 5.5 is used throughout this chapter. The influence of change in boundary conditions and the effects of inelasticity are illustrated in six cases, including:

Case I: Side-sway prevented during buckling (ideally braced)
Case II Side-sway permitted during buckling (no bracing)

Case III: Frame with base restraint
Case IV: Influence of inelastic behavior
Case V: Sway buckling of frame with a leaner column
Case VI: Buckling of a frame with semi-rigid joints

The slope-deflection functions, or stability functions derived in Chapter 4, are used frequently in this chapter, and are repeated here:

$$\phi^2 = \frac{PL_C^2}{EI_C} \tag{5.1a}$$

$$c = \frac{1}{\phi^2}\left(1 - \frac{\phi}{\tan\phi}\right) \tag{5.1b}$$

$$s = \frac{1}{\phi^2}\left(\frac{\phi}{\sin\phi} - 1\right) \tag{5.1c}$$

$$C = \frac{c}{c^2 - s^2} \tag{5.1d}$$

$$S = \frac{s}{c^2 - s^2} \tag{5.1e}$$

$$\frac{C^2 - S^2}{C} = \frac{1}{c} \tag{5.1f}$$

The geometric parameters in these equations are defined in Figure 5.5. The functions refer to the column members, each with an axial load P. The beams are without axial force, and thus $C = 4$ and $S = 2$.

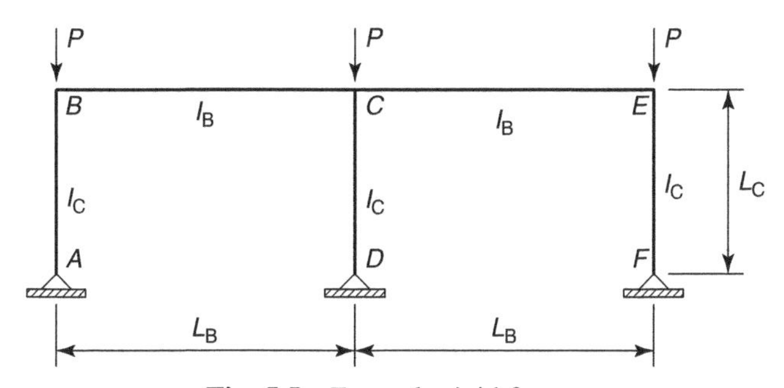

Fig. 5.5 Example rigid frame.

5.2.1 Case I: No Side-sway during Buckling

The assumed buckled shape of the frame is given in Figure 5.6. The top of the frame is prevented from moving laterally with respect to the base supports by a rigid brace. The rotation of each joint is shown clockwise. This orientation was assumed to be positive in the derivation of the slope-deflection equations in Chapter 4.

The slope-deflection equation for joint A in column AB is

$$M_{AB} = \frac{EI_C}{L_C}(C\theta_A + S\theta_B) = 0$$

From this equation

$$\theta_A = -\frac{S\theta_B}{C}$$

The corresponding slope-deflection equation for joint B is

$$M_{BA} = \frac{EI_C}{L_C}(S\theta_A + C\theta_B)$$

Substitution of θ_A and making use of the relationship in equation 5.1f, the following equation is obtained for M_{BA}.

$$M_{BA} = \frac{EI_C}{cL_C}\theta_B$$

Similarly

$$M_{CD} = \frac{EI_C}{cL_C}\theta_C$$

$$M_{EF} = \frac{EI_C}{cL_C}\theta_E$$

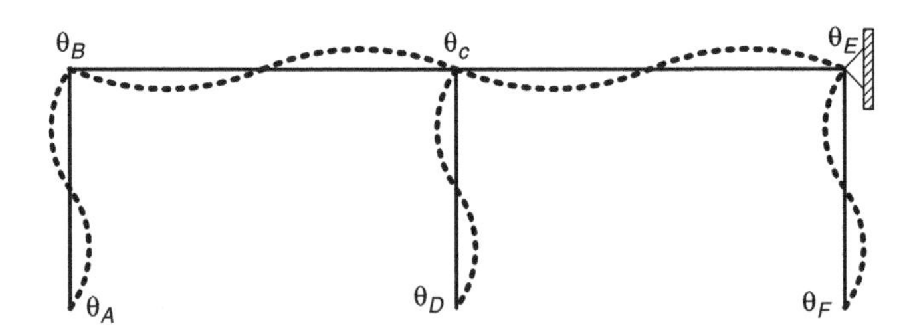

Fig. 5.6 The sway-prevented buckled shape.

The slope-deflection equations for the beams are given next:

$$M_{BC} = \frac{EI_B}{L_B}(4\theta_B + 2\theta_C)$$

$$M_{CB} = \frac{EI_B}{L_B}(2\theta_B + 4\theta_C)$$

$$M_{CE} = \frac{EI_B}{L_B}(4\theta_C + 2\theta_E)$$

$$M_{EC} = \frac{EI_B}{L_B}(2\theta_C + 4\theta_E)$$

There are three unknown slopes, θ_B, θ_C, θ_E, and so three equilibrium equations are needed. These are provided by summation of moments at joints B, C, and D:

$$M_{BA} + M_{BC} = 0$$
$$M_{CB} + M_{CD} + M_{CE} = 0$$
$$M_{EC} + M_{EF} = 0$$

Substitution of the slope-deflection equations into the equilibrium equations results in three homogeneous simultaneous equations. The buckling load is obtained by setting the determinant of the coefficients of the slopes equal to zero:

$$\begin{vmatrix} \left(\frac{1}{c} + 4\gamma\right) & 2\gamma & 0 \\ 2\gamma & \left(\frac{1}{c} + 8\gamma\right) & 2\gamma \\ 0 & 2\gamma & \left(\frac{1}{c} + 4\gamma\right) \end{vmatrix} = 0$$

where

$$\gamma = \frac{I_B L_C}{I_C L_B} \tag{5.2}$$

Performing the decomposition of the determinant results in the following characteristic equation, or *eigenfunction*:

$$\left[\frac{1}{c} + 4\gamma\right]\left[\frac{1}{c^2} + \frac{12\gamma}{c} + 24\gamma^2\right] = 0 \tag{5.3}$$

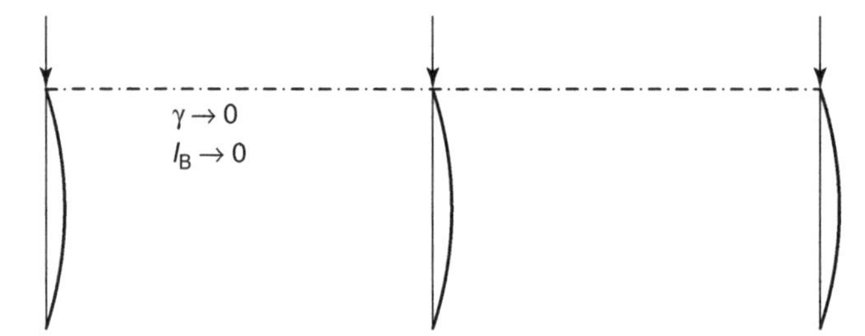

Fig. 5.7 Beam with zero stiffness.

Equation 5.3 has three roots:

$$\begin{aligned}\left(\frac{1}{c}\right)_1 &= -4\gamma \\ \left(\frac{1}{c}\right)_2 &= (-6+2\sqrt{3})\gamma = -2.536\gamma \\ \left(\frac{1}{c}\right)_3 &= (-6-2\sqrt{3})\gamma = -9.464\gamma\end{aligned} \tag{5.4}$$

The limits of equation 5.4 will next be checked to ascertain that no mistake was made in the derivation. The first limit to be considered is when the restraining beams have zero stiffness (Figure 5.7). In this case, $\gamma = \frac{I_B L_C}{I_C L_C} = 0$, and the buckling load should be $P_{cr} = \frac{\pi^2 EI_C}{L_C^2}$. From equations 5.4 we deduce that

$$\frac{1}{c} = 0 = \frac{\phi^2 \tan\phi}{\tan\phi - \phi} \rightarrow \tan\phi = 0 \rightarrow \phi^2 = \frac{PL_C^2}{EI_C} = \pi^2$$

Thus, the lower limit of the buckling load is correct. Next, we will ascertain the validity of the upper limit. This occurs when $\gamma = \frac{I_B L_C}{I_C L_C} = \infty$, (Figure 5.8)

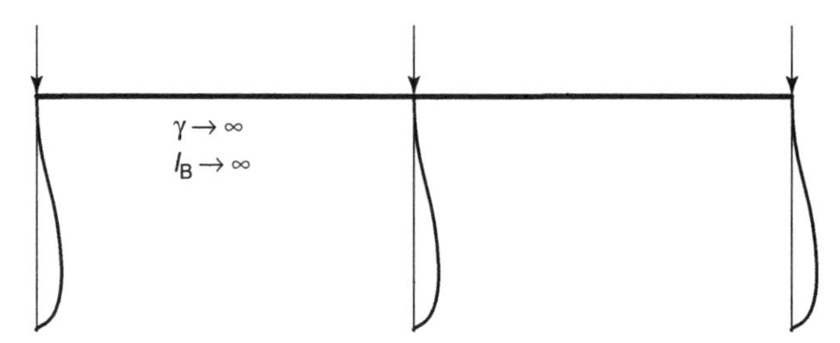

Fig. 5.8 Beams with infinite stiffness.

and the answer should be $P_{\text{cr}} = \frac{\pi^2 EI_C}{(0.70L_C)^2}$, or $\phi^2 = 20.20$. From equation 5.5, it is seen that

$$c = 0 = \frac{1}{\phi^2}\left[1 - \frac{\phi}{\tan\phi}\right] \rightarrow \tan\phi = \phi \rightarrow \phi = 4.49 \rightarrow \phi^2 = 20.20$$

The buckling load of the frame is, therefore

$$\pi^2 \leq \phi^2 = \frac{PL_C^2}{EI_C} \leq 20.20$$

The next question to be considered is, which of the equations 5.4 give the lowest buckling load? In order to obtain the answer, the curves for three equations are plotted in Figure 5.9. The lowest curve is equation 5.4b,

$$\gamma_{\text{cr}} = \frac{-1}{2.536c} = -\frac{0.394}{c} = -0.394\phi^2\left[\frac{\tan\phi}{\tan\phi - \phi}\right]$$

The mode-shape is obtained from the first equilibrium equation:

$$\theta_B = -\theta_C\left[\frac{2\gamma}{\frac{1}{c} + 4\gamma}\right] = -\theta_C\left[\frac{2\gamma}{-2.536\gamma + 4\gamma}\right] = -1.366\theta_C$$

Similarly, $\theta_E = -1.366\,\theta_C$. The buckled shape of the frame is shown in Figure 5.10.

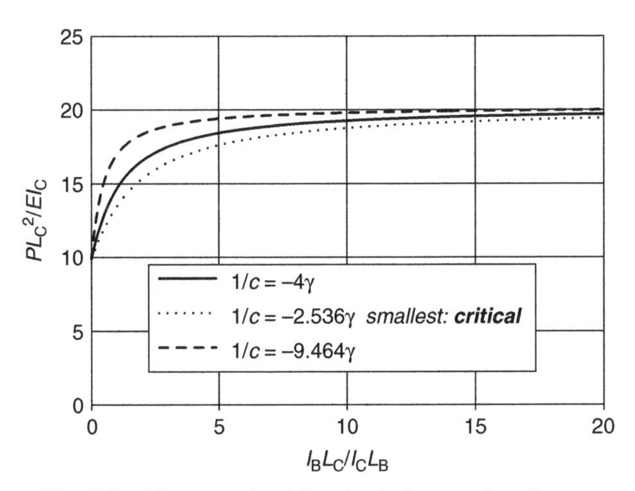

Fig. 5.9 Non-sway buckling loads for two-bay frame.

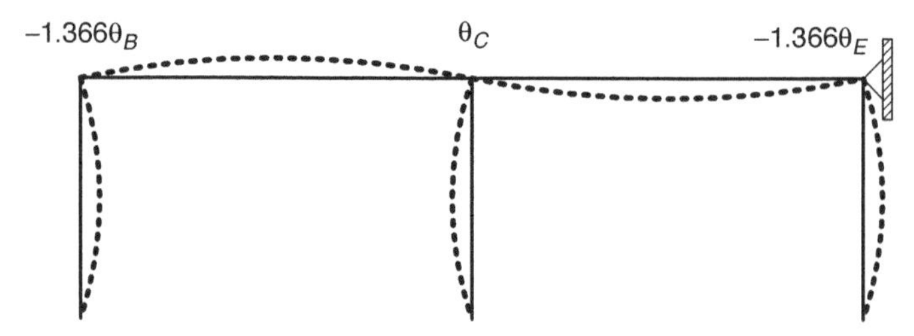

Fig. 5.10 Mode shape for lowest critical load.

The critical load can be determined using any solution software. Consider the following numerical example:

$$I_B = 2I_C$$

$$L_B = 3L_C$$

$$\gamma = \frac{I_B L_C}{I_C L_B} = \frac{2}{3}$$

$$\frac{1}{c} = \frac{\phi^2 \tan\phi}{\tan\phi - \phi} = \gamma(-6 + 2\sqrt{3}) = -1.691$$

Solving for the critical load gives $\phi^2 = \frac{PL_C^2}{EI_C} = 12.53$, or

$$P_{cr} = \frac{12.53EI_C}{L_C^2} = \frac{\pi^2 EI_C}{(0.89L_C)^2}$$

The effective length factor is thus 0.89.

Next we solve the problem with the AISC alignment chart method. The features of the solution are given in Figure 5.11.

For the story: $3P = 2\left[\frac{\pi^2 EI_C}{(0.90L_C)^2}\right] + \frac{\pi^2 EI_C}{(0.85L_C)^2} = \frac{\pi^2 EI_C}{(0.88L_C)^2}$. This compares with an effective length factor of 0.89 obtained by the analytical method.

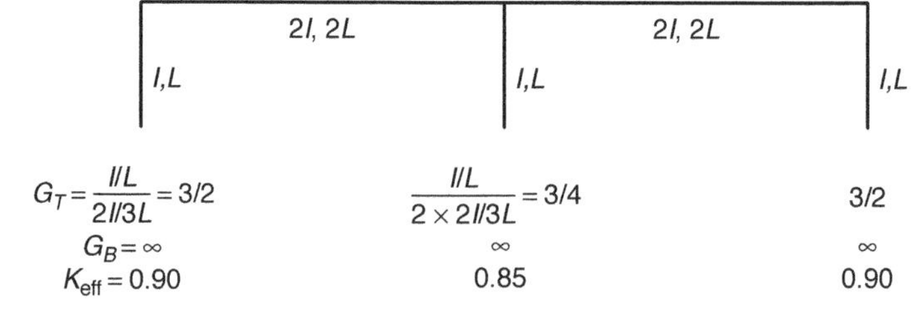

Fig. 5.11 Effective length factors using the alignment charts.

5.2.2 Case II: Side-sway Permitted during Buckling

The next case is illustrated in Figure 5.12. It differs from the case in Figure 5.6 in that the restraint preventing the lateral movement of the top of the frame is taken away and the frame is free to sway. Whereas in the non-sway case, the effective length factor could be between 0.5 and 1.0 depending on the stiffness of the beams, the effective length for the sway-buckling case can vary from 1.0 to infinity. In the previous case, it is conservative to simply use $K_{\text{eff}} = 1.0$. There is no such simple estimation of a safe, effective length factor for the frame in Figure 5.12.

The slope-deflection equations are derived as for the non-sway case. For the columns:

$$\rho = \frac{\Delta}{L_C}$$

$$M_{AB} = 0 = \frac{EI_C}{L_C}[C\theta_A + S\theta_B - (C+S)\rho]$$

$$\theta_A = -\frac{S}{C\theta_B} + \left(\frac{C+S}{C}\right)\rho$$

$$M_{BA} = \frac{EI_C}{L_C}[C\theta_B + S\theta_A - (C+S)\rho]$$

$$M_{BA} = \frac{EI_C}{L_C}\left(\frac{C^2 - S^2}{C}\right)(\theta_B - \rho)$$

$$M_{BA} = \frac{EI_C}{cL_C}(\theta_B - \rho)$$

$$M_{CD} = \frac{EI_C}{cL_C}(\theta_C - \rho)$$

$$M_{EF} = \frac{EI_C}{cL_C}(\theta_E - \rho)$$

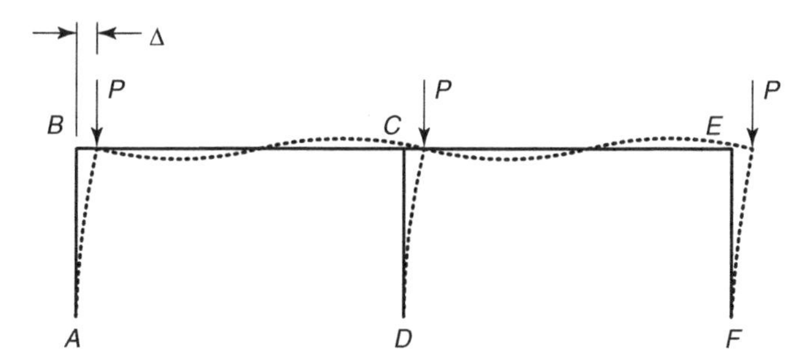

Fig. 5.12 Sway-permitted frame buckling mode shape.

For the beams:

$$M_{BC} = \frac{EI_B}{L_B}(4\theta_B + 2\theta_C)$$
$$M_{CB} = \frac{EI_B}{L_B}(2\theta_B + 4\theta_C)$$
$$M_{CE} = \frac{EI_B}{L_B}(4\theta_C + 2\theta_E)$$
$$M_{EC} = \frac{EI_B}{L_B}(2\theta_C + 4\theta_E)$$

Since there are four unknown displacements, θ_B, θ_C, θ_E, ρ, we need four equilibrium equations. The joint equilibrium equations are

$$M_{BA} + M_{BC} = 0$$
$$M_{CB} + M_{CD} + M_{CE} = 0$$
$$M_{EC} + M_{EF} = 0$$

Since there are only three equilibrium equations and four unknown displacements, a fourth condition is needed. That condition is a story equilibrium equation, which is obtained from the condition that the sum of the shear forces at the bases of the columns is zero at the instant of buckling. The forces are shown on Figure 5.13:

Taking moments about the top of each column, adding the three equations, and noting that the sum of the shears is zero, gives the fourth equilibrium equation:

$$\sum M_{AB} = PL_C\rho + M_{BA} + H_A L_C = 0$$
$$\sum M_{AC} = PL_C\rho + M_{CD} + H_D L_C = 0$$
$$\sum M_{AF} = PL_C\rho + M_{EF} + H_F L_C = 0$$

$$3PL_C\rho + M_{BA} + M_{CD} + M_{EF} + (H_A + H_D + H_F)L_C = 0$$
$$H_A + H_D + H_F = 0$$
$$3PL_C\rho + M_{BA} + M_{CD} + M_{EF} = 0$$

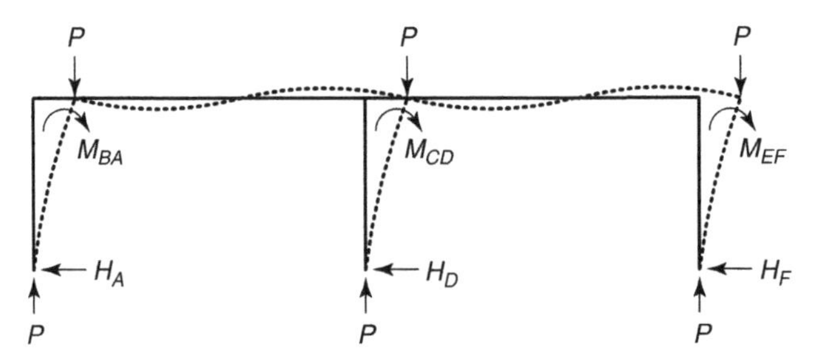

Fig. 5.13 Forces on the columns in the buckled frame.

Substitution of the slope-deflection equations into the four equilibrium equations results in four simultaneous homogeneous equations for the four unknowns. The determinant of the coefficients of the unknown deformations contains the critical load under which these deformations are possible (i.e., the sidesway buckling load of the frame).

Slope-deflection equations:

$$M_{BA} = \frac{EI_C}{cL_C}(\theta_B - \rho) \qquad M_{BC} = \frac{EI_B}{L_B}(4\theta_B + 2\theta_C)$$

$$M_{CB} = \frac{EI_B}{L_B}(2\theta_B + 4\theta_C)$$

$$M_{CD} = \frac{EI_C}{cL_C}(\theta_C - \rho) \qquad M_{CE} = \frac{EI_B}{L_B}(4\theta_C + 2\theta_E)$$

$$M_{EF} = \frac{EI_C}{cL_C}(\theta_E - \rho) \qquad M_{EC} = \frac{EI_B}{L_B}(2\theta_C + 4\theta_E)$$

Equilibrium equations:

$$\begin{aligned} M_{BA} + M_{BC} &= 0 \\ M_{CB} + M_{CD} + M_{CE} &= 0 \\ M_{EC} + M_{EF} &= 0 \\ M_{BA} + M_{CD} + M_{EF} + 3PL_C\rho &= 0 \end{aligned}$$

Determinant of the coefficients:

$$\begin{vmatrix} \frac{1}{c}+4\gamma & 2\gamma & 0 & \frac{-1}{c} \\ 2\gamma & \frac{1}{c}+8\gamma & 2\gamma & \frac{-1}{c} \\ 0 & 2\gamma & \frac{1}{c}+4\gamma & \frac{-1}{c} \\ \frac{1}{c} & \frac{1}{c} & \frac{1}{c} & 3\left(\phi^2 - \frac{1}{c}\right) \end{vmatrix} = 0$$

where

$$\phi = \sqrt{\frac{PL_C^2}{EI_C}} \tag{5.5}$$

$$c = \frac{1}{\phi^2}\left(1 - \frac{\phi}{\tan\phi}\right) \tag{5.6}$$

$$\gamma = \frac{I_B L_C}{I_C L_B} \tag{5.7}$$

Decomposition of the determinant results in the following characteristic equation:

$$\left(\frac{1}{c}+4\gamma\right)\left[\phi^2\left(\frac{1}{c^2}+\frac{12\gamma}{c}+24\gamma^2\right)-\frac{8\gamma}{c}\left(\frac{1}{c}+3\gamma\right)\right]=0 \tag{5.8}$$

The first bracket is identical to the first equation in equation 5.4. It corresponds to a higher mode of the nonsway buckling criterion. The buckling load appropriate to the sway-buckling mode is, therefore, the second bracket:

$$\phi^2\left(\frac{1}{c^2}+\frac{12\gamma}{c}+24\gamma^2\right)-\frac{8\gamma}{c}\left(\frac{1}{c}+3\gamma\right)=0 \tag{5.9}$$

Next we will check if this equation results in the correct upper and lower limits of the buckling load. When $\gamma = 0 \rightarrow I_B = 0$ the critical load should be zero, and when $\gamma = \infty \rightarrow I_B = \infty$ the critical load should be $P_{cr} = \frac{\pi^2 EI_C}{(2L_C)^2}$. This is indeed so, and thus the critical load is bounded by

$$0 \leq \frac{PL_C^2}{EI_C} \leq \frac{\pi^2}{4} \quad \text{as} \quad 0 \leq \frac{I_B L_C}{I_C L_B} < \infty \tag{5.10}$$

The relationship between $\frac{I_B L_C}{I_C L_B}$ and $\frac{PL_C^2}{EI_C}$ is shown in Figure 5.14. The critical load rises rapidly from zero as the beam stiffness ratio increase by small

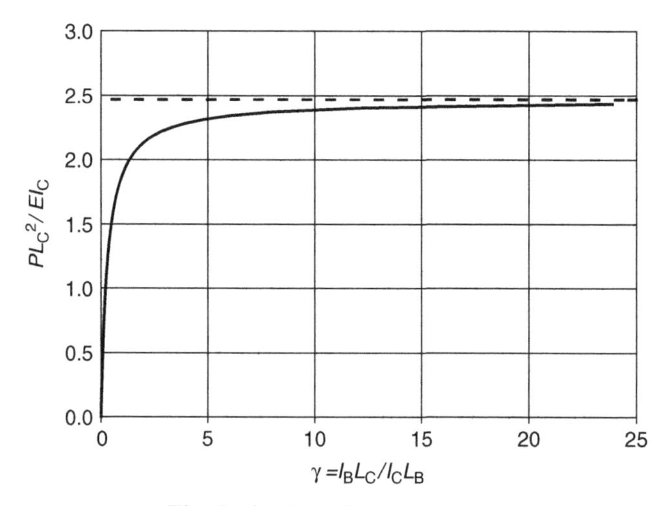

Fig. 5.14 Plot of equation 5.14.

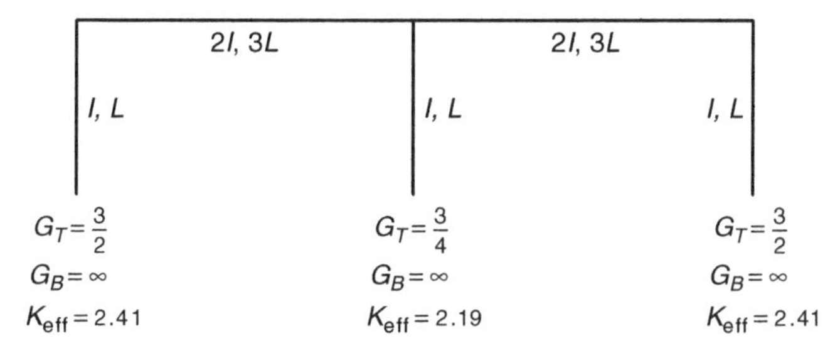

Fig. 5.15 Data for using the AISC alignment charts.

increments. For all practical purposes, $\frac{PL_C^2}{EI_C} = \frac{\pi^2}{4} = 2.47$, or the effective length factor is 2.0, when $\frac{I_B L_C}{I_C L_B} > 20$. The relationship between restraint, in this case from the beam, and the critical load as shown in Figure 5.14 is typical of stability problems: a little restraint helps a lot, but full rigidity is not necessary, or even practically attainable.

For $\gamma = \frac{2}{3}$ the MATHCAD equation solver gives

$$\phi = 1.33 \rightarrow \phi^2 = \frac{PL_C^2}{EI_C} = 1.77$$

This problem will next be solved with the AISC alignment charts. The flexibility factors and the effective lengths of each of the three columns are shown in Figure 5.15.

The buckling load for the structure is then the sum of the buckling loads of each column:

$$3P = \frac{\pi^2 EI_C}{L_C^2}\left[2 \times \frac{1}{2.41^2} + \frac{1}{2.19^2}\right] = 0.5525\frac{\pi^2 EI_C}{L_C^2}$$

$$\frac{PL_C^2}{EI_C} = 1.82$$

The two methods give comparable answers: 1.77 versus 1.82, or story effective length factors $K = 2.36$ versus 2.33, respectively, for the exact and the approximate method.

5.2.3 Case III: Frame with Base Restraints

Next we examine the effect of springs restraining the column bases. The frame is shown in Figure 5.16. The base of each column is restrained by a rotational spring with a spring constant α.

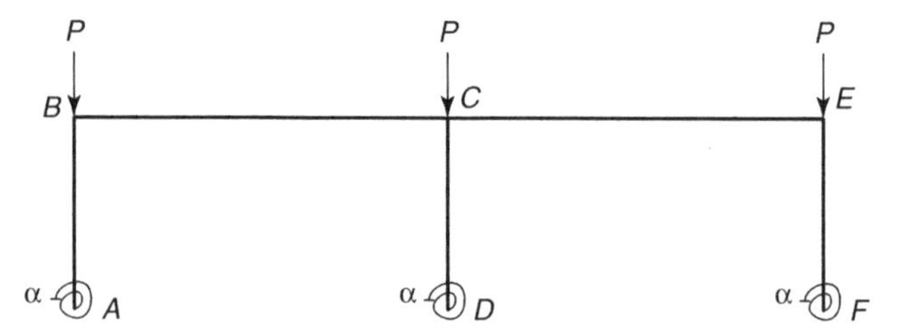

Fig. 5.16 Frame with base restraints.

From the previous solutions of this frame it can be assumed that rotations in the two outside columns are identical: $\theta_A = \theta_F$ and $\theta_B = \theta_E$. From this it follows that $M_{AB} = M_{FE}$ and $M_{BA} = M_{EF}$. The unknown displacements will then be θ_A, θ_B, θ_C, θ_D, ρ. The slope-deflections equations were given in the previous problem. There are five equilibrium equations:

$$
\begin{aligned}
M_{AB} + \alpha\theta_A &= 0 \\
M_{BA} + M_{BC} &= 0 \\
M_{CB} + M_{CD} + M_{CE} &= 0 \\
M_{DC} + \alpha\theta_D &= 0 \\
2M_{AB} + 2M_{BA} + M_{DC} + M_{CD} + 3PL_C\rho &= 0
\end{aligned}
$$

Substitution of the slope-deflection equations into the equilibrium equations will give five homogeneous simultaneous equations. The determinant of the coefficients of the unknown deflections is then equated to zero to obtain the critical load. Because of the excessive algebra that one would have to perform, the solution is obtained subsequently by numerical solution with the MATHCAD program.

$$
\begin{vmatrix}
C + \dfrac{\alpha L_C}{EI_C} & S & 0 & 0 & -(C+S) \\
S & C + 4\gamma & 2\gamma & 0 & -(C+S) \\
0 & 4\gamma & C + 8\gamma & S & -(C+S) \\
0 & 0 & S & C + \dfrac{\alpha L_C}{EI_C} & -(C+S) \\
2(C+S) & 2(C+S) & (C+S) & (C+S) & -6(C+S) + \dfrac{3PL_C^2}{EI_C}
\end{vmatrix} = 0 \tag{5.11}
$$

In the AISC alignment charts, the assumption is that all columns in the story fail at once. This approximation leads to the assumption that $\theta_B = \theta_C = \theta_E$

and $\theta_A = \theta_D = \theta_F$. The resulting buckling determinant is then a 3×3 matrix.

$$\begin{vmatrix} C+12\gamma & S & -(C+S) \\ S & C+\dfrac{\alpha L_C}{EI_C} & -(C+S) \\ (C+S) & (C+S) & -2(C+S)+\dfrac{PL_C^2}{EI_C} \end{vmatrix} = 0 \tag{5.12}$$

Table 5.1 lists results from the numerical solutions performed by MATHCAD for the case of $\gamma = \frac{I_B L_C}{I_C L_B} = 2.0$.

This table gives the solutions for the critical elastic buckling loads with the increase of the base restraint spring constant α.

$$P_{cr} = \frac{\phi^2 EI_C}{L_C^2} = \frac{\pi^2 EI_C}{\left(\dfrac{\pi}{\phi} L_C\right)^2} = \frac{\pi^2 EI_C}{(K_{eff} L_C)^2}$$

From this tabulation it can be observed that there is little difference between the results from the two determinants (equations 5.11 and 5.12), where the only difference is that in the second equation it is assumed, as for the AISC alignment charts, that the rotations of all joints in the story are identical. Essentially the same effective length factors can be calculated by using the

TABLE 5.1 Effective Length Factors for base Restrained Frames

$\frac{\alpha L_C}{EI_C}$	ϕ eq. 5.15	K_{eff} eq. 5.15	ϕ eq. 5.16	K_{eff} eq. 5.16	G_{base}
0	1.34	2.34	1.39	2.37	∞ (theoretical "pinned" base)
0.4	1.62	1.93	1.66	1.88	10 (nominally "pinned" base)
1.0	1.90	1.66	1.96	1.61	
2.0	2.17	1.45	2.24	1.41	
3.0	2.33	1.35	2.41	1.30	
4.0	2.45	1.28	2.53	1.24	1.0 (nominally "fixed" base)
10	2.74	1.15	2.84	1.11	
100	3.01	1.04	3.13	1.01	
1000	3.05	1.03	3.16	1.00	≈ 0 (theoretical "fixed" base)

AISC alignment charts as the values in column 5 of Table 5.1, with a lot less effort.

It is also evident by comparing the first and second lines in Table 5.1 that a small amount of restraint increases the strength substantially. By comparing the last two lines it is again demonstrated that very large increases of restraint yield only small increments of the buckling load, thus again showing that full rigidity is not worth striving for because one can get close enough with incomplete fixity. This observation was made by Galambos (Galambos 1960). The AISC Commentary (AISC 2005) has since 1963 recommended that a flexibility factor of $G = 10$ should be used instead of $G = \infty$ for nominally pinned bases, and $G = 1$ instead of $G = 0$ for nominally fixed bases. If information is available on the actual base conditions so that a base-restraint spring constant can be estimated, then a more precise effective length factor can be determined by the method just presented.

5.2.4 Case IV: Influence of Inelastic Behavior

The following problem is the same as Case II, the sway-buckling solution of the two-bay frame of Figure 5.5. The previous solution assumed that the material was elastic. The current problem expands the solution to the case where the columns can be yielded. The assumptions of the *tangent modulus theory*, as presented in Chapter 3, is used in calculating the frame buckling load.

Since the beams have no axial load, the slope-deflection equations will not be affected by yielding. From the previous elastic solution it is known that $\theta_B = \theta_E$, $\theta_A = \theta_F$ and, therefore

$$M_{BC} = M_{EC} = \frac{EI_{\text{B}}}{L_{\text{B}}}(4\theta_B + 2\theta_C)$$

$$M_{CB} = M_{CE} = \frac{EI_{\text{B}}}{L_{\text{B}}}(2\theta_B + 4\theta_C)$$

The columns each support the axial load P, and so their stiffness is governed by the tangent modulus $E_t = \tau E$. The slope-deflection equations are then

$$M_{BA} = M_{AB} = \frac{\tau EI_{\text{C}}}{cL_{\text{C}}}(\theta_B - \rho)$$

$$M_{CD} = \frac{\tau EI_{\text{C}}}{cL_{\text{C}}}(\theta_C - \rho)$$

With the definitions

$$\gamma = \frac{I_B L_C}{I_C L_B}$$
$$\phi^2 = \frac{P L_C^2}{\tau E I_C}$$
$$c = \frac{1}{\phi^2}\left(1 - \frac{\phi}{\tan\phi}\right)$$

and the equilibrium equations

$$M_{BA} + M_{BC} = 0$$
$$2M_{CB} + M_{CD} = 0$$
$$2M_{BA} + M_{CD} + 3PL_C\rho = 0$$

the buckling determinant becomes equal to

$$\begin{vmatrix} \frac{\tau}{c}+4\gamma & 2\gamma & -\frac{\tau}{c} \\ 4\gamma & \frac{\tau}{c}+8\gamma & -\frac{\tau}{c} \\ \frac{2\tau}{c} & \frac{\tau}{c} & \frac{3PL_C^2}{EI_C} - \frac{3\tau}{c} \end{vmatrix} = 0$$

The decomposition of this determinant results in the following buckling equation:

$$\frac{PL_C^2}{EI_C}\left[\left(\frac{\tau}{c}\right)^2 + \frac{12\tau\gamma}{c} + 24\gamma^2\right] - \left(\frac{\tau}{c} + 3\gamma\right)\frac{8\tau\gamma}{c} = 0 \qquad (5.13)$$

This equation is identical to equation 5.9 if the tangent modulus $\tau = 1.0$.

The tangent modulus that will be used for solving for the inelastic buckling load is τ equal to equation 5.14 (same as equations 3.103 and 3.104):

$$\begin{aligned} \tau &= 1.0 \quad \text{if} \quad 0 \leq p \leq 0.5 \\ \tau &= 4p(1-p) \quad \text{if} \quad 0.5 < p \leq 1.0 \end{aligned} \qquad (5.14)$$

where $p = \dfrac{P}{P_y}$

Now, we illustrate the application of the tangent modulus method with an example solved on the MATHCAD program. The beams are W24 × 94 shapes, and the columns are W14 × 99 shapes.

$$L_B = 40\text{ ft.}$$

$$L_C = 20\text{ ft.}$$

$$I_B = 2{,}700\text{ in}^4$$

$$I_C = 1{,}110\text{ in}^4 \qquad A = 29.1\text{ in}^2 \quad r_x = 6.17\text{ in.}$$

$$E = 29{,}000\text{ ksi} \quad F_y = 50\text{ ksi}$$

$$P_y = AF_y \qquad P_y = 1.455 \times 10^3\text{ kip}$$

$$\gamma = \frac{I_B L_C}{I_C L_B} \qquad \gamma = 1.216$$

$$\lambda = \frac{L_C}{\pi r_x}\sqrt{\frac{F_y}{E}} \qquad \lambda = 0.514$$

$$p = 0.6 \quad \text{(Assumed value to initiate numerical solution in the MATHCAD program)}$$

$$\tau(p) = \left| \begin{array}{ll} 1.0 & \text{if } p \le 0.5 \\ 4p(1-p) & \text{otherwise} \end{array} \right. \qquad \phi(p) = \pi\lambda\sqrt{\frac{p}{\tau(p)}} \qquad c(p) = \frac{1 - \frac{\phi(p)}{\tan(\phi(p))}}{\phi(p)^2}$$

$$F(p) = \phi(p)^2\left[\frac{1}{c(p)^2} + \frac{12\gamma}{c(p)\tau(p)} + 24\left(\frac{\gamma}{\tau(p)}\right)^2\right] - \frac{8\gamma}{c(p)\tau(p)} \times \left(\frac{1}{c(p)} + \frac{3\gamma}{\tau(p)}\right)$$

$$\text{Answer} = \text{root}(F(p),\ p) \qquad \text{Answer} = 0.676$$

$$P_{cr} = \text{Answer} \times P_y \qquad P_{cr} = 983.025\text{ kip}$$

For the condition of elastic buckling, $\tau = 1.0$, the critical load is $P_{cr} = 1,094$ kip. The tangent modulus ratio τ at the critical inelastic load of $p_{cr} = \frac{P_{cr}}{P_y} = 0.676$ is equal to $\tau_{cr} = 4p_{cr}(1 - p_{pr}) = 4 \times 0.676 \times (1 - 0.676) = 0.876$. As a result, the frame-buckling capacity of the structure was reduced by about 10 percent.

The inelastic buckling strength of the frame can be estimated by conventional design office methods using the AISC alignment charts. (Yura, 1971)

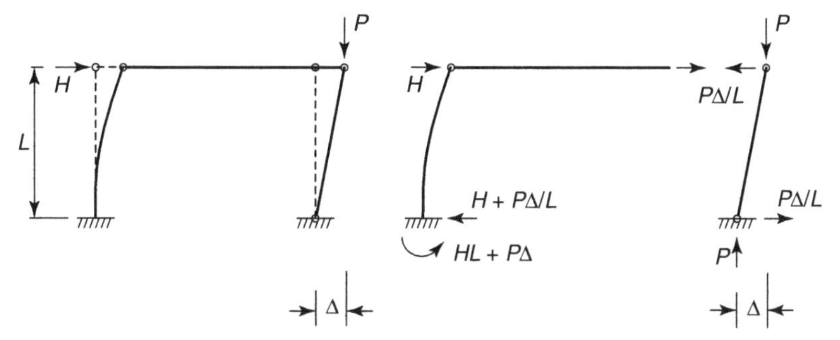

Fig. 5.17 Effects of leaning column on lateral resisting column.

5.2.5 Case V: Sway Buckling of Frame with a Leaner Column

The case to be considered next involves a frame that has, in addition to members with rigidly connected joints, columns that are pinned at their ends. Such members must support the load placed on them as pinned-end columns—that is, having an effective length factor of unity. They do not, however, contribute to the resistance of the frame to side-sway buckling, since they have no lateral stiffness. In a sense, the loads on these columns are an extra burden that is leaning on the lateral force-resisting system.

The frame shown in Figure 5.17 shows the effect of the leaning column on a lateral support column. The axial load on the leaning column decreases the capacity of the lateral support column by increasing the moment in the lateral resisting column. It is not uncommon in design practice to lean multiple pinned-end columns on a few lateral force resisting columns, and it is thus important to understand the effects of the leaning columns on the overall frame stability.

The two-bay single story example frame with a single leaning column in shown in Figure 5.18.

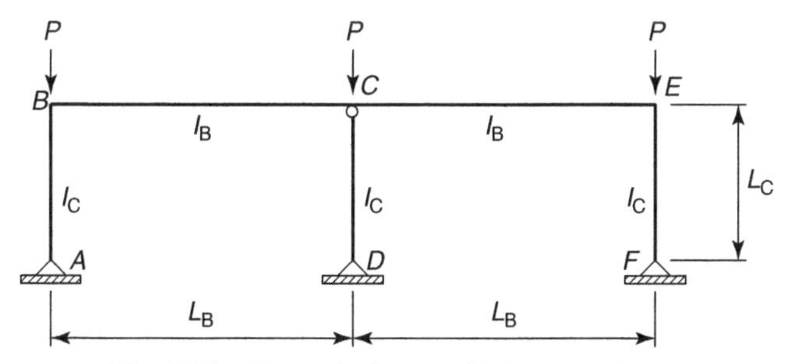

Fig. 5.18 Example frame with leaner column.

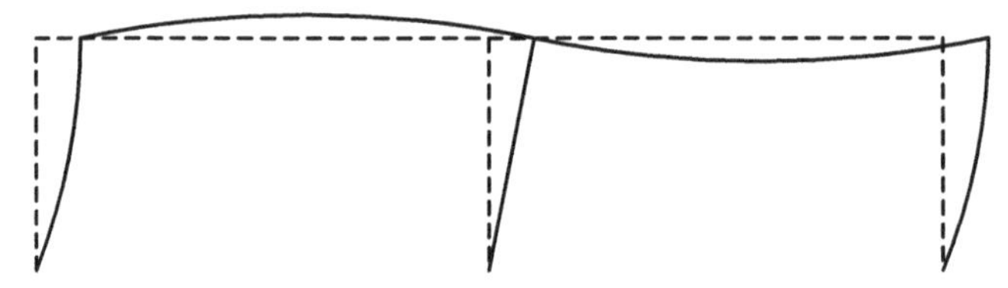

Fig. 5.19 The buckled shape.

Due to the anti-symmetry of the buckled shape, shown in Figure 5.19, $\theta_B = \theta_E$; therefore, $M_{BA} = M_{EF}$ and $M_{BC} = M_{EC}$. The moments and the equilibrium equations are

$$M_{BA} = M_{EF} = \frac{EI_{\mathrm{C}}}{cL_{\mathrm{C}}}(\theta_B - \rho)$$

$$M_{CB} = M_{CE} = \frac{EI_{\mathrm{B}}}{L_{\mathrm{B}}}(4\theta_C + 2\theta_B)$$

$$M_{EC} = M_{BC} = \frac{EI_{\mathrm{B}}}{L_{\mathrm{B}}}(4\theta_B + 2\theta_C)$$

$$M_{BA} + M_{BC} = 0$$
$$M_{CB} + M_{CE} = 0$$
$$M_{BA} + M_{EF} + 3PL_{\mathrm{C}}\rho = 0$$

Substitution of the moment expressions into the equilibrium equations yields three homogeneous simultaneous equations. The vanishing of the determinants of the coefficients of the unknown deformations will result in an equation that contains the critical sidesway buckling load.

$$\gamma = \frac{I_{\mathrm{B}}L_{\mathrm{C}}}{I_{\mathrm{C}}L_{\mathrm{B}}}, \ \phi^2 = \frac{PL_{\mathrm{C}}^2}{EI_{\mathrm{C}}}, \ c = \frac{1}{\phi^2}\left(1 - \frac{\phi}{\tan\phi}\right)$$

$$\begin{vmatrix} \frac{1}{c} + 4\gamma & 2\gamma & -\frac{1}{c} \\ 1 & 2 & 0 \\ \frac{2}{c} & 0 & 3\phi^2 - \frac{2}{c} \end{vmatrix} = 0$$

$$3\phi^2\left(\frac{2}{c} + 6\gamma\right) - \frac{16\gamma}{c} = 0$$

For $\gamma = 2$ the numerical solution gives $\phi = 1.12$ and the effective length then becomes equal to $K_{\mathrm{eff}} = \frac{\pi}{\phi} = 2.81$.

The same answer can be obtained with the AISC alignment charts, as illustrated in the following example.

For columns AB and EF,

$$G_{\mathrm{TOP}} = \frac{I_{\mathrm{C}}/L_{\mathrm{C}}}{0.5 I_{\mathrm{B}}/L_{\mathrm{B}}} = \frac{2}{\gamma} = 1.0,\ G_{\mathrm{Bottom}} = \infty,\ K_{\mathrm{eff}} = 2.3.$$

The 0.5 in the denominator is there because the far end of each beam has a slope in the opposite direction to the slope at the near end. In the derivation of the flexibility factor G it was assumed that the two end-slopes are in the same (clockwise) direction, as shown in Figure 5.6. The beams in Figure 5.19 have half the stiffness that they would have if they were bent in an S-shape.

The buckling load for the story is determined as follows:

$$3P_{\mathrm{cr}} = 2 \times \frac{\pi^2 E I_{\mathrm{C}}}{(2.3 L_{\mathrm{C}})^2}$$

$$P_{\mathrm{cr}} = \frac{2}{3} \times \frac{\pi^2 E I_{\mathrm{C}}}{(2.3 L_{\mathrm{C}})^2} = \frac{\pi^2 E I_{\mathrm{C}}}{(2.82 L_{\mathrm{C}})^2}$$

$$K_{\mathrm{eff}} = 2.82$$

The presence of the load on the leaner member adds to the destabilizing second-order effect (the $P\Delta$ effect shown in Figure 5.17) by reducing the elastic critical load by the ratio $(\frac{2.30}{2.82})^2 = 0.67$. In this case, this error would basically wipe out the factor of safety.

Frames with leaning columns will be discussed in more depth in Chapters 7 and 8.

5.2.6 Case VI: Buckling of a Frame with Partially Restrained (PR) Joints

In this case, the problem of the stability of a frame with partially restrained (PR) beam-to-column connections, shown in Figure 5.20, is considered. The stiffness of the connection is idealized as a rotational spring.

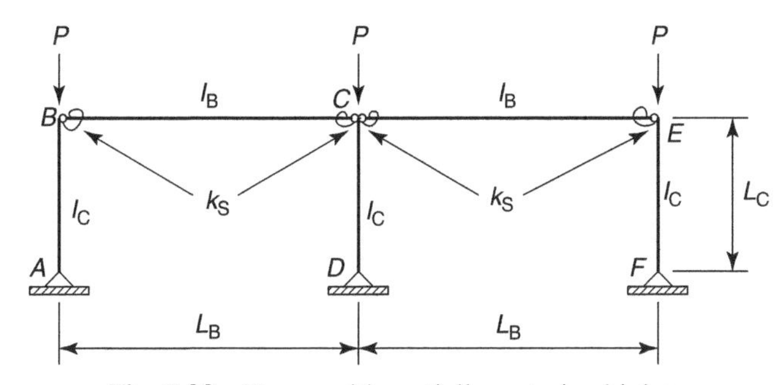

Fig. 5.20 Frame with partially restrained joints.

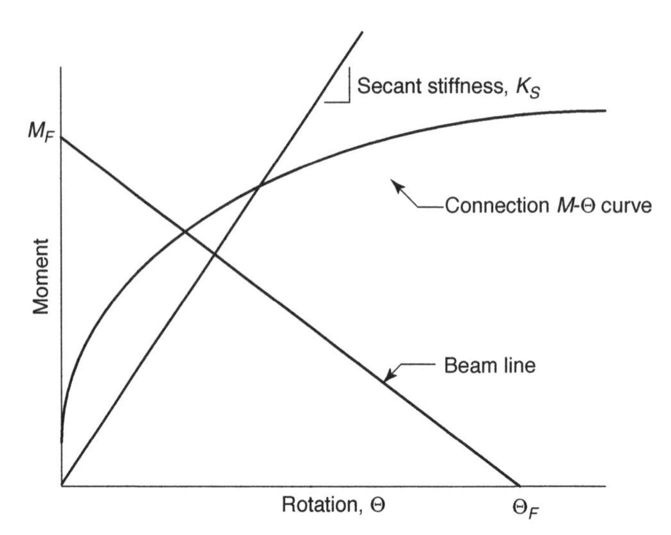

Fig. 5.21 Beam line diagram to determine secant stiffness of the connection.

PR connections are those that fall between the idealizations of pinned and fixed connections. Typical moment rotation curves for all three classes of connection are shown in Figure 5.21. There has been significant research into the behavior of PR connections, as well as methods for analysis and design. The stiffness of the connection is dependent on the load level and is given as the tangent stiffness of the moment rotation curve. For design purposes, this may be estimated by the secant stiffness determined by a beam line analysis, as shown in Figure 5.21. The choice of design stiffness or approach is beyond the scope of this book. The reader is referred to any of the references presented at the back of the chapter for insight into how to model a PR frame.

For purposes of considering stability, a general rotational spring stiffness k_s is used. The influence of the magnitude of the spring constants on the buckling load is also examined. In order to perform the frame buckling analysis it is necessary to derive the relationship between the end-slopes of a beam and the corresponding end moments. In the interests of simplicity it is assumed that the spring constants at each end of a beam are the same, and that there is no axial force in the beams. For most practical purposes these assumptions are appropriate. The free-body diagram of the beam with end-restraints is shown on Figure 5.22.

The slopes at the two ends are given by the following expressions:

$$\begin{aligned}\theta_A &= \frac{M_A L}{3EI} - \frac{M_B L}{6EI} + \frac{M_A}{k_s} \\ \theta_B &= -\frac{M_A L}{6EI} + \frac{M_B L}{3EI} + \frac{M_B}{k_s}\end{aligned} \tag{5.15}$$

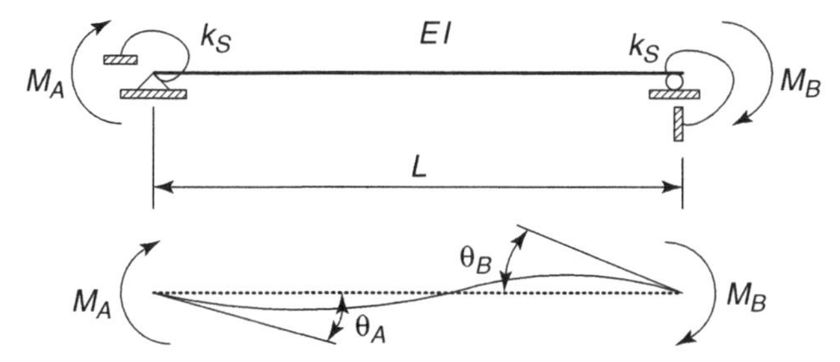

Fig. 5.22 Beam with semi-rigid connection at its ends.

The first two terms represent the end-slopes due to the end moments, the third term is the contribution of the restraint. The two simultaneous equations above are solved for the end moments to obtain the slope-deflection equations. After some algebraic manipulations and the introduction of the following abbreviations, the slope-deflection equations for beams with end-restraints are given next.

$$\alpha = \frac{k_s L}{EI} \tag{5.16}$$

$$\begin{aligned} B &= \left(\frac{1}{\alpha} + \frac{1}{3}\right)^2 - \frac{1}{36} \\ C_S &= \left(\frac{1}{\alpha} + \frac{1}{3}\right)\frac{1}{B} \end{aligned} \tag{5.17}$$

$$S_S = \frac{1}{6B} \tag{5.18}$$

$$\begin{aligned} M_A &= \frac{EI}{L}(C_S\theta_A + S_S\theta_B) \\ M_A &= \frac{EI}{L}(S_S\theta_A + C_S\theta_B) \end{aligned} \tag{5.19}$$

When $k_s = \alpha \to \infty$ it can be shown that $C_S = 4$ and $S_S = 2$. These are the values used in all the previous cases where the joints between the columns and the beams were entirely rigid.

In the following derivation the method of obtaining the buckling load of the two-bay frame of Figure 5.20 is illustrated. From the sway-buckling solution of the frame with rigid joints, Case II, it is known that the buckling shape of the frame is antisymmetric and, therefore, $\theta_B = \theta_E$. The slope-deflection equations are then the following:

$$M_{BC} = M_{EC} = \frac{EI_B}{L_B}(C_S\theta_B + S_S\theta_C)$$

$$M_{CB} = M_{CE} = \frac{EI_B}{L_B}(S_S\theta_B + C_S\theta_C)$$

$$M_{BA} = M_{EF} = \frac{\tau EI_C}{cL_C}(\theta_B - \rho) \tag{5.20}$$

$$M_{CD} = \frac{\tau EI_C}{cL_C}(\theta_C - \rho)$$

In the last two of these equations, the incorporation of the tangent modulus ratio τ is included so that an inelastic analysis can also be performed (as in Case V).

The equilibrium equations are the following expressions:

$$\begin{aligned} &M_{BA} + M_{BC} = 0 \\ &M_{CB} + M_{CD} + M_{CE} = 0 \\ &M_{BA} + M_{CD} + M_{EF} + 3PL_C\rho = 0 \end{aligned} \tag{5.21}$$

Substitution of the slope-deflection equations into the equilibrium equations will give three simultaneous homogeneous equations. The vanishing of the determinant of the coefficients of the unknown deformations θ_B, θ_C and ρ provides the answer to the critical load for the frame.

$$\begin{vmatrix} \frac{\tau}{c}+\gamma C_S & \gamma S_S & -\frac{\tau}{c} \\ 2\gamma S_S & \frac{\tau}{c}+2\gamma C_S & -\frac{\tau}{c} \\ \frac{2\tau}{c} & \frac{\tau}{c} & \frac{3PL_C^2}{EI_C} - \frac{3\tau}{c} \end{vmatrix} = 0 \tag{5.22}$$

Following is an example problem solved by MATHCAD:

$L_B = 40\,\text{ft.}$ $\quad E = 29{,}000\,\text{ksi}$ $\quad A = 29.1\,\text{in}^2$

$L_C = 20\,\text{ft.}$ $\quad F_y = 50\,\text{ksi}$ $\quad r_x = 6.17\,\text{in.}$

$k_S = 10^5\,\text{in}\dfrac{\text{kip}}{\text{rad}}$ $\quad \alpha = \dfrac{k_S L_B}{EI_B}$ $\quad I_B = 2{,}700\,\text{in}^4$

$I_C = 1{,}110\,\text{in}^4$

$P_y = AF_y$ $\quad P_y = 1.455 \times 10^3\,\text{kip}$

$\gamma = \dfrac{I_B L_C}{I_C L_B}$ $\quad \gamma = 1.216$

$$\lambda = \frac{L_C}{\pi r_x}\sqrt{\frac{F_y}{E}} \qquad \lambda = 0.514$$

$$B = \left(\frac{1}{\alpha}+\frac{1}{3}\right)^2 - \frac{1}{36} \quad C_S = \left(\frac{1}{\alpha}+\frac{1}{3}\right)\frac{1}{B} \quad S_S = \frac{1}{6B}$$

$p = 0.5$ (Assumed value to initiate numerical solution in the MATHCAD program)

$$\tau(p) = \begin{vmatrix} 1.0 \text{ if } p \le 0.5 \\ 4p(1-p) \text{ otherwise} \end{vmatrix} \quad \phi(p) = \pi\lambda\sqrt{\frac{p}{\tau(p)}} \quad c(p) = \frac{1 - \dfrac{\phi(p)}{\tan(\phi(p))}}{\phi(p)^2}$$

$$F(p) = \begin{bmatrix} \dfrac{\tau(p)}{c(p)}+\gamma C_S & \gamma S_S & -\dfrac{\tau(p)}{c(p)} \\ 2(\gamma S_S) & \dfrac{\tau(p)}{c(p)}+2\gamma C_S & -\dfrac{\tau(p)}{c(p)} \\ 2\dfrac{\tau(p)}{c(p)} & \dfrac{\tau(p)}{c(p)} & 3\pi^2\lambda^2 p - 3\dfrac{\tau(p)}{c(p)} \end{bmatrix}$$

$\text{Answer}(p) = \text{root}(|F(p)|, p) \quad \text{Answer}(p) = 0.257$
$\text{Answer}(p)P_y = 373.933 \text{ kip}$

The problem above is for a frame with W24 $\times$ 94 beams and W14 $\times$ 99 columns with yield strength of 50 ksi and a spring constant of 10^9 inch-kip/radian. This is essentially full rigidity. The same problem was solved with fully rigid connections in Case V and the answers are almost identical: $p = \frac{P_{cr}}{P_y} = 0.676$ or $P_{cr} = 0.676P_y = 0.676AF_y = 983$ kip. The effect of changes in the spring constant can be seen in the following table:

Spring constant k_s	$\dfrac{P_{cr}}{P_y}$	P_{cr}
inch-kips/radian		kip
10^9	0.676	983
10^7	0.670	975
10^6	0.613	892
10^5	0.257	374

The following data will put the information in this table into perspective: Chen et al. (1996) use an example of a semi-rigid frame where a W12 $\times$ 45 beam is connected to a W8 $\times$ 40 column with a L6X4 $\times$ 3/3 top and seat angle connection. The initial stiffness of the joint is given as

$k_s = 2.25 \times 10^6$ inch-kips/radian. The treatment herein of the stability of frames with semi-rigid joints is but an introduction to an important topic of frame design, and at the end of this chapter there is a list of references for further study.

5.3 SUMMARY

Concepts of frame stability were presented in this chapter. The vehicle for demonstrating these concepts was a two-bay, one-story rectangular frame (Figure 5.5). The bay-widths were equal, the two beams were equal in size, and the three columns were also equal to each other. The complicating effects of lateral-torsional and out-of-plane buckling, as well as the effects of shear deformation and member shortening and bowing, were not included so that the derivations could remain straightforward. Six cases were examined. The conditions included the frames where all three beam-to-column joints were fully rigid (Cases I, II, III, IV,), partially restrained (Case VI), and where the center column was pinned at both ends (Case V). The column bases were pinned except for Case III, for which the bases were restrained by elastic springs simulating the full range from fully pinned to fully fixed ends. The material was assumed to be elastic in all but Cases IV and VI, where the tangent modulus concept was employed to obtain the inelastic solution to frame buckling.

In each of the cases that were examined, the method of approach was to first develop the expressions for the analytically exact equations that contained the buckling load. These equations were then solved by numerical routines built into commercial mathematical software programs. The particular program used in this instance was MATHCAD. Any other program, such as EXCEL, MATLAB, or MATHEMATICA can also be used with equal success. Such programs permit the exploration of the effects of different geometries, material properties and restraint conditions. Once the program is established it is easy to keep solving problems. For each of the cases, the next step was to solve the problem by using the AISC alignment chart method. This method was introduced and derived in Chapter 2. It is a powerful tool that can be very helpful for solving individual problems. It is also a tool that is familiar to structural engineers in the design office, and the method is taught to undergraduate students.

The analytically exact derivations in this chapter were all based on utilizing the force method of structural analysis. The method is also familiarly known as the slope-deflection method. This method uses the definitions of the end-moment on any member in the frame expressed in terms of joint rotations and bar rotations, as well as member stiffness expressions and the stability functions. These are dependent on the axial force in the member. Once the moments at the ends of each member are expressed, these are then substituted into the equilibrium equations. The number of these is equal to the unknown deformations. They are formulated on the deformed, or buckled, shape of the members, including the moments due to the axial force times the deflection. The small deflection assumptions apply, since in the formulation it is stipulated that there are no deflections before buckling. At buckling, an infinitesimally small adjacent buckled configuration defines the bifurcation of the equilibrium. The slope-deflection method was used in this chapter because it is an "exact" method within the assumptions of small deflection buckling and the tangent modulus theory. The approximation from this method is inherent only in the algorithm of the mathematical solution routine. The finite element method could just as well have been used. In daily design office practice this would be the preferred method because the analyst would not have to formulate moment equations and equilibrium relationships. The justification for the approach used in this chapter is purely for the purpose of demonstrating the underlying principles in their most basic form.

5.4 SELECTED REFERENCES ON FRAMES WITH PARTIALLY RESTRAINED JOINTS

AISC-SSRC Ad Hoc Committee on Frame Stability. 2003. "Background and Illustrative Examples on Proposed Direct Analysis Method for Stability Design of Moment Frames." Technical White Paper, Technical Committee 10, AISC, 17 pgs.

Ammerman and R. T. Leon. 1989. "Unbraced Frames with Semi-Rigid Composite Connections." *Engineering Journal*, AISC, 27(1): 1–11.

Barakat, M., and W. F. Chen. 1991. "Design Analysis of Semi-Rigid Frames: Evaluation and Implementation." *Engineering Journal*, AISC, 28(2): 55–64.

Christopher, J. E., and R. Bjorhovde. 1999. "Semi-Rigid Frame Design Methods for Practicing Engineers." *Engineering Journal*, AISC, 36(1): 12–28.

Goverdhan, A.V. 1983. "A Collection of Experimental Moment-Rotation Curves and Evaluation of Prediction Equations for Semi-Rigid Connections." M.Sc. Thesis, Vanderbilt University, Nashville, TN.

Goverdhan, A.V., and S. D. Lindsey. 1996. "PR Connections in Design Practice." In *Connections in Steel Structures III: Behavior, Strength and Design, Proceedings of the Third International Workshop, Trento, Italy,* ed. R. Bjorhovde, A. Colson, and J. Stark, Elsevier Science, 505–514.

Kishi, N., and W. F. Chen. 1986. *Database of Steel Beam-to-Column Connections*. Structural Engineering Report No. CE-STR-86-26, 2 Vols., School of Civil Engineering, Purdue University, West Lafayette, IN.

Leon, R. T., J. J. Hoffman, and T. Staeger. 1996. *Partially Restrained Composite Connections*. Steel Design Guide Series 8, AISC, 59 pp.

Surovek, A. E., D. W. White, and R. T. Leon. 2005. "Direct Analysis for Design Evaluation and Design of Partially-Restrained Steel Framing Systems." *Journal of Structural Engineering*. ASCE, September.

Weynand, K., M. Huter, P. A. Kirby, L. A. P. da Silva, and P. J. S. Cruz. 1998. "SERICON—A Databank for Tests on Semi-Rigid Joints." *COST C1: Control of the Semi-Rigid Behavior of Civil Engineering Structural Connections, Proceedings of the International Conference*. Liege, Belgium, September, pp. 217–228.

PROBLEMS

5.1. Find the elastic effective length factor for each column if the top and center beams are W27 × 84 sections, the bottom restraining beam is a W18 × 35 section, and the columns are W14 × 109 sections. Consider only in-plane buckling. Assume that the buckling deformation is about the major axis of the sections. The bay width is 240 in. and the story height is 180 in.

Make an exact frame stability analysis using the slope-deflection method and compare with the answers obtained by the AISC alignment charts for sway-permitted frames. If you have access to a finite element second-order analysis program (such as MASTAN), check your work by this program also.

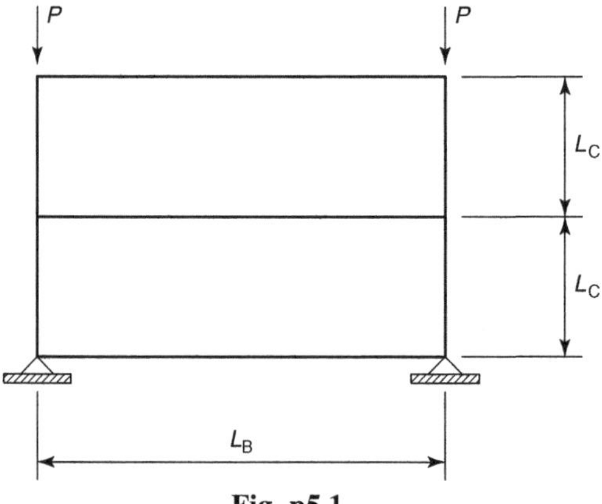

Fig. p5.1

5.2. Determine the design strength of this frame according to Chapter E of the 2005 AISC Specification. Calculate the effective length using the alignment charts. Check the critical load with the slope-deflection method. $L = 200$ in. and $F_y = 50$ ksi. The beams are W21 × 62 sections, the pinned columns are W8 × 31 sections, the right frame column is a W8 × 58 section, and the two left columns are W8 × 48 sections. All deflections are about the major axes of the members.

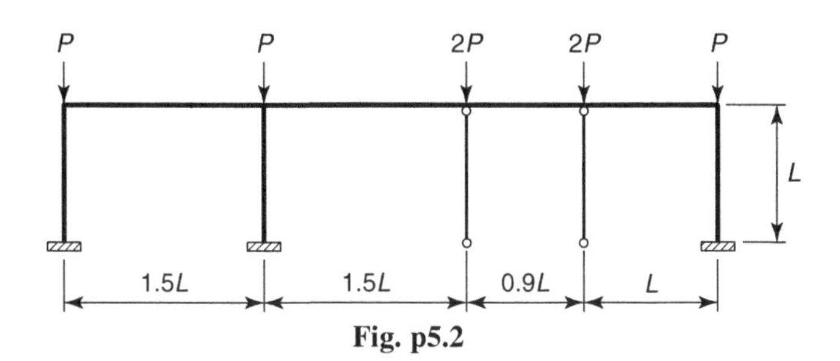

Fig. p5.2

5.3. Determine the elastic side-sway buckling load of this frame with stepped columns and unsymmetric loading, using the slope-deflection method. Check your answer by a finite element program, if you have one available. If you do not, think about how you would prove to yourself that your answer is reasonable? Make such an evaluation. (Think upper bounds and lower bounds to the answer!)

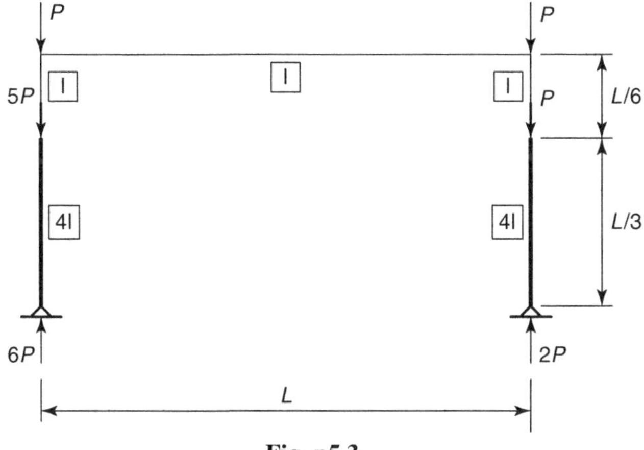

Fig. p5.3

5.4. Determine the elastic critical load of this rigid frame.

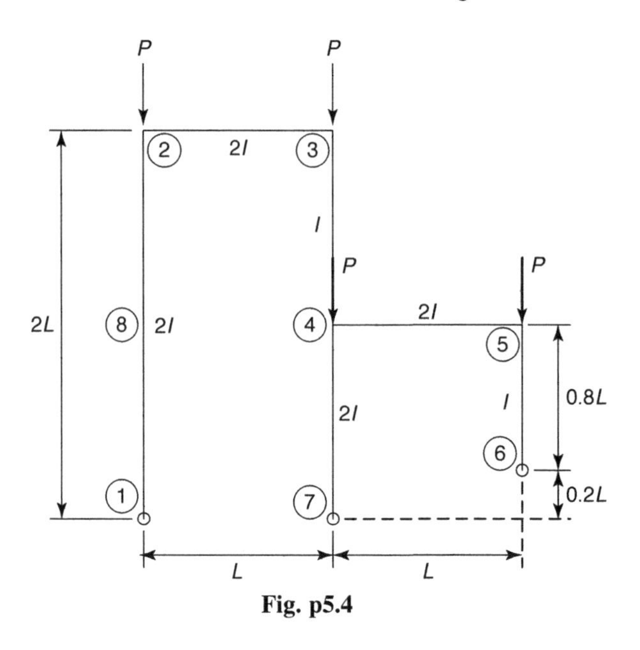

Fig. p5.4

5.5. Determine the elastic critical load of this frame.

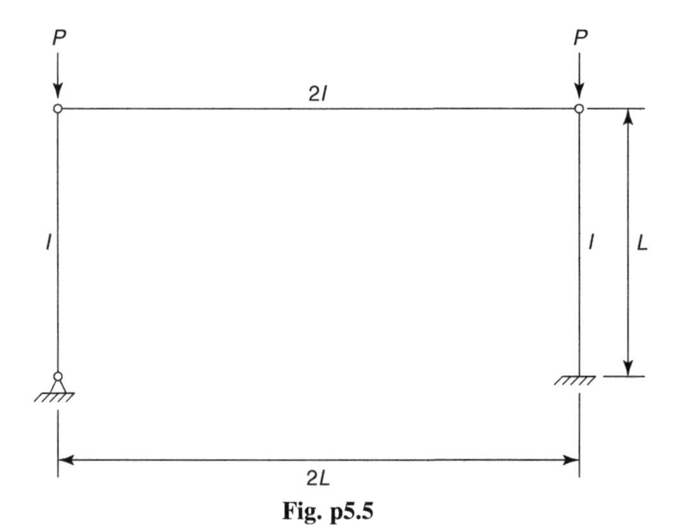

Fig. p5.5

5.6. Determine the elastic buckling load of this frame.

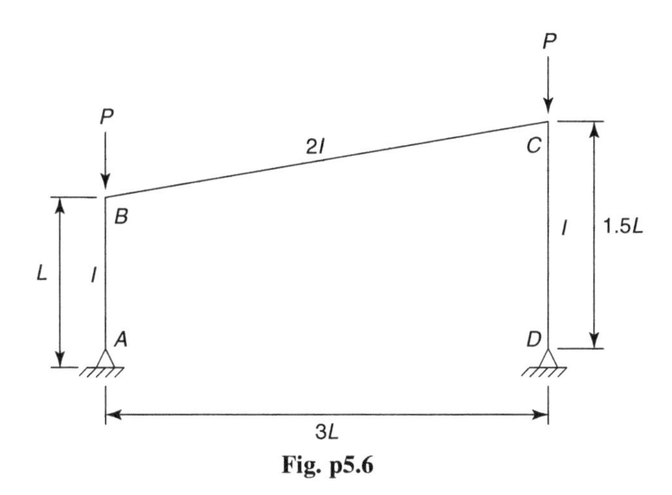

Fig. p5.6

5.7. Interaction buckling exercise. Determine the elastic buckling load of this frame for $P_2 = 0$, $P_1/2$ and P_1.

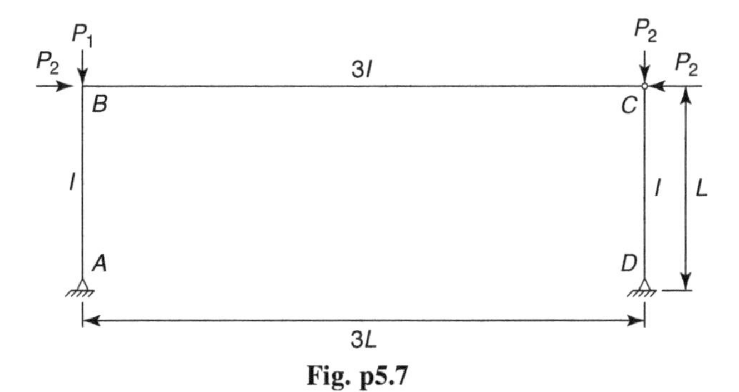

Fig. p5.7

CHAPTER SIX

LATERAL-TORSIONAL BUCKLING

6.1 INTRODUCTION

The preceding chapters dealt with the stability of members and frames where the loading, the resulting deformation, and the eventual buckled configuration, were all restricted to the same plane, thus the definition *in-plane* behavior. This chapter addresses a type of unstable behavior that is called *lateral-torsional buckling*. In this type of instability the member is singly or doubly symmetric, it is loaded by forces in the plane of symmetry, and it deforms in this plane until, at a critical loading, the member both deflects out of its plane of symmetry and twists. This behavior is typical of wide-flange beams that are loaded so that the resulting bending moments act about the x-axis. Figure 6.1 shows lateral-torsional buckling of a wide-flange beam, exaggerated as a result of a fire, in a flour-mill building in Minneapolis, Minnesota.

The emphasis in this chapter is on elastic buckling. Inelastic buckling and its implications on design rules are covered in the final section of the chapter. Doubly and singly symmetric wide-flange beams and beam-columns are also considered. Lateral-torsional instability is of particular importance during erection, before the lateral bracing system is fully installed. In fact, it is during the act of installing the braces that many fatal accidents have occurred. Figure 6.2 illustrates the bracing of a curved girder bridge during its erection.

Fig. 6.1 Lateral torsional beam buckling exaggerated by fire.

6.2 BASIC CASE: BEAMS SUBJECTED TO UNIFORM MOMENT

The reference case for lateral-torsional buckling is shown in Figure 6.3.

The basic case is a beam with the following attributes:

- The beam behaves elastically.
- It is simply supported.
- It is subjected to uniform moment about the major principal axis.

Fig. 6.2 Curved girder bridge erection bracing.

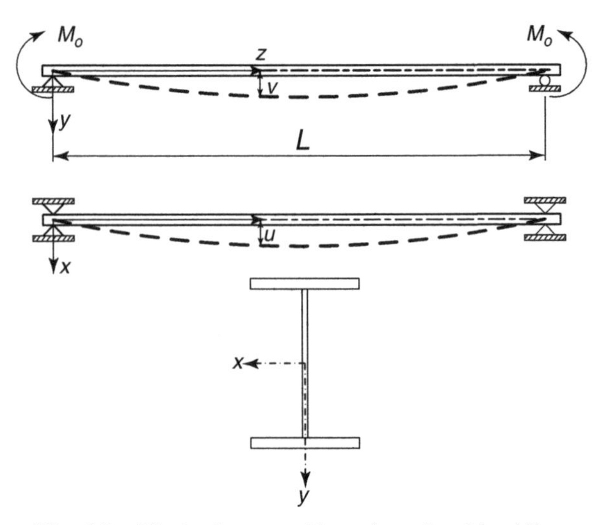

Fig. 6.3 The basic case of lateral-torsional buckling.

- The cross-section is double symmetric (and thus the centroid and the shear center coincide).
- Bending occurs about the major axis.

This concept is analogous to the pinned-end column that was the reference case for axially loaded columns in Chapter 2. The sketch in Figure 6.3 shows the orientation of the x- and y- axes. The z-axis is at the centroid and it originates at the left end of the beam that has a length L. The end moments M_o act in the y-z plane. This is also the plane of the web, and thus bending is about the major axis of the cross section. The moment M_o deforms the beam in the direction of the y-axis through a deflection v.

The member continues to deflect as M_o is increased from zero, without any out-of-plane movement, until a critical moment M_{ocr} is reached. This moment represents the point of bifurcation when equilibrium is possible in both an unbuckled and in a buckled configuration. Conceptually, this condition is the same as the buckling of the ideal Euler column presented in Chapter 2. In case of the pinned-end column, bifurcation is the transition from a perfectly straight to a laterally deflected geometry. For the beam, the buckled deflection changes from an in-plane configuration at the critical moment M_{ocr} to an out-of-plane mode represented by a lateral deflection u and an angle of twist ϕ. Because the buckled shape includes both a *lateral deflection* and an *angle of twist*, this case of instability is called *lateral-torsional buckling.*

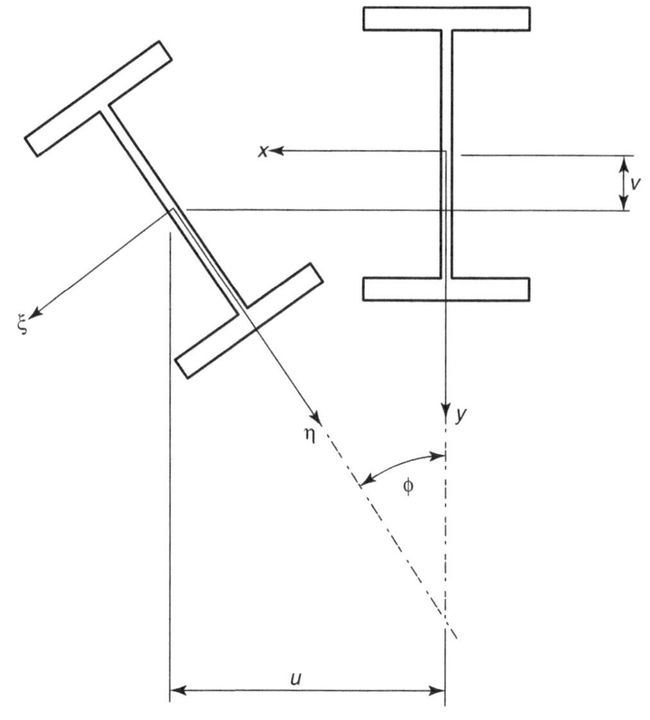

Fig. 6.4 The cross-section before and after buckling.

The lateral deflection is shown on Figure 6.3. The simply supported end of the beam is the idealized condition when lateral deflection and twist are prevented, the end is free to rotate laterally, and the end cross-section is free to warp.

Figure 6.4 illustrates the transition from the unbuckled configuration to the buckled location of the cross-section at the critical moment. The translated and rotated cross-section has moved from the pre-buckling state with a transverse deflection v, to a new location with a lateral deflection u, and an angle of twist ϕ. The original centroidal coordinates x and y of the cross-section in the buckled configuration are designated as ξ and η.

In the following derivation, these conditions are assumed:

- The deflections and angles of twist are small.
- The material is elastic, homogeneous and isotropic.
- There is no local buckling of the flanges and the web.
- There is no distortion of the cross section during buckling.

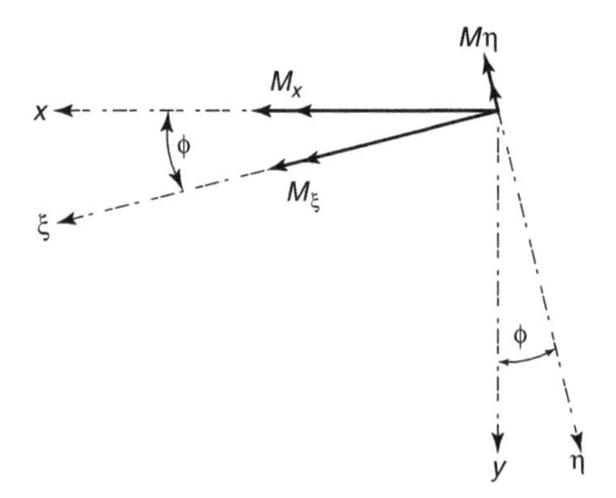

Fig. 6.5 Decomposition of M_x.

The sketch in Figure 6.5 shows the decomposition of the moment M_x at location z into components in the direction ξ and η. For the assumed small displacements $\cos\phi \approx 1$ and $\sin\phi \approx \phi$.

$$M_\xi = M_x \cos\phi \approx M_x$$
$$M_\eta = -M_x \sin\phi \approx -M_x\phi$$

From Figure 6.6, it follows that $M_x = -M_o$, and thus

$$M_\xi = M_x = -M_o$$
$$M_\eta = -M_x\phi = M_o\phi$$

Equilibrium requires that these moments are equal to the internal moments, resulting in the following two differential equations for bending about the x and y axes, respectively:

$$M_\xi = -EI_x \frac{d^2v}{dz^2} = -EI_x v'' = -M_o$$
$$EI_x v'' - M_o = 0$$

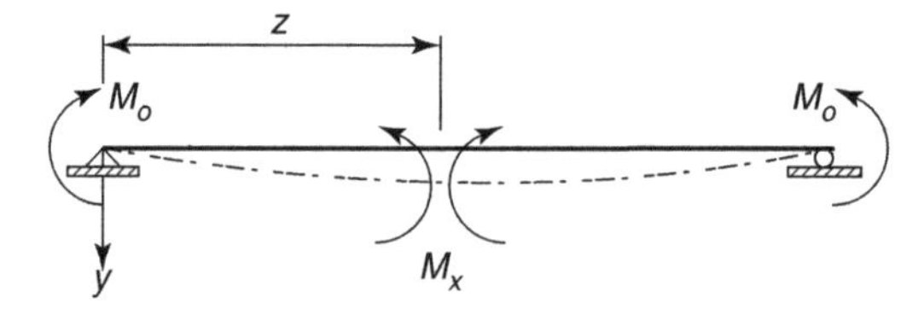

Fig. 6.6 The moment M_x at location z.

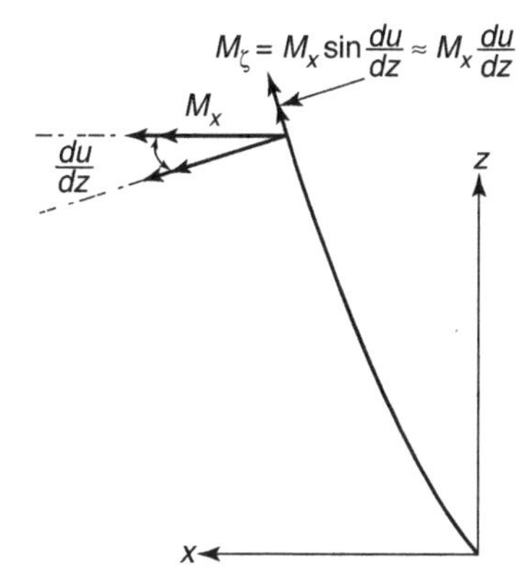

Fig. 6.7 Torsional component.

In this formula, E is the modulus of elasticity and I_x is the moment of inertia about the x-axis. This equation involves only the in-plane deflection v. It is of no further interest in the buckling problem.

The second equation defines out-of-plane bending:

$$M_\eta = -EI_y u'' = M_o\phi$$
$$EI_y u'' + M_o\phi = 0$$

I_y is the moment of inertia about the y-axis. This differential equation contains the two buckling deformations u and ϕ. An additional differential equation is derived from the torsional component of M_x along the displaced z-axis, shown as ζ on the sketch in Figure 6.7.

$$M_\zeta = M_x \sin\frac{du}{dz} = -M_o \sin u' \approx -M_o u'$$

The internal moment of torsion consists of a warping and a uniform torsion component. G is the shear modulus, C_w is the warping constant, and J is the St. Venant's torsion constant.

$$EC_w\phi''' - GJ\phi' = -M_o u'$$
$$EC_w\phi''' - GJ\phi' + M_o u' = 0$$

The two differential equations involving the lateral-torsional displacements u and ϕ are then equal to

$$\begin{aligned} EI_y u'' + M_o\phi &= 0 \\ EC_w\phi''' - GJ\phi' + M_o u' &= 0 \end{aligned} \tag{6.1}$$

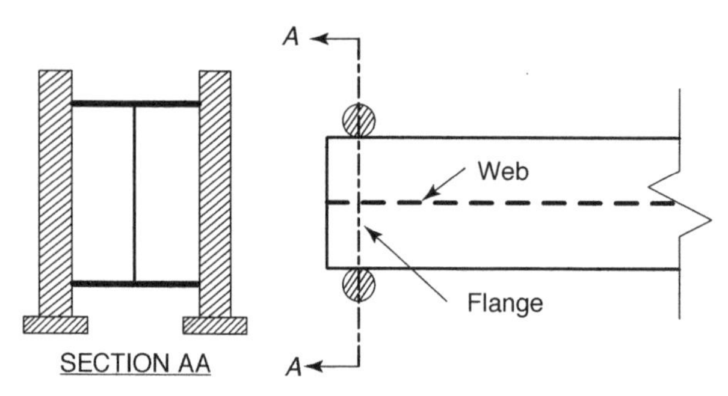

Fig. 6.8 Definition of a simple support.

From the first of these equations

$$u'' = \frac{-M_o\phi}{EI_y}$$

Differentiation of the second equation with respect to z and substitution results in the following differential equation for the elastic lateral-torsional buckling of a simply supported prismatic wide-flange beam under a uniform moment:

$$EC_w\phi^{iv} - GJ\phi'' - \frac{M_o^2\phi}{EI_y} = 0 \tag{6.2}$$

The boundary conditions defining a simply supported end are shown in Figure 6.8. These are expressed in terms of the deformations u and ϕ as follows:

$$\begin{aligned} u(0) = u(L) = \phi(0) = \phi(L) = 0 \\ u''(0) = u''(L) = \phi''(0) = \phi''(L) = 0 \end{aligned} \tag{6.3}$$

The first line in equation 6.3 states that the lateral deflection and the angle of twist equals zero at each end. The first two conditions in the second line say that there is no moment about the y-axis at the ends, and the third and fourth conditions mean that the ends of the wide-flange shape are free to warp.

Once the boundary conditions have been determined, the differential equations can be solved. The resulting set of homogeneous equations yields the expression for the critical lateral-torsional buckling moment. Equation 6.2 can be written as

$$\phi^{iv} - \frac{GJ}{EC_w}\phi'' - \frac{M_o^2}{E^2C_wI_y}\phi = 0$$

Let

$$\lambda_1 = \frac{GJ}{EC_w}$$
$$\lambda_2 = \frac{M_o^2}{E^2 C_w I_y} \tag{6.4}$$

The differential equation then becomes equal to

$$\phi^{iv} - \lambda_1 \phi'' - \lambda_2 \phi = 0$$

The roots of the differential equation are

$$\phi[r^4 - \lambda_1 r^2 - \lambda_2] = 0$$

where

$$r = \begin{bmatrix} \alpha_1 \\ -\alpha_1 \\ i\alpha_2 \\ -i\alpha_2 \end{bmatrix}$$

$$\alpha_1 = \sqrt{\frac{\lambda_1 + \sqrt{\lambda_1^2 + 4\lambda_2}}{2}} \tag{6.5}$$

$$\alpha_2 = \sqrt{\frac{-\lambda_1 + \sqrt{\lambda_1^2 + 4\lambda_2}}{2}} \tag{6.6}$$

The expression for the angle of twist can be written in either of the following two ways:

$$\phi = A_1 e^{\alpha_1 z} + A_2 e^{-\alpha_1 z} + A_3 e^{i\alpha_2 z} + A_4 e^{-i\alpha_2 z} \tag{6.7}$$

$$\phi = C_1 \cosh \alpha_1 z + C_2 \sinh \alpha_1 z + C_3 \sin \alpha_2 z + C_4 \cos \alpha_2 z \tag{6.8}$$

In these equations, A_1, A_2, A_3, A_4 and C_1, C_2, C_3, C_4 are constants of integration dependent on the boundary conditions, and $i = \sqrt{-1}$. It is more convenient to work with the second equation that uses trigonometric and hyperbolic functions. If the boundary conditions $\phi(0) = \phi(L) = \phi''(0) = \phi''(L) = 0$ are substituted into equation 6.8, then the following four homogeneous simultaneous equations are obtained:

$$\begin{bmatrix} 1 & 0 & 0 & 1 \\ \alpha_1^2 & 0 & 0 & -\alpha_2^2 \\ \cosh \alpha_1 L & \sinh \alpha_1 L & \sin \alpha_2 L & \cos \alpha_2 L \\ \alpha_1^2 \cosh \alpha_1 L & \alpha_1^2 \sinh \alpha_1 L & -\alpha_2^2 \sin \alpha_2 L & -\alpha_2^2 \cos \alpha_2 \end{bmatrix} \begin{Bmatrix} C_1 \\ C_2 \\ C_3 \\ C_4 \end{Bmatrix} = 0$$

A nontrivial solution is obtained by setting the determinant of the coefficients equal to zero. Decomposition of the determinant leads to the following equation:

$$(\alpha_1^2 + \alpha_2^2)\sinh\alpha_1 L \times \sin\alpha_2 L = 0$$

Since the expression before the multiplication sign is not equal to zero, the eigenvalue must reside in the equation

$$\sin\alpha_2 L = 0 \tag{6.9}$$

Equation 6.9 is true if $\alpha_2 L = n\pi$, where $n = 1, 2, 3, \ldots$

Substitution of $\alpha_2 = \sqrt{\frac{-\lambda_1 \pm \sqrt{\lambda_1^2 + 4\lambda_2}}{2}}$ gives the following equation:

$$\lambda_2 = \frac{n^2\pi^2}{L^2}\left[\lambda_1 + \frac{n^2\pi^2}{L^2}\right]$$

The lowest eigenvalue is when $n = 1$. Substitution of $\lambda_1 = \frac{GJ}{EC_w}$; $\lambda_2 = \frac{M_o^2}{E^2 C_w I_y}$ into the above expression leads to the following equation for the critical lateral-torsional buckling moment:

$$M_{ocr} = \frac{\pi}{L}\sqrt{EI_y GJ}\sqrt{1 + \frac{\pi^2 EC_w}{GJL^2}} \tag{6.10}$$

Equation 6.10 is the basic equation that is the anchor point of further generalizations of the lateral-torsional buckling solutions. This equation is, to repeat, strictly the *elastic buckling moment of a simply supported prismatic wide-flange beam subject to uniform moment about its major axis.* It is analogous to the Euler buckling load of an ideal axially loaded column, $P_{cr} = \frac{\pi^2 EI}{L^2}$. The graph relating the critical elastic buckling moment M_{cr} to the unbraced length L of the beam between the two end supports is very similar to the curve relating the buckling load to the length for the pinned-end column: at a large length the strength tends toward zero, and at very short length it tends toward infinity. Such beam-buckling curves are shown in Figure 6.9 for two wide-flange shapes, the W8 $\times$ 31 having a *column-type* cross-section (flange width and depth of section are equal), and the W16 $\times$ 26 with a flange width much smaller than its depth (*beam-type* cross-section). The curves have the ratio M_{cr}/M_y as the ordinate and the

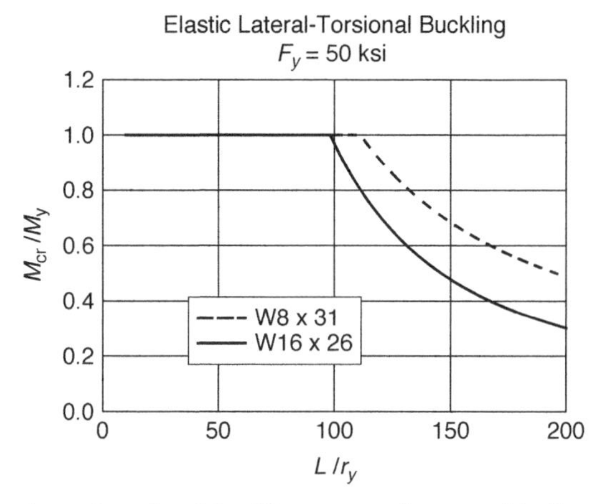

Fig. 6.9 Lateral-torsional buckling curves for two wide-flange beams.

minor-axis slenderness ratio L/r_y as the abscissa. $M_y = S_x F_y$, the yield moment, and r_y is the minor-axis moment of inertia.

Even though the calculation of the elastic critical moment is a somewhat more complicated process than the determination of the buckling load of a column, the formula can be easily programmed on a spreadsheet, as illustrated in the printout in Figure 6.10.

The equation for the critical buckling moment of equation 6.10 is valid for the prediction of the elastic lateral-torsional instability limit state of a laterally and torsionally simply supported beam with a doubly symmetric wide-flange cross-section under uniform bending about the major axis. This basic case is expanded next to include other boundary conditions, other

$$M_{cr} = \frac{\pi}{L}\sqrt{EI_y GJ}\sqrt{1 + \frac{\pi^2 EC_w}{GJL^2}}$$

W16 x 26

E =	29,000	ksi
G =	11,165	ksi
Fy =	50	ksi
L =	360	in
Iy =	9.59	in^4
J =	0.262	in^4
Cw =	565	in^6
Sx =	38.4	in^3
M_{cr} =	297	in-kip
F_{cr} =	7.74	ksi

Fig. 6.10 Example calculation.

loading cases, other cross-sections, and it will be generalized to incorporate buckling in the inelastic range.

6.3 THE EFFECT OF BOUNDARY CONDITIONS

End restraints are considered in the model by the boundary conditions used in solving the differential equations. By changing the boundary conditions, we can include the influence of end restraints on the critical moment. First, it is assumed that the lateral boundary conditions remain pinned,

$$u(0) = u(L) = u''(0) = u''(L) = 0,$$

but that the torsional boundary conditions are fixed:

$$\phi(0) = \phi(L) = \phi'(0) = \phi'(L) = 0.$$

This set of boundary conditions implies that the ends of the beam are prevented from freely warping by a thick end plate, or by a channel stiffener. An example of this type of restraint is illustrated in Figure 6.11 (Ojalvo and Chambers 1977). Substitution of the torsional boundary conditions into equation 6.6 results in four homogeneous simultaneous equations. The determinant of the matrix of the coefficients is given as

$$\begin{vmatrix} 1 & 0 & 0 & 1 \\ 0 & \alpha_1 & \alpha_2 & 0 \\ \cosh \alpha_1 L & \sinh \alpha_1 L & \sin \alpha_2 L & \cos \alpha_2 L \\ \alpha_1 \sinh \alpha_1 & \alpha_1 \cosh \alpha_1 L & \alpha_2 \cos \alpha_2 L & -\alpha_2 \sin \alpha_2 L \end{vmatrix} = 0$$

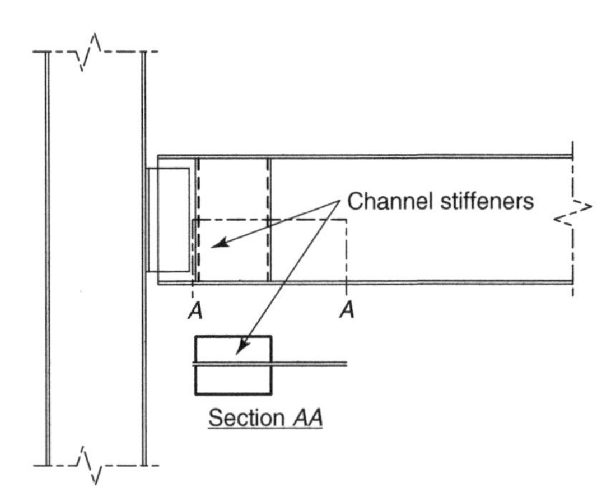

Fig. 6.11 Stiffener arrangement to prevent warping (Figure 5.6, Galambos 1998.)

Decomposition of this determinant leads to the following transcendental equation that can only be solved by trial and error to obtain the critical moment:

$$\cosh \alpha_1 L \times \cos \alpha_2 L - 1 + \left(\frac{\alpha_2^2 - \alpha_1^2}{2\alpha_1 \alpha_2} \right) \sinh \alpha_1 L \times \sin \alpha_2 L = 0 \quad (6.11)$$

For example, the critical moment for a 224-inch-long W16 $\times$ 26 beam has a critical moment of 1050 kip-inches. For this shape

$$J = 0.262 \text{ in.}^4; \quad C_w = 565 \text{ in.}^6; \quad I_y = 9.59 \text{ in.}^4$$

From this example, it can be seen that the solution to this problem is not easy, even though a closed-form analytical solution can be obtained. The need for a computationally difficult trial-and-error solution makes this type of approach, at best, cumbersome in design. For most other boundary conditions, obtaining a closed-form analytic solution is either much more difficult or even impossible. There is a substantial amount of literature on solving lateral-torsional buckling problems by various numerical methods (Galambos 1998, Trahair 1993) or using finite element programs. Approximate formulas based on fitting equations to numerically obtained values from parametric studies are presented in Chapter 6 of the SSRC *Guide to Stability Design Criteria for Metal Structures* (Galambos 1998). For design office use, however, an extension of the effective length method is suggested in the Commentary of the 2005 AISC Specification, as illustrated in the following modification of the basic lateral-torsional buckling equation (equation 6.10):

$$M_{cr} = \frac{\pi}{k_y L} \sqrt{E I_y G J} \sqrt{1 + \frac{\pi^2 E C_w}{G J (k_z L)^2}} \quad (6.12)$$

In this equation k_y is the effective length factor for the lateral buckling mode, and k_z is effective length factor for the torsional buckling mode. For both ends pinned, $k = 1.0$, for both ends fixed $k = 0.5$, and for one end fixed and the other end pinned, $k = 0.7$. It should be recognized that for the case of lateral-torsional buckling, these values are only good approximations. They are, however, sufficient for the design office. When precise values are needed, the engineer must consult the literature to see if the answer is readily available, otherwise a finite element or numerical analysis procedure can be employed. The effective length procedure is illustrated in the graphs of Figure 6.12 for a W16 $\times$ 26 beam.

The solid curve represents the basic case, where both ends are simply supported. This is the lowest buckling strength. The highest buckling strength is when both ends are fixed. The two intermediate curves are for the cases where the ends are laterally pinned and torsionally fixed, and where they are laterally fixed and torsionally pinned, respectively. The black square is the buckling load that was obtained numerically from equation 6.11:

$$L/r_y = \frac{224}{1.12} = 200$$

$$M_{cr}/M_y = \frac{1050}{S_x F_y} = \frac{1055}{38.4 \times 50} = 0.547$$

Since this value is for the case of torsionally fixed and laterally pinned ends, the solid rectangular point compares with the dashed curve in Figure 6.12, indicating that the approximation is conservative.

Further topics related to the effect of the boundary conditions will be considered after the effect of loading on the elastic lateral-torsional buckling strength, which is presented next.

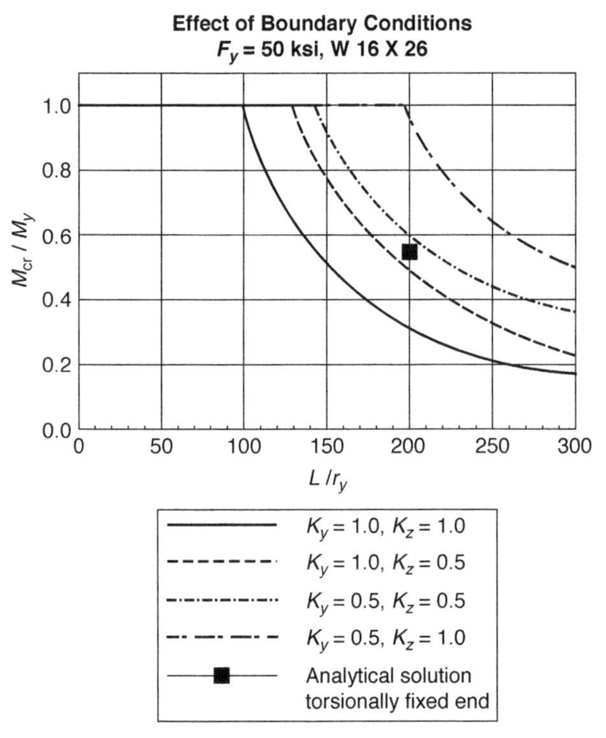

Fig. 6.12 Illustration of the effect of boundary conditions on the elastic critical moment.

6.4 THE EFFECT OF LOADING CONDITIONS

The basic case of a laterally and torsionally simply supported wide-flange beam subjected to a uniform moment about its major axis was amenable to an analytical solution that resulted in a simple formula, equation 6.10, relating the span L to the critical moment M_{cr}. This formula will produce a conservative result in most cases. However, most beams in a structure are not subject to uniform moment, and most supports are not simple. The loading and boundary conditions that are of practical importance have, unfortunately, differential equations that are either too complicated or even impossible to be solved analytically. For the research engineer there are a multitude of approximate and numerical solutions available, such as a variety of different energy methods, the finite difference method, the finite integral method, as well as many modern finite element codes (Trahair 1993). There are hundreds of papers in the structural mechanics literature that give results for a great variety of lateral-torsional buckling problems (Galambos 1998). Practicing engineers require more straightforward approaches than complicated analytical solutions, however. Design standards have dealt with this dilemma by a multiplier C_b (equation 6.13) of the basic formula to account for the various possible loading conditions:

$$M_{cr} = \frac{C_b \pi}{L} \sqrt{EI_y GJ} \sqrt{1 + \frac{\pi^2 EC_w}{GJL^2}} \tag{6.13}$$

The first formula for C_b to find its way into structural design codes is the result of work presented by Salvadori (1955). The formula, and its interpretation, is given in Figure 6.13. The fundamental assumption for this formula is that the bending moment diagram between points of lateral bracing is linear. In the United States, the formula was first used in the AISC specification for steel buildings (AISC 1963). In the 1993 *Load and Resistance Factor Design* (LRFD) edition of this specification (AISC 1993), this equation was replaced by one based on the maximum moment and on the values of the moments at the one-fourth points between the lateral brace locations. This formula is based on the work of Kirby and Nethercot (1979) and it is presented in Figure 6.14.

Other modern structural steel design standards in the world consider the effect of loading in essentially the same way: The basic formula is multiplied by a factor that is dependent on the shape of the moment diagram.

There is one omission in either of the two formulas just presented: They do not account for the position of the load on the y-axis of the cross-section.

The situation is illustrated in Figure 6.15. The load at the centroid (Figure 6.15a) is the basis of the original derivation for M_{ocr} and in the development

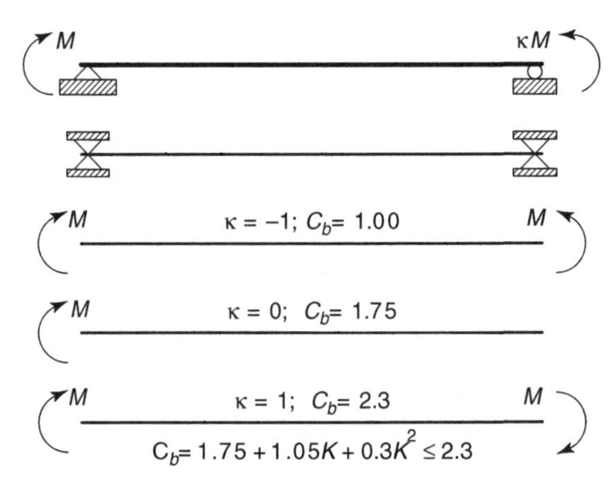

Fig. 6.13 Salvadori formula for C_b.

of C_b. A load applied at the tension flange, as shown in 6.15b, will help stabilize the beam, and it increases the buckling load. A load placed on the compression flange will destabilize the beam, and then the buckling load is reduced. The literature on lateral-torsional buckling provides many ways of dealing with this case.

One approach takes into account both end restraint and load position. The procedure was developed by Nethercot and Rockey (1972) and Nethercot (1983) by fitting equations to numerically obtained values of the critical moment, and it is reproduced here from Chapter 5 in the SSRC Guide (Galambos 1998):

The base critical moment is defined by equation 6.10, rewritten here in the following abbreviated form:

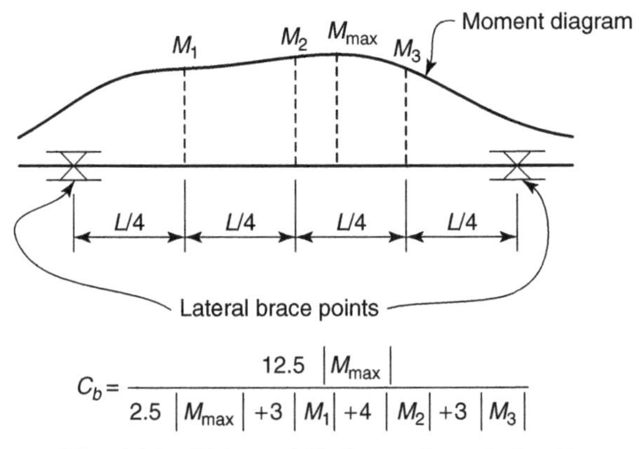

$$C_b = \frac{12.5\,|M_{max}|}{2.5\,|M_{max}| + 3\,|M_1| + 4\,|M_2| + 3\,|M_3|}$$

Fig. 6.14 Kirby and Nethercot formula for C_b.

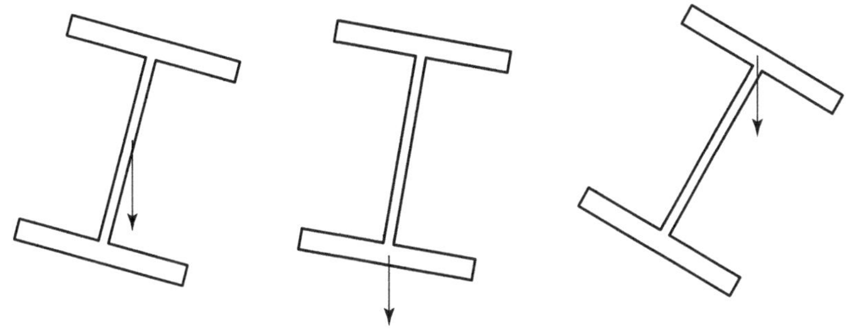

Fig. 6.15 Effect of the location of the load.

$$M_{ocr} = \frac{\pi}{L}\sqrt{EI_yGJ}\sqrt{1+W^2} \tag{6.14}$$

where

$$W = \frac{\pi}{L}\sqrt{\frac{EC_w}{GJ}} \tag{6.15}$$

The load condition modifier C_b is equal to:

$$C_b = AB^{2y/d} \tag{6.16}$$

The parameters A and B are given by the formulas in Figure 6.16; d is the depth of the cross-section and y is the distance from the centroid to the location where the load is applied. It is positive if the load is applied below the mid-height, and negative if it is applied above the mid-height. Thus, if

Loading	Moment Diagram	M	A	B
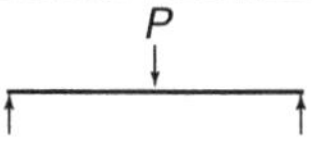 P		$\frac{PL}{4}$	1.35	$1-0.180W^2+0.649W$
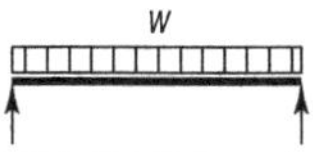 w		$\frac{wL^2}{8}$	1.12	$1-0.154W^2+0.535W$
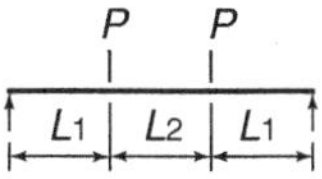 P P; L_1 L_2 L_1	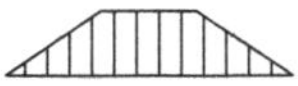	PL_1	$1+\left(\frac{L_1}{2L_1+L_2}\right)^2$	$\frac{(1-0.465W^2+1.636W)L_1}{2L_1+L_2}$

Fig. 6.16 Coefficients for beams with simple supports (Figure 5.3, SSRC Guide).

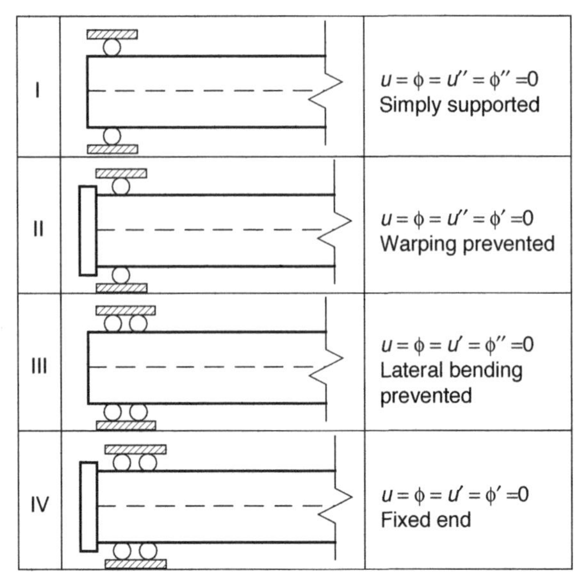

Fig. 6.17 Idealized end restraints (Figure 5.4, SSRC Guide).

the load is applied at the top flange, $y = -d/2$ and $C_b = A/B$; if it is located at the center then $y = 0$ and $C_b = A$; and if it is at the bottom flange then is $y = d/2$ and $C_b = AB$. The additional effect of end-restraint on the coefficients A and B can be obtained from Figures 6.17 and 6.18.

The following three example problems illustrate methods of calculating elastic lateral-torsional buckling loads for beams. In particular they show the effects of the load location on the magnitudes of the critical moments.

Loading	Restraint	A	B
(concentrated load at midspan)	*I*	1.35	$1 - 0.180W^2 + 0.649W$
	II	$1.43 + 0.485W^2 + 0.463W$	$1 - 0.317W^2 + 0.619W$
	III	$2.0 - 0.074W^2 + 0.304W$	$1 - 0.207W^2 + 1.047W$
	IV	$1.916 - 0.424W^2 + 1.851W$	$1 - 0.466W^2 + 0.923W$
(uniformly distributed load)	*I*	1.13	$1 - 0.225W^2 + 0.571W$
	II	$1.2 + 0.416W^2 + 0.402W$	$1 - 0.154W^2 + 0.535W$
	III	$1.9 - 0.120W^2 + 0.006W$	$1 - 0.100W^2 + 0.806W$
	IV	$1.643 - 0.405W^2 + 0.771W$	$1 - 0.339W^2 + 0.625W$

Fig. 6.18 Restraint categories (Figure 5.5, SSRC Guide).

6.4.1 Example 6.1 Effect of Load Location

Calculate the critical load for different load positions and boundary conditions for a 40 ft. long W24 × 55 wide-flange beam subjected to a concentrated load P at its center.

Initial data

$$\begin{array}{ll} E = 29,000\,\text{ksi} & I_y = 29.1\,\text{in.}^4 \\ G = 0.385\,\text{E} & C_w = 3870\,\text{in.}^6 \\ F_y = 50\,\text{ksi} & J = 1.18\,\text{in.}^4 \\ L = 40\,\text{ft.} & S_x = 114\,\text{in.}^3 \end{array}$$

$$W = \frac{\pi}{L}\sqrt{\frac{EC_w}{GJ}} = 0.604$$

$$M_{ocr} = \frac{\pi}{L}\sqrt{EI_yGJ}\sqrt{1+W^2} = 806''\,\text{k}$$

Case I: Lateral and torsional simple supports (Figure 6.16).

1. P is applied at the top flange:

$$A = 1.35$$
$$B = 1 - 0.180W^2 + 0.649W = 1.32$$
$$C_b = \frac{A}{B}$$
$$M_{cr} = C_b M_{ocr} = 824''\,\text{k}$$
$$P_{cr} = \frac{4M_{cr}}{L} = 6.8\,\text{kip}$$

2. P is applied at the bottom flange:

$$A = 1.35$$
$$B = 1 - 0.180W^2 + 0.649W = 1.32$$
$$C_b = AB = 1.78$$
$$M_{cr} = C_b M_{ocr} = 1438''\,\text{k}$$
$$P_{cr} = \frac{4M_{cr}}{L} = 11.9\,\text{kip}$$

Case II: Laterally simply supported ends; warping is prevented at the supports (Case II, Figures 6.17 and 6.18)

1. P is applied at the top flange.

$$A = 1.43 + 0.485W^2 + 0.463W = 1.89$$
$$B = 1 - 0.317W^2 + 0.619W = 1.26$$
$$C_b = \frac{A}{B} = 1.50$$
$$M_{\text{cr}} = C_b M_{o\text{cr}} = 1209'' \text{k}$$
$$P_{\text{cr}} = \frac{4M_{\text{cr}}}{L} = 10.0 \, \text{kip}$$

2. P is applied at the bottom flange.

$$A = 1.43 + 0.485W^2 + 0.463W = 1.89$$
$$B = 1 - 0.317W^2 + 0.619W = 1.26$$
$$C_b = AB = 2.37$$
$$M_{\text{cr}} = C_b M_{o\text{cr}} = 1914'' \text{k}$$
$$P_{\text{cr}} = \frac{4M_{\text{cr}}}{L} = 15.9 \, \text{kip}$$

6.4.2 Example 6.2 Stability of a Beam with An Erector at Its Center

An erector weighing 250 lb. must install a lateral brace at the center during the erection of a 50 ft. long $W16 \times 26$ beam, as shown in Figure 6.19. The erector sits on the top flange of the beam. Derive a formula for the lateral-torsional buckling of a wide-flange beam loaded at the centroid by its self-weight w and by an erector whose weight P is applied at the top flange. Use

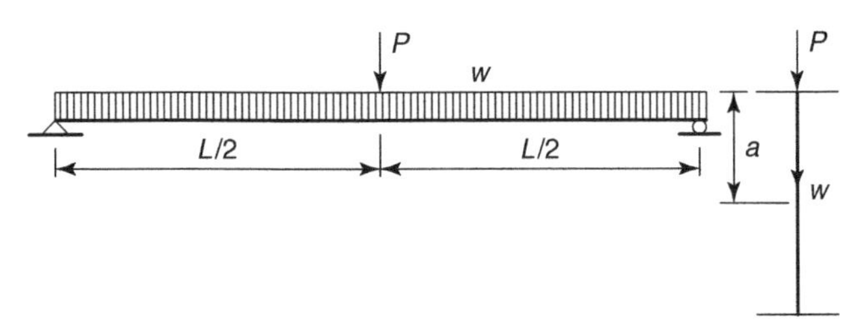

Fig. 6.19 Example 6.2 beam.

an approximate energy method in the derivation. Is it safe for the erector to perform the required job?

1. *Derivation of the formula for the elastic critical moment:*
   The total potential of the virtual work due to the lateral deflection u and the torsional twist ϕ is equal to

$$U + V_{\mathrm{p}}$$

where U is the elastic strain energy

$$U = \frac{1}{2}\int_0^L \left[EI_y(u'')^2 + GJ(\phi')^2 + EC_w(\phi'')^2\right] dz$$

and V_{p} is the potential of the external loads

$$V_{\mathrm{p}} = \int_0^L M_x u'' \phi dz - \frac{Pa}{2}\left[\phi\left(\frac{L}{2}\right)\right]^2$$

The coordinate z has its origin at the left support, and the moment about the x-axis M_x is equal to

$$M_x^w = \frac{wL^2}{2}\left(\frac{z}{L}\right)\left(1 - \frac{z}{L}\right)$$

$$M_x^P = \frac{PL}{2} \times \begin{Bmatrix} \dfrac{z}{L} & \text{for } 0 \le \dfrac{z}{L} \le \dfrac{L}{2} \\ 1 - \dfrac{z}{L} & \text{for } \dfrac{L}{2} \le \dfrac{z}{L} \le L \end{Bmatrix}$$

The deformations are assumed to be sinusoidal. The coefficients A and B are the maximum amplitudes at the center of the beam. The distance a defines the location of P from the centroid. For top flange loading, it is equal to $d/2$, where d is the total depth of the section. Following are the expressions of the assumed deformations u and ϕ, and their first and second derivatives with respect to the coordinate z:

$$u = A\sin\left(\frac{\pi z}{L}\right) \qquad \phi = B\sin\left(\frac{\pi z}{L}\right)$$

$$u' = A\left(\frac{\pi}{L}\right)\cos\left(\frac{\pi z}{L}\right) \qquad \phi' = B\left(\frac{\pi}{L}\right)\cos\left(\frac{\pi z}{L}\right)$$

$$u'' = A\left(-\frac{\pi^2}{L^2}\right)\sin\left(\frac{\pi z}{L}\right) \qquad \phi'' = B\left(-\frac{\pi^2}{L^2}\right)\sin\left(\frac{\pi z}{L}\right)$$

The following integrals are needed for the evaluation of the total potential:

$$\int_0^L \sin^2\left(\frac{\pi z}{L}\right)dz = \int_0^L \cos^2\left(\frac{\pi z}{L}\right)dz = \frac{L}{2}$$

$$\int_0^L z\sin^2\left(\frac{\pi z}{L}\right)dz = \frac{L^2}{4}$$

$$\int_0^L z^2\sin^2\left(\frac{\pi z}{L}\right)dz = \left(\frac{1}{6}-\frac{1}{4\pi^2}\right)L^3 = 0.14134L^3$$

$$\int_0^{\frac{L}{2}} z\sin^2\left(\frac{\pi z}{L}\right)dz = \left(\frac{1}{16}+\frac{1}{4\pi^2}\right)L^2 = 0.08783L^2$$

$$\int_{\frac{L}{2}}^{L} z\sin^2\left(\frac{\pi z}{L}\right)dz = \left(\frac{3}{16}-\frac{1}{4\pi^2}\right)L^2 = 0.16217L^2$$

After some algebra, we arrive at the following expression for the total potential:

$$U + V_p = A^2\left[\frac{\pi^4 EI_y}{4L^3}\right] + B^2\left[\frac{\pi^4 EC_w}{4L^3}+\frac{\pi^2 GJ}{4L}-\frac{Pa}{2}\right] - AB[0.5362wL + 0.8668P]$$

According to the virtual work theory, the equilibrium of the buckled shape is obtained by setting the derivatives of the total potential with respect to the coefficients A and B equal to zero:

$$\frac{\partial(U+V_p)}{\partial A} = 0 = 2A\left[\frac{\pi^4 EI_y}{4L^3}\right] - B[0.5362wL + 0.8668P]$$

$$\frac{\partial(U+V_p)}{\partial B} = 0 = 2B\left[\frac{\pi^4 EC_w}{4L^3}+\frac{\pi^2 GJ}{4L}-\frac{Pa}{2}\right] - A[0.5362wL + 0.8668P]$$

These are two homogeneous simultaneous equations that will give the buckling load when the determinant of the coefficients of A and B is equated to zero:

$$\begin{vmatrix} \dfrac{\pi^4 EI_y}{2L^3} & -(0.5362wL + 0.8668P) \\ -(0.5362wL + 0.8668P) & \dfrac{\pi^4 EC_w}{2L^3}+\dfrac{\pi^2 GJ}{2L} - Pa \end{vmatrix} = 0$$

Decomposition of the determinant gives the following quadratic equation for the critical load combination of w and P:

$$\frac{\pi^4 EI_y}{2L^3}\left[\frac{\pi^4 EC_w}{2L^3}+\frac{\pi^2 GJ}{2L}-Pa\right]=(0.5362wL+0.8668P)^2$$

Substitution of the basic critical moment expression (equation 6.10)

$$M_{ocr}=\frac{\pi}{L}\sqrt{EI_yGJ}\sqrt{1+\frac{\pi^2 EC_w}{GJL^2}}$$

will result, after some algebra, in the following equation:

$$\frac{\pi^4 M_{ocr}^2}{4L^2}-Pa\frac{\pi^4 EI_y}{2L^3}=(0.5362wL+0.8668P)^2 \tag{6.17}$$

If $a=0$ this equation can be checked for accuracy on problems for which there are answers in Figure 6.16. When $P=0$,

$$\frac{\pi^2 M_{ocr}}{2L}=0.5362wL\rightarrow M_{cr}=\frac{wL^2}{8}=\frac{\pi^2 M_{ocr}}{2\times 8\times 0.5362}=1.15M_{ocr}$$

That is, $C_b=1.15$; this compares with $C_b=1.12$ given in Figure 6.16. Similarly, when $w=0$

$$\frac{\pi^2 M_{ocr}}{2L}=0.8668P\rightarrow M_{cr}=\frac{PL}{4}=\frac{\pi^2 M_{ocr}}{2\times 4\times 0.8668}=1.42M_{ocr}$$

The value of $C_b=1.42$ compares with $C_b=1.35$ in Figure 6.16. The approximate derivation thus gives an acceptably accurate equation for the case of distributed load with a concentrated load at the top of the flange.

We now return to the problem of the erector at the center of the $W16\times26$ beam. For this beam, the self-weight $w=26$ lb./ft. The basic critical moment is found to be $M_{ocr}=160.40$ in-kips from equation 6.10. The location of the erector above the centroid is $a=d/2=15.7/2=7.85$ in. The span is $L=50$ ft. $=600$ in. Substitution of these values into equation 6.17 gives the critical load $P_{cr}=586$ lb. that, when applied at the top of the beam flange, will cause elastic lateral-torsional buckling. The safety factor of the 250 lb. erector is thus $586/250=2.3$.

6.4.1 Example 6.3 Effect of Restraints from Adjacent Spans on Beam Stability

A simply supported wide-flange beam of length L is under a distributed load w, as shown in Figure 6.20. There are lateral braces at each third point. The central braced segment is approximately under uniform moment. The two other braced segments act as restraining members. Determine the effective length factor to be used in the equation for the critical moment, equation 6.12. Assume that the lateral and the torsional restraints are the same. The central segment is the critical length. The approximate procedure is from Chapter 5 of the SSRC Guide (Galambos 1998), and the original idea is due to Trahair (1993).

Analogous to the effective length determination for frames in Chapter 5, the effective length can be obtained for beams also using the AISC alignment charts for the sway-prohibited case. The restraint factor G includes an approximation for the reduction in the stiffness of the restraining beams.

$$G = \frac{\alpha_m}{\alpha_r}$$

$$\alpha_m = \frac{2EI_y}{L_m} = \frac{6EI_y}{L}$$

$$\alpha_r = \frac{nEI_y}{L_r}\left[1 - \frac{M_{cr}^m}{M_{cr}^r}\right] = \frac{3EI_y}{L/3}\left[1 - \frac{C_b^m M_{ocr}}{C_b^r M_{ocr}}\right] = \frac{9EI_y}{L}\left[1 - \frac{1}{1.75}\right]$$

$$G = \frac{6}{9\left[1 - \frac{1}{1.75}\right]} = 1.56$$

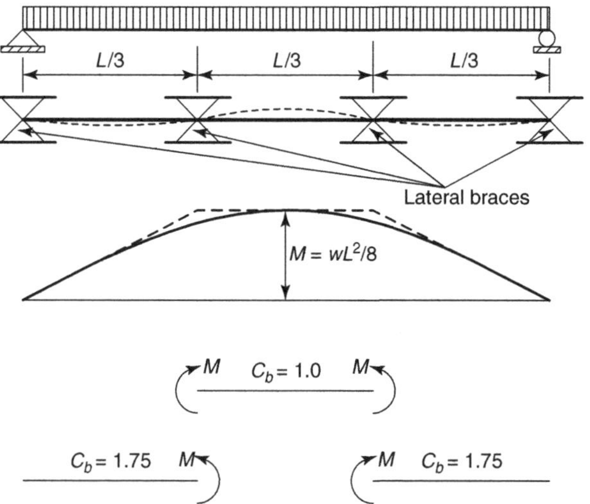

Fig. 6.20 Beam braced at third points in Example 6.3.

The letters m and r refer to the restrained and the restraining segments, respectively. The coefficient $n = 3$ if the far end of the restraining segment is pinned, $n = 4$ if it is fixed, and $n = 2$ if it is continuous. Here $n = 3$ applies. The effective length factor from the AISC alignment chart for $G_A = G_B = G = 1.56$ is $k = 0.83$. The critical lateral-torsional moment is thus equal to:

$$M_{cr} = \frac{\pi}{0.83(L/3)} \sqrt{EI_y GJ} \sqrt{1 - \frac{\pi^2 EC_w}{GJ[0.83(L/3)]^2}}$$

Summary of Important Points—Elastic LTB of Symmetric Cross-sections

- As the name suggests, lateral torsional buckling is a phenomenon including both out-of plane buckling and twisting of the cross-section; this leads to a beam displacement that includes contributions from lateral displacement and the angle of twist.
- The twisting of the cross-section includes components of warping torsion and St. Venant's torsion, and the resistance of the cross-section to this twisting is dependant on C_w, the warping constant, and J, the St. Venant's torsion constant and the polar moment of inertia.
- The critical moment of the cross-section will vary based on the end restraint (both lateral and torsional), the moment gradient, and the placement of the load.

6.5 LATERAL-TORSIONAL BUCKLING OF SINGLY SYMMETRIC CROSS-SECTIONS

In the previous three sections we considered the lateral-torsional buckling behavior of wide-flange beams with a doubly symmetric cross-section. In this part of Chapter 6, we consider the case of the lateral-torsional buckling of wide-flange beams with a singly symmetric cross-section. (see Figure 6.21).

The differential equations for the simply supported beam under uniform moment, as shown in Figure 6.3, are equal to (Timoshenko and Gere 1961, Galambos 1968, Vlasov 1961):

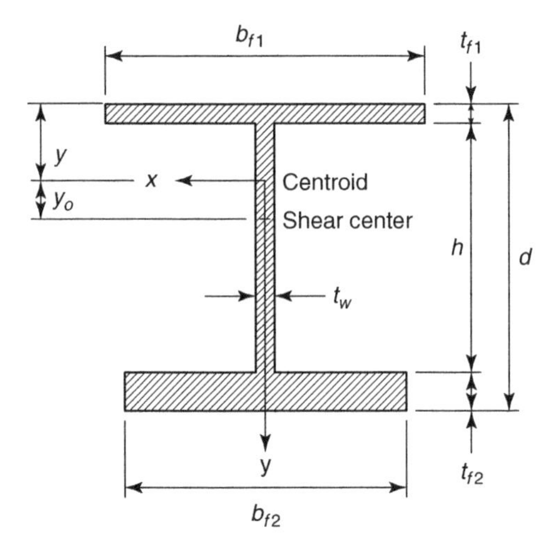

Fig. 6.21 The singly symmetric cross-section geometry.

$$EI_y u'' + M_o \phi = 0 \tag{6.18}$$

$$EC_w \phi''' - (GJ + M_o \beta_x)\phi + M_o u = 0 \tag{6.19}$$

The first equation is equal to the first equation in 6.1, the second equation has an additional term $M_o \beta_x$. This term derives from the additional twisting moment caused by the normal stresses on each of the two differently warped surfaces on the differential elements dz along the z-axis of the beam.

$$\beta_x = \frac{1}{I_x} \int_{\text{Area}} y(x^2 + y^2)dA - 2y_o \tag{6.20}$$

is a cross-section property. I_x is the major-axis moment of inertia, x and y are principal axes, and y_o is the distance of the shear center from the centroid, as shown in Figure 6.21.

For simply supported boundary conditions, $u = u'' = 0$, equation 6.18 can be rearranged to give

$$u'' = -\frac{M_o \phi}{EI_y}$$

Differentiating equation 6.19 once and substituting u'' will result in the following differential equation for the angle of twist ϕ:

$$EC_w \phi^{iv} - (GJ + M_o \beta_x)\phi'' - \frac{M_o \phi}{EI_y} = 0 \tag{6.21}$$

The solution of this differential equation follows the same course as the solution of equation 6.2, resulting finally in the same eigenfunction $\sin \alpha_2 L = 0$ (equation 6.9). This expression can only be true if $\alpha_2 L = n\pi$. For the lowest eigenvalue $n = 1$. From equation 6.6, we get

$$\alpha_2 = \sqrt{\frac{\sqrt{\lambda_1^2 + 4\lambda_2} - \lambda_1}{2}}$$

$$\lambda_1 = \frac{GJ + M_o\beta_x}{EC_w} \tag{6.22}$$

$$\lambda_2 = \frac{M_o^2}{E^2 I_y C_w}$$

The equations for α_2 and λ_2 are identical to the previous equations for the doubly symmetric wide-flange cross-section; only λ_1 is different, since it includes the extra term $M_o\beta_x$.

$$\alpha_2^2 L^2 = \pi^2 = \frac{1}{2}\left(\sqrt{\lambda_1^2 + 4\lambda_2} - \lambda_1\right)$$

$$\lambda_2 = \frac{\pi^2}{L^2}\left(\lambda_1 + \frac{\pi^2}{L^2}\right)$$

$$\frac{M_o^2}{E^2 I_y C_w} = \frac{\pi^2}{L^2}\left[\frac{GJ + M_o\beta_x}{EC_w} + \frac{\pi^2}{L^2}\right]$$

Solving the quadratic equation for M_{ocr} results in one or the other of the following two alternate formulas for the critical moment:

$$M_{ocr} = \frac{\pi^2 EI_y\beta_x}{2L^2}\left[1 \pm \sqrt{1 + \frac{4}{\beta_x^2}\left(\frac{GJL^2}{\pi^2 EI_y} + \frac{C_w}{I_y}\right)}\right] \tag{6.23}$$

$$M_{ocr} = \frac{\pi}{L}\sqrt{EI_y GJ}\left[\frac{\pi\beta_x}{2L}\sqrt{\frac{EI_y}{GJ}} \pm \sqrt{1 + \frac{\pi^2 EC_w}{GJL^2} + \frac{\pi^2\beta_x^2}{4L^2}\left(\frac{EI_y}{GJ}\right)}\right] \tag{6.24}$$

The plus sign is for the top flange in compression and the minus sign is for the top flange in tension. For a doubly symmetric wide-flange section $\beta_x = 0$.

Equation 6.24 then becomes equal to

$$M_{ocr} = \frac{\pi}{L}\sqrt{EI_yGJ}\left[\pm\sqrt{1+\frac{\pi^2 EC_w}{GJL^2}}\right]$$

This formula is identical to equation 6.10 that was previously derived for the doubly symmetric cross-section. Equations for the applicable cross section properties for the singly symmetric wide-flange shape are listed in Table 6.1.

One of the most commonly used singly symmetric shapes is the T-section. For this section, (see Figure 6.21 and Table 6.1)

$$b_{f2} = \alpha = C_w = 0.$$

The critical lateral-torsional buckling moment is then (from equation 6.24):

$$M_{ocr} = \frac{\pi}{L}\sqrt{EI_yGJ}\left[\frac{\pi\beta_x}{2L}\sqrt{\frac{EI_y}{GJ}} \pm \sqrt{1+\frac{\pi^2\beta_x^2}{4L^2}\left(\frac{EI_y}{GJ}\right)}\right] \tag{6.25}$$

or from equation 6.23:

$$M_{ocr} = \frac{\pi^2 EI_y\beta_x}{2L^2}\left[1 \pm \sqrt{1+\frac{4}{\beta_x^2}\left(\frac{GJL^2}{\pi^2 EI_y}\right)}\right] \tag{6.26}$$

If we introduce the cross-section constants

$$A_1 = \frac{\pi^2 EI_y\beta_x}{2} \tag{6.27}$$

$$A_2 = \frac{4GJ}{\pi^2\beta_x^2 EI_y} \tag{6.28}$$

then the critical moment for the T-beam can be written as

$$M_{ocr} = \frac{A_1}{L^2}\left(1 \pm \sqrt{1 + A_2L^2}\right) \tag{6.29}$$

The curves in Figure 6.22 illustrate several aspects of the lateral-torsional stability of T-beams. The cross-section used in the example is the rolled WT18 × 67.5 shape. The cross-sectional properties from the AISC Manual are given below in Table 6.2.

TABLE 6.1 Formulas for Cross-section Properties for the Shape in Figure 6.21

Area of top flange	$A_{f1} = b_{f1} t_{f1}$
Area of bottom flange	$A_{f2} = b_{f2} t_{f2}$
Area of web	$A_w = h t_w$
Area	$A = A_{f1} + A_{f2} + A_w$
Distance of centroid from top flange	$y = \frac{1}{A}\left[\frac{A_{f1} t_{f1}}{2} + A_w\left(t_{f1} + \frac{h}{2}\right) + A_{f2}\left(d - \frac{t_{f2}}{2}\right)\right]$
Moment of inertia about x-axis	$I_x = \frac{b_{f1} t_{f1}^3 + h^3 t_w + b_{f2} t_{f2}^3}{12} + A_{f1}\left(y - \frac{t_{f1}}{2}\right)^2 + A_w\left(t_{f1} + \frac{h}{2} - y\right)^2 + A_{f2}\left(d - \frac{t_{f2}}{2} - y\right)^2$
Moment of inertia about y-axis	$I_y = \frac{b_{f1}^3 t_{f1} + b_{f2}^3 t_{f2} + h t_w^3}{12}$
Distance from centroid to center of top flange	$\overline{y} = y - \frac{t_{f1}}{2}$
Distance between flange centers	$h_o = d - \frac{t_{f1} + t_{f2}}{2}$
α	$\alpha = \frac{1}{1 + \left(\frac{b_{f1}}{b_{f2}}\right)^3 \left(\frac{t_{f1}}{t_{f2}}\right)}$
Distance between centroid and shear center	$y_o = \alpha h_o - \overline{y}$
β_x	$\beta_x = \frac{1}{I_x}\left\{ \left(h_o - \overline{y}\right)\left[\frac{b_{f2}^3 t_{f2}}{12} + b_{f2} t_{f2}\left(h_o - \overline{y}\right)^2 + \frac{t_w\left(h_o - \overline{y}\right)^3}{4}\right] - \overline{y}\left(\frac{b_{f1}^3 t_{f1}}{12} + b_{f1} t_{f1} \overline{y}^2 + \frac{t_w \overline{y}^3}{4}\right)\right\} - 2y_o$
Warping constant	$C_w = \frac{h_o^2 b_{f1}^3 t_{f1} \alpha}{12}$

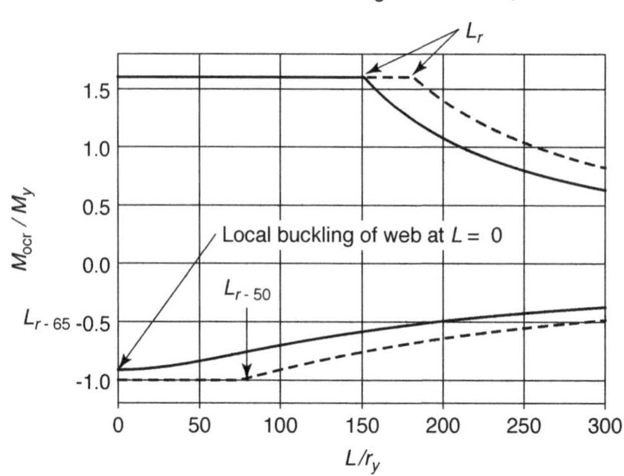

Fig. 6.22 T-beam stability curves.

The relationship between the critical moment and the length of the member is presented in Figure 6.22 in nondimensional form: the critical moment is the ordinate, and it is divided by the yield moment M_y; the span length is the abscissa, and it is divided by the minor-axis slenderness ratio r_y. The usual condition is when the flange of the T-beam is in compression. In this case, the positive sign in equation 6.29 applies. According to the AISC Specification Section F9, the nominal moment M_n shall not exceed the lesser value of M_p, the plastic moment, and 1.6 times the yield moment, M_y, when the flange of the beam is in compression. The length at which the

TABLE 6.2 Data for WT18 × 67.5 T-beam

Handbook Values	Material Data	Computed Values
$b_f = 11.95$ in.	$E = 29,000$ ksi	$\beta_x = 13.17$ in.
$t_f = 0.79$ in.	$G = 0.385E = 11,185$ ksi	$A_1 = 213 \times 10^6$ kip − in.^{3}
$d = 17.78$ in.	$F_y = 65$ ksi	$A_2 = 2.766 \times 10^{-5} \times \frac{1}{\text{in.}^2}$
$t_w = 0.6$ in.	$M_y = S_x F_y = 3230.5''$ k	$\frac{A_1 A_2}{2} = 2949''$ k
$A = 19.9$ in.^{2}	$M_p = Z_x F_y = 5856.5''$ k	
$I_x = 636$ in.^{4}	$M_n = \min(1.6M_y, M_p) = 5168.8''$ k	
$S_x = 49.7$ in.^{3}		
$I_y = 113$ in.^{4}		
$r_y = 2.38$ in.		
$J = 3.48$ in.^{4}		
$y = 4.96$ in.		

maximum permissible nominal moment is reached is the point L_r indicated in Figure 6.22. This length is obtained by setting $M_{ocr} = M_n$ in equation 6.29, resulting in the following equation:

$$L_r = \frac{A_1}{M_n}\sqrt{\frac{2M_n}{A_1} + A_2} \tag{6.30}$$

For $F_y = 65\,\text{ksi}$, $L_r = 360\,\text{in.} = 30\,\text{ft.}$ or $\frac{L}{r_y} = 151$. Also shown in Figure 6.22 in dotted lines is the critical moment curve for a beam with $F_y = 50\,\text{ksi}$.

The relations between the critical moment and the length in the negative region of Figure 6.22 are obtained from equation 6.29 when the negative sign is used. In this case, the tip of the stem of the T-shape is in compression. Such a situation is not very common in practice, but it needs to be also investigated. The AISC (2005) Specification Section F9 prescribes the upper limit of the nominal moment to be the yield moment, M_y. Two types of behavior are possible: (1) the yield moment is reached, as shown by the dashed line for $F_y = 50\,\text{ksi}$, at a length L_r that can be determined from equation 6.30 by setting $M_{ocr} = -M_y$ into equation 6.29 and solving for

$$L_r = \frac{A_1}{M_y}\sqrt{A_2 - \frac{2M_y}{A_1}} \tag{6.31}$$

and (2) the yield moment is not achieved. In this latter case, equation 6.29 is solved for the critical moment for a zero length member. Substitution results in $M_{ocr} = \frac{0}{0}$, and L'Hospital's rule from calculus must be used to obtain an answer, as shown in the next few lines of derivation:

$$\lim M_{ocr}\; as\; L \to 0 = \frac{\frac{d\left[A_1\left[1-(A_2L^2+1)^{\frac{1}{2}}\right]\right]}{dL}}{\frac{d(L^2)}{dL}}$$

$$= -\frac{A_1}{2}\left\{\frac{\frac{d\left[A_2L(A_2L^2+1)^{-\frac{1}{2}}\right]}{dL}}{\frac{d(L)}{dL}}\right\}$$

$$= -\frac{A_1}{2}\left\{\frac{d[A_2(A_2L^2+1)^{-\frac{1}{2}} - \frac{1}{2}A_2L(A_2L^2+1)^{-\frac{3}{2}} \times 2A_2L]}{dL}\right\}$$

$$= -\frac{A_1A_2}{2}$$

Thus

TABLE 6.3 Summary of Elastic Lateral-torsional Buckling Strength of a T-beam

Flange in compression	$M_{ocr} = \frac{A_1}{L^2}\left(1 + \sqrt{A_2 L^2 + 1}\right)$
	$L_r = \frac{A_1}{M_n}\sqrt{\frac{2M_n}{A_1} + A_2}$
Web tip in compression	$M_{ocr} = \frac{A_1}{L^2}\left(1 - \sqrt{A_2 L^2 + 1}\right)$
	$L_r = \frac{A_1}{M_y}\sqrt{\frac{-2M_y}{A_1} + A_2} \quad \text{if } M_y \leq \frac{GJ}{\beta_x}$
	$\text{if } M_y > \frac{GJ}{\beta_x}, \text{ then } \sigma_{cr}^{\text{web local buckling}} = \frac{GJ}{\beta_x S_x}$

$$M_{ocr}^{L \to 0} = -\frac{A_1 A_2}{2} = -\frac{GJ}{\beta_x} \tag{6.32}$$

For the case of the WT18 × 67($Fy = 65$ ksi) the critical moment at zero length is equal to 2949 in-kip, or 0.913 M_y. Converting this moment into a stress acting at the tip of the stem gives $\sigma_{cr} = \frac{M_{ocr}}{S_x} = \frac{2949}{49.7} = 59\,\text{ksi} < F_y = 65\,\text{ksi}$. This stress can be thought of as the elastic local buckling stress of the stem, and it is designated as such in section F9 of the AISC (2005) Specification. Table 6.3 summarizes the equations applicable to T-beam lateral-torsional stability.

In order to avoid the cumbersome calculation of β_x with equation 6.20, Kitipornchai and Trahair (1980) performed an extensive numerical parametric study, and they proposed the following simple approximate formula:

$$\beta_x^{\text{Appr.}} = h_o\left\{0.9(2\rho - 1)\left[1 - \left(\frac{I_y}{I_x}\right)^2\right]\right\} \tag{6.33}$$

where

$$\rho = \frac{I_{yc}}{I_y} = \frac{t_{fc} b_{fc}^3}{12 I_y} \tag{6.34}$$

The subscript c refers to the compression flange. This formula is quite accurate as long as $\frac{I_y}{I_x} < 0.5$. We will next consider equation 6.24:

$$M_{ocr} = \frac{\pi}{L}\sqrt{EI_y GJ}\left[\frac{\pi\beta_x}{2L}\sqrt{\frac{EI_y}{GJ}} \pm \sqrt{1 + \frac{\pi^2 EC_w}{GJL^2} + \frac{\pi^2\beta_x^2}{4L^2}\left(\frac{EI_y}{GJ}\right)}\right]$$

Consider the first term inside the bracket, and let it be designated as B_1

$$B_1 = \frac{\pi\beta_x}{2L}\sqrt{\frac{EI_y}{GJ}} = \frac{\pi h_o}{2L}\sqrt{\frac{EI_y}{GJ}} \times 0.9(2\rho - 1)\left[1 - \left(\frac{I_y}{I_x}\right)^2\right]$$

$$\left[1 - \left(\frac{I_y}{I_x}\right)^2\right] \approx 1.0$$

$$\frac{0.9\pi}{2}\sqrt{\frac{1}{0.385}} \approx 2.25$$

$$B_1 = 2.25\left(2\frac{I_{yc}}{I_y} - 1\right)\frac{h_o}{L}\sqrt{\frac{I_y}{L}}$$

Next let

$$B_2 = \frac{\pi^2 EC_w}{GJL^2}$$

$$C_w = I_{yc}\alpha h_o^2 = \rho(1 - \rho)I_y h_o^2$$

where α is from Table 6.1 with $\frac{b_1^3 t_1}{12} = I_{yc}$, the y-axis moment of inertia of the compression flange

$$\frac{\pi^2 E}{G} = \frac{\pi^2}{0.385} \approx 25$$

$$B_2 = 25\left(\frac{h_o}{L}\right)^2\left(\frac{I_{yc}}{J}\right)\left(1 - \frac{I_{yc}}{I_y}\right)$$

The approximate expression for the elastic lateral-torsional buckling moment of singly symmetric wide-flange beam is

$$M_{ocr} = \frac{\pi}{L}\sqrt{EI_y GJ}\left(B_1 \pm \sqrt{1 + B_2 + B_1^2}\right) \tag{6.35}$$

The exact and the approximate method will be illustrated in Example 6.4.

For the T-beam $C_w = B_2 = 0$; $\frac{I_{yc}}{I_y} = 1.0$; $1 - \left(\frac{I_y}{I_x}\right)^2 \approx 1$; $\beta_x \approx 0.9h_o$; $h_o \approx 0.98d$, and $B = B_1 = 2.3\frac{d}{L}\sqrt{\frac{I_y}{J}}$. The approximate elastic lateral-torsional buckling moment for a T-beam is thus

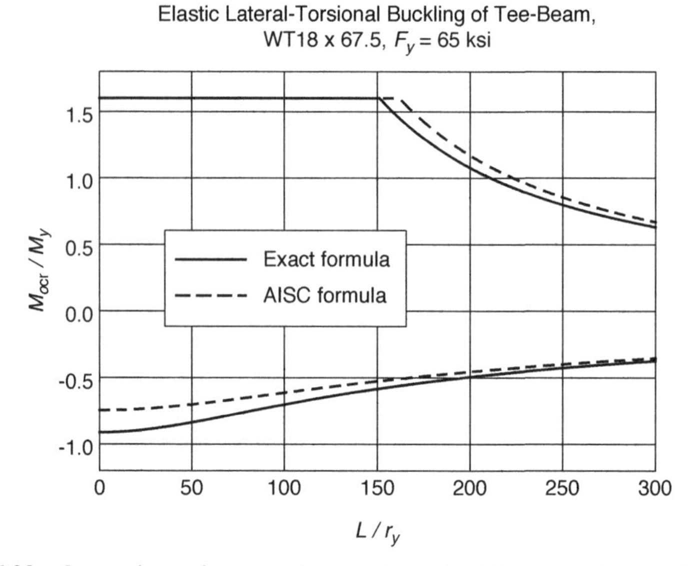

Fig. 6.23 Comparison of exact and approximate buckling strength for a T-beam.

$$M_{ocr} = \frac{\pi}{L}\sqrt{EI_yGJ}\left(B \pm \sqrt{1+B^2}\right) \tag{6.36}$$

This is the equation used in section F9 of the 2005 AISC Specification. The resulting curves for the WT18 × 67.5 beam are shown as dashed lines in Figure 6.23. Equation 6.36 overestimates the critical moment when the flange is in compression, and it underestimates it when the tip of the stem is in compression.

6.5.1 Example 6.4 Stability of Beam with Unequal Flanges

Calculate the elastic lateral-torsional buckling strength of a singly-symmetric wide-flange beam shown in Figure 6.21. The following dimensions are given:

Dimensions:

$$E = 29,000\,\text{ksi}$$
$$G = 0.385\text{E}$$
$$F_y = 50\,\text{ksi}$$
$$L = 15\,\text{ft.}$$
$$b_{f1} = 10\,\text{in.}$$
$$b_{f2} = 5\,\text{in.}$$
$$h = 25\,\text{in.}$$

$$t_{f1} = 0.75 \text{ in.}$$
$$t_{f2} = 0.75 \text{ in.}$$
$$t_w = 0.5 \text{ in.}$$
$$d = 26.5 \text{ in.}$$

Here are the calculated properties:

$$\begin{array}{ll} A = 23.75 \text{ in.}^2 & \\ y = 11.22 \text{ in.} & I_y = 70.6 \text{ in.}^4 \\ I_x = 2418 \text{ in.}^4 & y_o = -7.98 \text{ in.} \\ \bar{y} = 10.84 \text{ in.} & C_w = 4605 \text{ in.}^6 \\ h_o = 25.75 \text{ in.} & J = 3.18 \text{ in.}^4 \\ \alpha = 0.111 & \beta_x = 18.75 \text{ in.} \end{array}$$

Elastic Critical Moments (equation 6.23)

Case I: top flange in compression

$$M_{ocr} = \frac{\pi^2 E I_y \beta_x}{2L^2}\left[1 + \sqrt{1 + \frac{4}{\beta_x^2}\left(\frac{C_w}{I_y} + \frac{GJL^2}{\pi^2 E I_y}\right)}\right] = 14,880'' \text{k}$$

$$\sigma_{top} = \frac{M_{ocr} y}{I_x} = 69 \text{ ksi(compression)}$$

$$\sigma_{bottom} = \frac{M_{ocr}(d - y)}{I_x} = 94 \text{ ksi(tension)}$$

The beam will not fail by elastic lateral-torsional buckling if the top flange is in compression.

Case II: bottom flange in compression

$$M_{ocr} = \frac{\pi^2 E I_y \beta_x}{2L^2}\left[1 - \sqrt{1 + \frac{4}{\beta_x^2}\left(\frac{C_w}{I_y} + \frac{GJL^2}{\pi^2 E I_y}\right)}\right] = -3192'' \text{k}$$

$$\sigma_{top} = \frac{M_{ocr} y}{I_x} = 14.8 \text{ ksi(tension)}$$

$$\sigma_{bottom} = \frac{M_{ocr}(d - y)}{I_x} = 20.2 \text{ ksi(compression)}$$

The beam will fail by elastic lateral-torsional buckling if the bottom flange is in compression.

Elastic Critical Moment (equation 6.35):

Case I: top flange in compression

$$I_{yc} = \frac{b_{f1}^3 t_{f1}}{12} = 62.5 \text{ in.}^4$$

$$B_1 = 2.25\left(2\frac{I_{yc}}{I_y} - 1\right)\frac{h_o}{L}\sqrt{\frac{I_y}{J}} = 1.17$$

$$B_2 = 25\left(\frac{h_o}{L}\right)^2 \frac{I_{yc}}{J}\left(1 - \frac{I_{yc}}{I_y}\right) = 1.15$$

$$M_{ocr} = \frac{\pi}{L}\sqrt{EI_y GJ}\left(B_1 + \sqrt{1 + B_2 + B_1^2}\right) = 14,330'' \text{k} (\text{vs. } 14880'' \text{k})$$

Case II: bottom flange in compression

$$I_{yc} = \frac{b_{f2}^3 t_{f2}}{12} = 7.8 \text{ in.}^4$$

$$B_1 = 2.25\left(2\frac{I_{yc}}{I_y} - 1\right)\frac{h_o}{L}\sqrt{\frac{I_y}{J}} = -1.18$$

$$B_2 = 25\left(\frac{h_o}{L}\right)^2 \frac{I_{yc}}{J}\left(1 - \frac{I_{yc}}{I_y}\right) = 1.12$$

$$M_{ocr} = \frac{\pi}{L}\sqrt{EI_y GJ}\left(B_1 + \sqrt{1 + B_2 + B_1^2}\right) = 3263'' \text{k} (\text{vs. } 3192'' \text{k})$$

6.6 BEAM-COLUMNS AND COLUMNS

The lateral-torsional stability of doubly and singly symmetric beams was examined in the previous sections of this chapter. This section will describe the extension of the theory to include beam-columns and columns. A general formula that will incorporate the lateral-torsional stability of beams, columns, and beam-columns will be derived. The loading and geometry are shown in Figure 6.24. This Figure is the same as Figure 6.3, except that an axial load that acts through the centroid of the cross-section is added.

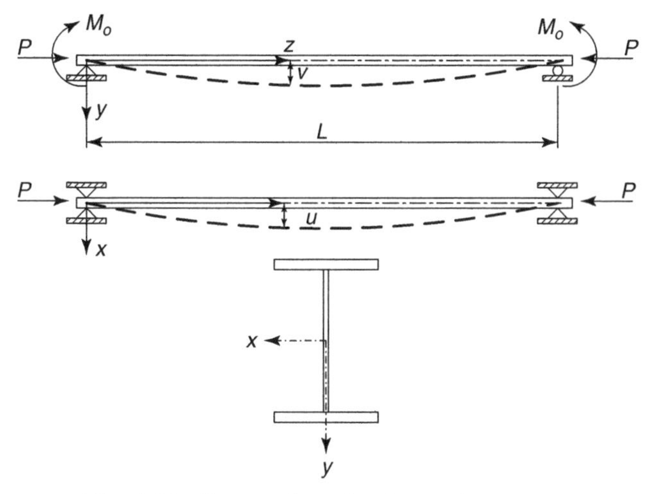

Fig. 6.24 Beam-column loading and geometry.

With the addition of the compressive axial load P the differential equations of lateral-torsional buckling, as previously presented in equations 6.1 for the doubly symmetric cross section and equations 6.18 and 6.19 for the singly symmetric section, become equal to (Galambos 1968)

$$EI_y u^{iv} + Pu'' + (M_o + Py_o)\phi'' = 0 \tag{6.37}$$

$$EC_w \phi^{iv} - (GJ - P\bar{r}_o^2 + M_o \beta_x)\phi'' + (M_o + Py_o)u'' = 0 \tag{6.38}$$

For the case of "simple" supports,

$$u(0) = u(L) = \phi(0) = \phi(L) = 0$$
$$u''(0) = u''(L) = \phi''(0) = \phi''(L) = 0$$

the assumed displacements and their derivatives are

$$u = C_1 \sin\frac{\pi z}{L} \qquad \phi = C_2 \sin\frac{\pi z}{L}$$

$$u'' = -C_1\left(\frac{\pi}{L}\right)^2 \sin\frac{\pi z}{L} \quad \phi'' = -C_2\left(\frac{\pi}{L}\right)^2 \sin\frac{\pi z}{L}$$

$$u^{iv} = C_1\left(\frac{\pi}{L}\right)^4 \sin\frac{\pi z}{L} \quad \phi^{iv} = C_2\left(\frac{\pi}{L}\right)^4 \sin\frac{\pi z}{L}$$

Substitution of these expressions into the previous differential equations, noting that each term is multiplied by $\sin\frac{\pi z}{L} \neq 0$, which can therefore be canceled out, results in the two simultaneous equations:

$$\begin{bmatrix} \left[\dfrac{\pi^2 EI_y}{L^2} - P\right] & [-(M_o + Py_o)] \\ [-(M_o + Py_o)] & \left[\dfrac{\pi^2 EC_w}{L^2} + GJ - P\bar{r}_o^2 + M_o\beta_x\right] \end{bmatrix} \begin{bmatrix} C_1 \\ C_2 \end{bmatrix} = 0 \quad (6.39)$$

In these equations

$$\beta_x = \frac{1}{I_x} \int_{\text{Area}} y(x^2 + y^2)dA - 2y_o \quad (6.20)$$

introduced in the previous section, and

$$\bar{r}_o^2 = y_o^2 + \frac{I_x + I_y}{A} \quad (6.40)$$

where y_o is the shear center distance from the centroid, as shown in Figure 6.21. Two further definitions will now be given:

The *y-axis elastic buckling load* of the column

$$P_{ey} = \frac{\pi^2 EI_y}{L^2} \quad (6.41)$$

and the *torsional buckling load* of the column

$$P_{ez} = \left(\frac{\pi^2 EC_w}{L^2} + GJ\right)\frac{1}{\bar{r}_o^2} \quad (6.42)$$

The two simultaneous equations 6.39 can now be written as

$$\begin{bmatrix} [P_{ey} - P] & [-(M_o + Py_o)] \\ [-(M_o + Py_o)] & [\bar{r}_o^2 P_{ez} - P\bar{r}_o^2 + M_o\beta_x] \end{bmatrix} \begin{bmatrix} C_1 \\ C_2 \end{bmatrix} = 0 \quad (6.43)$$

Since at buckling $C_1 \neq 0$ and $C_2 \neq 0$, the determinant of the coefficients in equation 6.43 must be equal to zero. Thus

$$\begin{vmatrix} [P_{ey} - P] & [-(M_o + Py_o)] \\ [-(M_o + Py_o)] & [\bar{r}_o^2 P_{ez} - P\bar{r}_o^2 + M_o\beta_x] \end{vmatrix} = 0 \quad (6.44)$$

Decomposition of the determinant results in the following characteristic equation that contains the combination of critical forces that define elastic lateral-torsional instability of singly or doubly symmetric prismatic simply supported members loaded by a concentric axial force and/or a uniform moment acting in the plane of symmetry.

$$\left(P_{ey} - P\right)\left(\bar{r}_o^2 P_{ez} - \bar{r}_o^2 P + M_o \beta_x\right) - \left(M_o + P y_o\right) = 0 \tag{6.45}$$

When $P = 0$, $M_o^2 - P_{ey}\beta_x M_o - P_{ey}P_{ez}\bar{r}_o^2 = 0$. From this equation, one can derive equation 6.23 or 6.24 with a few algebraic manipulations.

$$M_{ocr} = \frac{\pi^2 E I_y \beta_x}{2L^2}\left[1 \pm \sqrt{1 + \frac{4}{\beta_x^2}\left(\frac{GJL^2}{\pi^2 E I_y} + \frac{C_w}{I_y}\right)}\right] \tag{6.23}$$

$$M_{ocr} = \frac{\pi}{L}\sqrt{EI_y GJ}\left[\frac{\pi \beta_x}{2L}\sqrt{\frac{EI_y}{GJ}} \pm \sqrt{1 + \frac{\pi^2 E C_w}{GJL^2} + \frac{\pi^2 \beta_x^2}{4L^2}\left(\frac{EI_y}{GJ}\right)}\right] \tag{6.24}$$

When $M_o = 0$, then the following equation is obtained for the *flexural-torsional buckling load* of singly symmetric pinned-end column:

$$P_{cr} = \frac{P_{ey} + P_{ez}}{2H}\left[1 - \sqrt{1 - \frac{4P_{ey}P_{ez}H}{\left(P_{ey} + P_{ez}\right)^2}}\right] \tag{6.46}$$

where

$$H = 1 - \frac{y_o^2}{\bar{r}_o^2} \tag{6.47}$$

The term *flexural-torsional buckling load* and the definition H are the notations used in the AISC (2005) Specification.

In the case of a doubly symmetric cross-section, equations 6.24 and 6.46 simplify to

$$M_{ocr}^{P=0} = \frac{\pi}{L}\sqrt{EI_y GJ}\sqrt{1 + \frac{\pi^2 E C_w}{GJL^2}} \tag{6.10}$$

$$P_{cr}^{M_o=0} = \min\left(P_{ey}, P_{ez}\right) \tag{6.48}$$

because $y_o = \beta_x = 0$. From equation 6.45, one can derive the following interaction equation:

$$M_{ocr} = \sqrt{\left(\frac{I_x + I_y}{A}\right)(P_{ey} - P)(P_{ez} - P)} \tag{6.49}$$

Since

$$M_{ocr}^{P=0} = \bar{r}_o^2 P_{ey} P_{ez} = \frac{\pi}{L}\sqrt{EI_y GJ}\sqrt{1 + \frac{\pi^2 EC_w}{GJL^2}} \tag{6.50}$$

another variant of the interaction equation can be derived:

$$\frac{P}{P_{ey}} + \left(\frac{M_{ocr}}{M_{ocr}^{P=0}}\right)^2 \times \frac{1}{1 - \frac{P}{P_{ez}}} = 1 \tag{6.51}$$

Equation H1-2 in the AISC (2005) Specification simplifies equation 6.51 by assuming

$$\frac{1}{1 - \frac{P}{P_{ez}}} \approx 1$$

$$\frac{P}{P_{ey}} + \left(\frac{M_{ocr}}{M_{ocr}^{P=0}}\right)^2 = 1 \tag{6.52}$$

A conservative formula variant is used in many steel design standards by conservatively omitting the square in the second term on the left of equation 6.52:

$$\frac{P}{P_{ey}} + \frac{M_{ocr}}{M_{ocr}^{P=0}} = 1 \tag{6.53}$$

The curves relating the interaction between the critical axial load and the critical moment are illustrated in Figures 6.25 and 6.26.

The curves in Figures 6.25 and 6.26 relate the end-moment M_o (abscissa) and the axial force P (ordinate) of the beam-column shown in Figure 6.24. M_o and P are divided by the yield moment M_y and the squash load P_y. The heavy solid line represents the limit state of yielding

$$\frac{P}{P_y} + \frac{M_o}{M_y} = 1 \tag{6.54}$$

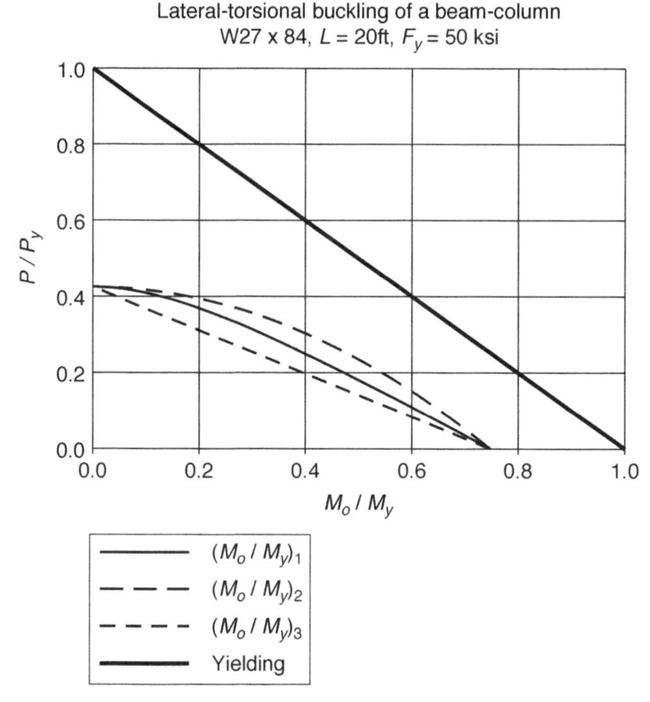

Fig. 6.25 Lateral-torsional buckling is the only limit state.

where $P_y = AF_y$ and $M_y = S_xF_y$. The curves with the thinner lines are for the limit state of lateral-torsional buckling. The middle curve (solid line) is labeled $\left(\frac{M_o}{M_y}\right)_1$ and it represents equation 6.51. This is the lateral-torsional buckling strength that includes the effect of both the y-axis and the torsional buckling loads, P_{ey} and P_{ez}. The upper curve (long dashed line) is labeled $\left(\frac{M_o}{M_y}\right)_2$ and it is the approximation for which the effect of P_{ez} is ignored (equation 6.52). It is somewhat on the unconservative side with respect to equation 6.51. The lowest curve (short dashed line) is labeled $\left(\frac{M_o}{M_y}\right)_3$ and it is the straight line approximation of equation 6.53. The curves in Figure 6.25 are for a 20 ft. long W27 × 84 beam. The governing limit state is lateral-torsional buckling, well below the yield limit. The curves in Figure 6.26 are for a 15 ft. long beam. In this case, the limit state is a mixed one between yielding and lateral-torsional buckling.

The representations in Figures 6.25 and 6.26 are for purely elastic behavior. In the next section of this chapter, the modifications made in structural design standards for the inelastic behavior will be presented.

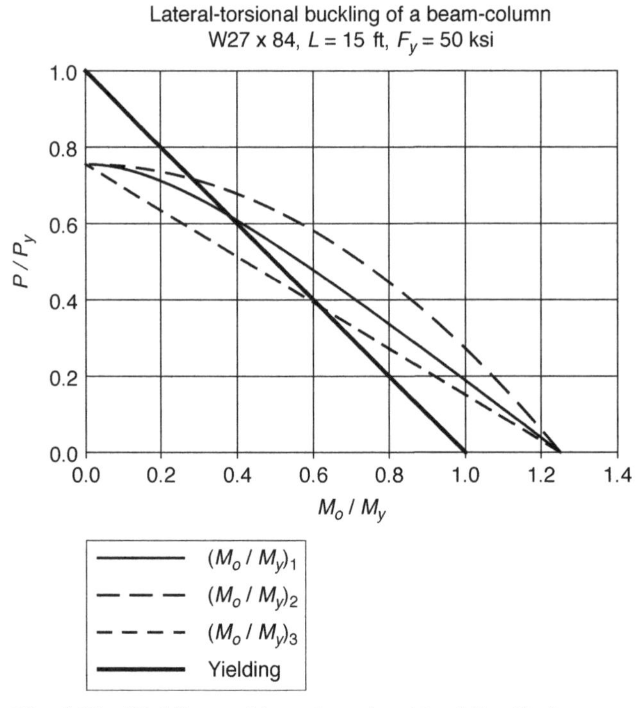

Fig. 6.26 Yielding and lateral-torsional buckling limit states.

We present one more case in this section: flexural-torsional buckling of an axially loaded T-column. The curves in Figure 6.27 represent the buckling strength of a pinned-end axially loaded WT5 × 15 column. The abscissa is the length of the column and the ordinate is the critical stress $F_{\text{cr}} = \frac{P_{\text{cr}}}{A}$. The solid curve consists actually of two lines: one for the x-axis and one for the y-axis buckling stress. The latter is obtained from equation 6.41. The reason for the overlap is because for this section I_x and I_y are nearly identical. The dashed curve is the flexural-torsional buckling stress from equation 6.46, and it controls the strength of this member. While the x- and y-axis elastic buckling curves approach infinity as the length of the column nears zero, the flexural-torsional curve approaches the torsional buckling stress obtained from equation 6.42, since the warping constant $C_w = 0$ for a T-shape. Since the buckling stress is above the yield stress, inelastic behavior will need to be considered. The critical flexural buckling stresses F_{ex} and F_{ey} are replaced by the AISC column curve equations (discussed in Chapter 3). The equations are shown next, and the resulting buckling curves are shown in Figure 6.28. It can be seen that if inelastic behavior is included, the differences between the flexural and flexural-torsional behavior are

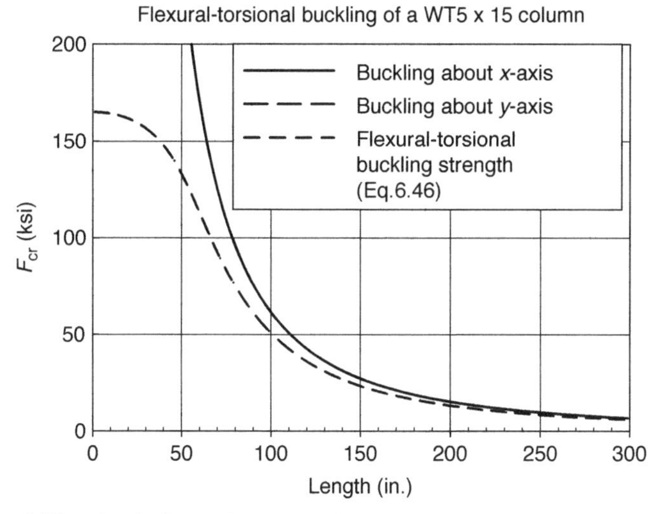

Fig. 6.27 Elastic flexural-torsional buckling of a pinned-end T-column.

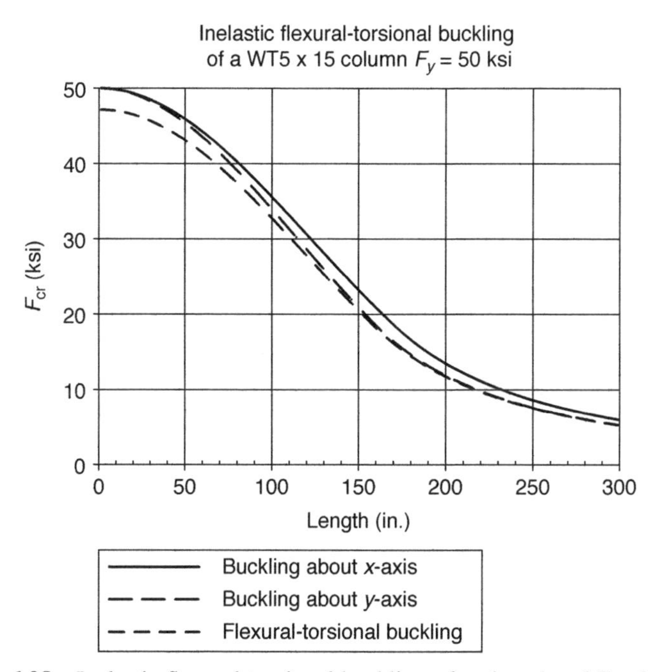

Fig. 6.28 Inelastic flexural-torsional buckling of a pinned-end T-column.

greatly reduced. Following are the details of the formulas from the AISC Specification used in the construction of Figure 6.28

$$F_{ex} = \frac{\pi^2 EI_x}{AL^2};\ F_{xcr} = \begin{Bmatrix} F_y \times 0.658^{\frac{F_y}{F_{ex}}} & \text{if } F_{ex} \geq 0.44F_y \\ 0.877F_{ex} & \text{if } F_{ex} \leq 0.44F_y \end{Bmatrix}$$

$$F_{ey} = \frac{\pi^2 EI_y}{AL^2};\ F_{ycr} = \begin{Bmatrix} F_y \times 0.658^{\frac{F_y}{F_{ey}}} & \text{if } F_{ey} \geq 0.44F_y \\ 0.877F_{ey} & \text{if } F_{ey} \leq 0.44F_y \end{Bmatrix}$$

$$F_{zcr} = F_{ez} = \frac{GJ}{A\bar{r}_o^2}$$

$$F_{cr} = \text{Min}\left\{ F_{xcr}, \frac{F_{ycr} + F_{zcr}}{2H}\left[1 - \sqrt{1 - \frac{4F_{ycr}F_{zcr}H}{\left(F_{ycr} + F_{zcr}\right)^2}}\right]\right\}$$

6.7 INELASTIC LATERAL-TORSIONAL BUCKLING

In Chapter 3, the subject of the elastic buckling theory for axially loaded members was extended to the cases where portions of the column cross section were yielded. The same concepts apply also to the lateral-torsional buckling of beams. The theoretical relationships between the tangent modulus, the reduced modulus, and the ultimate loads are valid also for perfectly straight beams, as are the effects of initial imperfections of geometry and of residual stresses. Because of the more complex geometry and the presence of both lateral and torsional deformations, the solutions for the inelastic lateral buckling problems for beams are far more difficult than they were for columns. Nevertheless, there are an abundance of numerical solutions, reports and papers of carefully conducted tests available in the literature. The solution of this problem is outside the scope of this book, and the reader is directed for an introduction to inelastic lateral-torsional buckling in the books and papers by Galambos (1968, 1998) and Trahair (1983, 1993). In the following, the method of solving the inelastic lateral-torsional buckling problem will be illustrated on a very simple example.

A simply supported thin rectangular beam of length L is subject to uniform bending (Figure 6.3) about its x-axis. The cross-section is shown in Figure 6.29.

At any cross-section the beam is subject to a bending moment M, as illustrated in Figure 6.30. Two stress distributions are shown at the right of this figure. The first is the stress distribution at the instant when the stress at the extreme fiber reaches the yield stress F_y. The moment causing this stress

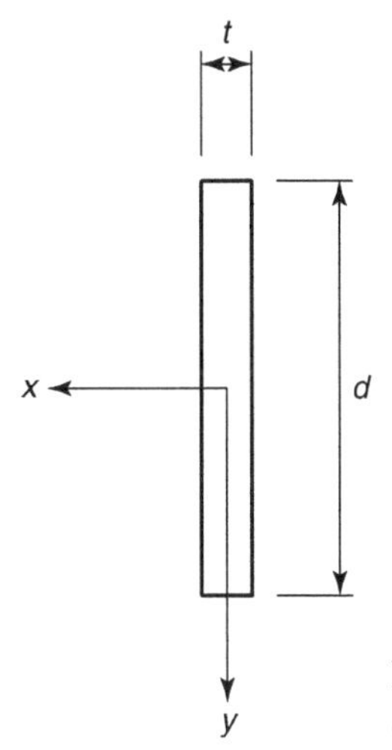

Fig. 6.29 Thin rectangular cross-section of beam that is bent about the x-axis.

distribution is the yield moment, M_y. It is obtained by taking moments of the stress blocks about the neutral axis:

$$M_y = 2t\left(\frac{1}{2} \times \frac{d}{2} \times F_y\right)\frac{2}{3} \times \frac{d}{2} = \frac{d^2 t F_y}{6} \tag{6.55}$$

The right stress block represents the case when part of the cross-section is yielded. It is assumed that the stress–strain diagram is ideally elastic-plastic (see Figure 3.18). The extent of yielding from the extreme fiber is γd. Taking moments about the neutral axis results in the following expression for the inelastic moment:

$$M = 2t\left[\gamma d\left(\frac{d-\gamma d}{2}\right) + \frac{1}{2}\left(\frac{d-2\gamma d}{2}\right) \times \frac{2}{3}\left(\frac{d-2\gamma d}{2}\right)\right] F_y$$

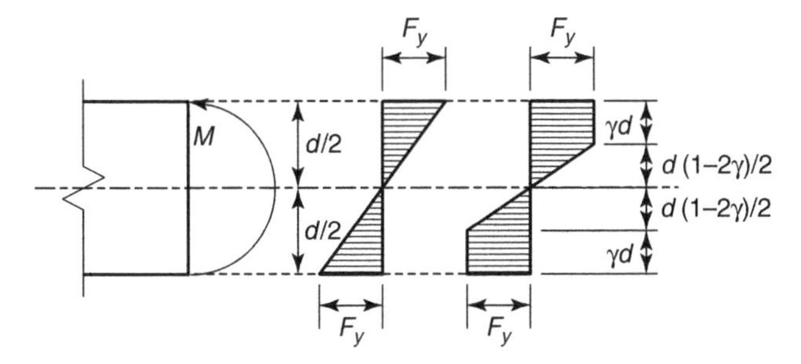

Fig. 6.30 Stress distributions.

After some algebra, we obtain this equation:

$$M = d^2 t \left(\frac{1 + 2\gamma - 2\gamma^2}{6} \right) F_y \tag{6.56}$$

When $\gamma = 0$, $M = M_y = \frac{d^2 t F_y}{6}$, the yield moment, and when $\gamma = \frac{1}{2}$ the moment becomes equal to the plastic moment

$$M_p = \frac{d^2 t F_y}{4} \tag{6.57}$$

Division of equation 6.56 by the plastic moment will result in equation 6.58:

$$\frac{M}{M_p} = \frac{2}{3} [1 + 2\gamma(1 - \gamma)] \tag{6.58}$$

The cross-section available for resisting lateral-torsional buckling will be the elastic core, as defined by the part of Figure 6.31 that is without cross-hatching. In this region, the material modulus is E, the elastic modulus. The cross-hatched region has no material stiffness according to the assumed stress–strain curve; thus, $E = 0$. This is also the assumption made in Chapter 3, where the *tangent modulus theory* was introduced.

The lateral-torsional elastic buckling moment for a rectangular beam is, from equation 6.10 because the warping constant $C_w \approx 0$ for a narrow rectangle, equal to

$$M_{cr} = \frac{\pi}{L} \sqrt{E I_y G J} \tag{6.59}$$

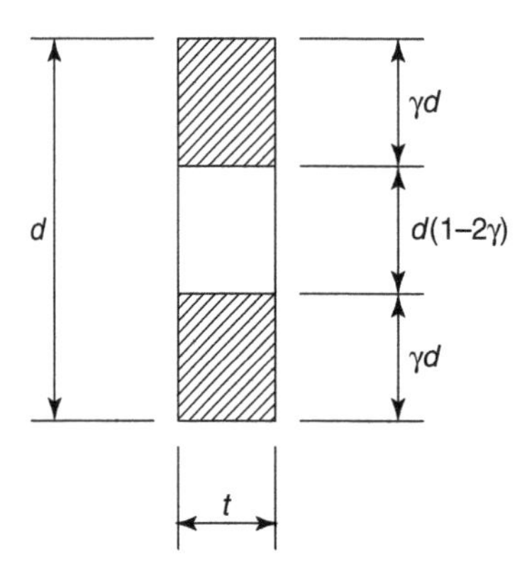

Fig. 6.31 Elastic core of the cross-section.

In this equation $G = 0.385E$, and

$$I_y = \frac{t^3 d(1-2\gamma)}{12}; \text{ and } J = \frac{t^3 d(1-2\gamma)}{3}$$

Substitution of the appropriate formulas into equation 6.59 leads to the following equation for the lateral-torsional buckling moment:

$$\frac{M_{cr}}{M_p} = 1.30\left(\frac{E}{F_y}\right)\left(\frac{t}{d}\right)\frac{1-2\gamma}{\left(\frac{L}{t}\right)} \tag{6.60}$$

The ratio L/t at the limit of elastic buckling is computed from setting $\frac{M_{cr}}{M_p} = \frac{M_y}{M_p} = \frac{2}{3}$:

$$\left(\frac{L}{t}\right)_y = 1.95\left(\frac{E}{F_y}\right)\left(\frac{t}{d}\right) \tag{6.61}$$

The critical moment expressions for the rectangular section are defined in the domain of the nondimensional parameters $\frac{M_{cr}}{M_p}$ and $\left(\frac{L}{t}\right)$ are the following:

$$\frac{M_{cr}}{M_p} = 1.30\left(\frac{F_y}{E}\right)\left(\frac{t}{d}\right)\frac{1}{\left(\frac{L}{t}\right)} \quad \text{if } \frac{L}{t} \geq 1.95\left(\frac{F_y}{E}\right)\left(\frac{t}{d}\right) \tag{6.62}$$

$$\frac{M_{cr}}{M_p} = 1.30\left(\frac{F_y}{E}\right)\left(\frac{t}{d}\right)\frac{1-2\gamma}{\left(\frac{L}{t}\right)} \quad \text{if } \frac{L}{t} < 1.95\left(\frac{F_y}{E}\right)\left(\frac{t}{d}\right) \tag{6.63}$$

The value of γ to be used in these equations is evaluated from equation 6.58, and it is equal to

$$\gamma = \frac{1}{2}\left(1 - \sqrt{3 - 3\frac{M_{cr}}{M_p}}\right) \tag{6.64}$$

Equations 6.64 can then be solved for L/t as shown next:

$$\frac{L}{t} = 1.30\left(\frac{F_y}{E}\right)\left(\frac{t}{d}\right)\frac{1}{\left(\frac{M_{cr}}{M_p}\right)} \quad \text{if } \frac{M_{cr}}{M_p} \leq \frac{2}{3}$$

$$\frac{L}{t} = 1.30\left(\frac{F_y}{E}\right)\left(\frac{t}{d}\right)\frac{\sqrt{3 - 3\frac{M_{cr}}{M_p}}}{\left(\frac{M_{cr}}{M_p}\right)} \quad \text{if } 1.0 \geq \frac{M_{cr}}{M_p} > \frac{2}{3} \tag{6.65}$$

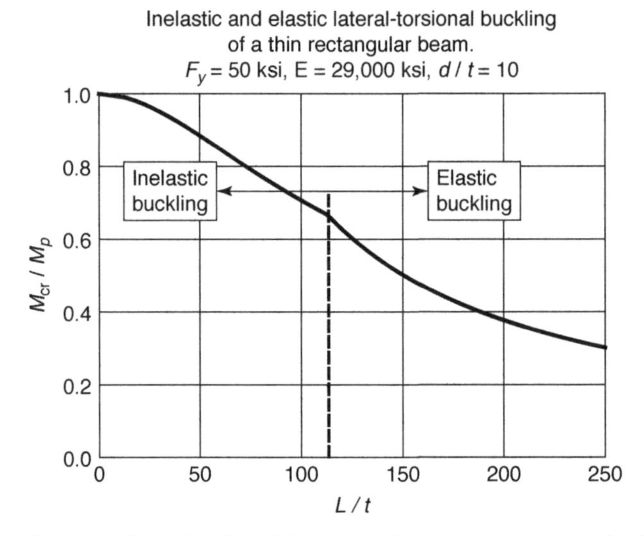

Fig. 6.32 Lateral-torsional buckling curve for a narrow rectangular beam.

For example, if $\frac{E}{F_y} = \frac{29{,}000}{50}$ and $\frac{d}{t} = 10$, then

$$
\begin{aligned}
\frac{L}{t} &= \frac{75.4}{\left(\frac{M_{cr}}{M_p}\right)} \quad \text{if } \frac{M_{cr}}{M_p} \leq \frac{2}{3} \\
\frac{L}{t} &= \frac{75.4\sqrt{3 - 3\frac{M_{cr}}{M_p}}}{\left(\frac{M_{cr}}{M_p}\right)} \quad \text{if } 1.0 \geq \frac{M_{cr}}{M_p} > \frac{2}{3}
\end{aligned}
\tag{6.66}
$$

The resulting curve is given in Figure 6.32. In the inelastic region there is a curve that ends at $M_{cr} = M_p$ when the length approaches zero, as one should expect.

The example of the lateral-torsional buckling of the beam with a thin rectangular cross-section was an easy problem. The inelastic lateral-torsional buckling of a wide-flange beam with residual stresses and having different loading and boundary conditions, is a much more difficult problem. Nevertheless, many cases have been solved, and accompanying experiments have provided confidence in the solutions. It is not expected that such calculations, or even the direct use of the numerical and/or experimental data, will be used in the daily work in design offices. Structural design codes have criteria that specify some form of a transition from the end of the elastic

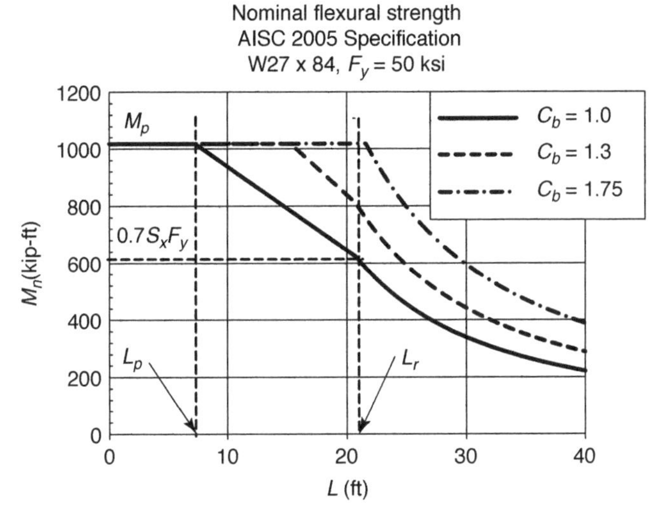

Fig. 6.33 AISC rules for lateral-torsional buckling strength of wide-flange beams.

region to the plastic moment. One such criterion is the straight-line transition in the AISC Specification (AISC 1999 and 2005). The method is illustrated in Figure 6.33.

The base curve is for the beam under uniform bending, $C_b = 1.0$, the lowest curve (solid line) in the Figure. Elastic buckling governs for longer beams. It terminates at the length designated as L_r. At this length the elastic critical moment equals the value when the maximum compressive stress due to the applied load, plus the maximum compressive residual stress equals the yield stress. The maximum residual stress is assumed to be 30 percent of the yield stress, and thus the moment corresponding to L_r equals $0.7S_xF_y$. S_x is the elastic section modulus. It is a well demonstrated experimental and analytical fact that a full plastic moment M_p can be attained for a finite length L_p. In the AISC Specification a straight line is prescribed between the points L_r, $0.7S_xF_y$ and L_p, M_p. For the case of nonuniform bending, $C_b > 1.0$, the transition from elastic to inelastic buckling is still at the same L_r that was established for the case of uniform bending. The straight line then extends to a point L_p, C_bM_p. This point is larger than the plastic moment, and the cross section obviously cannot attain it. Therefore, the maximum moment is curtailed to be equal to the plastic moment M_p. With this construct it is possible that beams will be in the elastic range of lateral-torsional buckling right up to the plastic moment. This assumed behavior can be justified by the fact that the plastic moment will occur at the end of the unbraced segment that is under reversed moment while the entire length of the beam is still elastic.

Structural design standards in the modern codes of other nations do not employ this straight-line method, using instead nonlinear transition

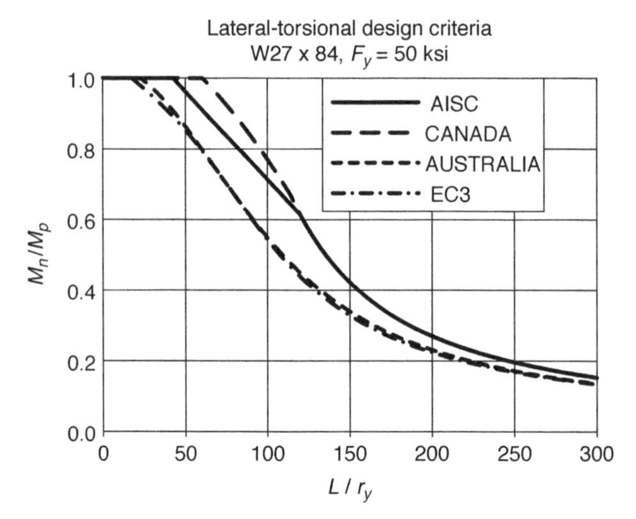

Fig. 6.34 Lateral-torsional buckling around the world.

equations. The curves in Figure 6.34 compare the AISC approach with the design curves of Canada, Australia, and the Eurocode EC3. The applicable formulas are listed in Table 6.4. Following the table there will be an example problem that compares the required moments for the four standards.

6.7.1 Example 6.5 Comparative Designs from Around the World

Calculate the nominal lateral-torsional buckling moment using the design criteria from the United States, Canada, Australia, and the European Community.

The following information is given:

Rolled shape: W27 × 84

$$F_y = 50\,\text{ksi}$$
$$E = 29,000\,\text{ksi}$$
$$G = 0.385\,E = 11,165\,\text{ksi}$$
$$\text{Unbraced length}\,L = 15\,\text{ft.} = 180\,\text{in.}$$

$$I_y = 106\,\text{in.}^4$$
$$J = 2.81\,\text{in.}^4$$
$$C_w = 17,900\,\text{in.}^6 \qquad M_p = 12,200''\,\text{k}$$
$$S_x = 213\,\text{in.}^3 \qquad M_{ocr} = 13,300''\,\text{k}$$
$$Z_x = 244\,\text{in.}^4$$

TABLE 6.4 Lateral-torsional Design Rules around the World

Design Standard	Design Formulas
AISC/05, USA	$M_n = \begin{Bmatrix} M_p & \text{if } L \le L_p \\ M_p - (M_p - 0.7F_yS_x)\left(\frac{L - L_p}{L_r - L_p}\right) & \text{if } L_p < L \le L_r \\ M_{ocr} & \text{if } L > L_r \end{Bmatrix}$
CSA S16, Canada	$M_n = \begin{Bmatrix} 1.15M_p\left(1 - \frac{0.28M_p}{M_{ocr}}\right) & \text{if } M_{ocr} \ge 0.67M_p \\ M_{ocr} & \text{if } M_{ocr} \le 0.67M_p \end{Bmatrix}$
AS9100, Australia	$M_n = 0.6M_p\left[\sqrt{\left(\frac{M_p}{M_{ocr}}\right)^2 + 3} - \frac{M_p}{M_{ocr}}\right]$
EC3, European Community	$M_n = \frac{M_p}{\Phi_{LT} + \sqrt{\Phi_{LT}^2 - \bar{\lambda}_{LT}^2}} \le M_p$ $\Phi_{LT} = 0.5\left[1 + \alpha_{LT}(\bar{\lambda}_{LT} - 0.2) + \bar{\lambda}_{LT}^2\right]$ $\bar{\lambda}_{LT} = \sqrt{\frac{M_p}{M_{ocr}}};\ \alpha_{LT} = \begin{Bmatrix} 0.21 & \text{if } \frac{d}{b_f} \le 2 \\ 0.34 & \text{if } \frac{d}{b_f} \ge 2 \end{Bmatrix}$
Definitions These formulas apply to simply supported rolled wide-flange beams subjected to uniform moment.	$M_p = F_yZ_x$; tabulated in AISC Handbook $M_{ocr} = \frac{\pi}{L}\sqrt{EI_yGJ}\sqrt{1 + \frac{\pi^2 EC_w}{GJL^2}}$ $L_p = 1.76r_y\sqrt{\frac{E}{F_y}}$; tabulated $L_r = \frac{r_yX_1}{0.7F_y}\sqrt{1 + \sqrt{1 + X_2(0.7F_y)^2}}$; tabulated $X_1 = \frac{\pi}{S_x}\sqrt{\frac{EGJA}{2}};\quad X_2 = \frac{4C_w}{I_y}\left(\frac{S_x}{GJ}\right)^2;$

AISC/05 Specification

$$L_p = 88\,\text{in.}; \quad L_r = 250\,\text{in.}; \quad L = 180\,\text{in.}$$

Since $L_p \leq L \leq L_r$

$$M_n = M_p - \left(M_p - 0.7F_yS_x\right)\left(\frac{L - L_p}{L_r - L_p}\right) = 9,500''\,\text{k}$$

Canadian Specification

$$0.67M_p = 8174''\,\text{k}$$

Since $0.67M_p > M_{ocr} = 13,300''\,\text{k}$

$$M_n = 1.15M_p\left(1 - \frac{0.28M_p}{M_{ocr}}\right) = 10,427''\,\text{k}$$

Australian Specification

$$M_n = 0.6M_p\left[\sqrt{\left(\frac{M_p}{M_{ocr}}\right)^2 + 3} - \frac{M_p}{M_{ocr}}\right] = 7,632''\,\text{k}$$

European Specification

Since for the W27 × 84 $\dfrac{d}{b_f} > 2$, $\alpha_{LT} = 0.34$

$$\bar{\lambda}_{LT} = \sqrt{\frac{M_p}{M_{ocr}}} = 0.958$$

$$\Phi_{LT} = 0.5\left[1 + \alpha_{LT}\left(\bar{\lambda}_{LT} - 0.2\right) + \bar{\lambda}_{LT}^2\right] = 1.088$$

$$M_n = \frac{M_p}{\Phi_{LT} + \sqrt{\Phi_{LT}^2 + \bar{\lambda}_{LT}^2}} = 7,613''\,\text{k}$$

$$< M_p = 12,200''\,\text{k} \rightarrow M_n = 7,613''\,\text{k}$$

Country	Nominal lateral-torsional strength, inch-kips	Nominal lateral-torsional stress, ksi, $f = M_n/S_x$
USA	9,500	44.6
Canada	10,430	49.0
Australia	7630	35.8
Europe	7610	35,7

The summary table and the curves Figure 6.34 show a considerable difference in the required lateral-torsional buckling strength between the four regions. In all four jurisdictions the elastic critical moment is the same. The difference is in the type of transition from the elastic to the inelastic domain. The two North American standards implicitly assume that the beam has no initial out-of-straightness. The Australian and the European standards allow a substantial penalty for the geometric imperfections. In this comparison it must be mentioned that the four standards have also different load recurrence intervals, load factors, and resistance factors (in the Eurocode these factors are called partial factors). This means that the codes do not necessarily have the same intended reliability of not exceeding a design limit state. Thus, comparisons have perhaps a limited merit. Example 6.5 was introduced mainly to acquaint the student with the lateral-torsional buckling formulas used in other parts of the world.

One additional item needs to be mentioned here. The formula we have been using throughout this chapter for the basic elastic critical moment M_{ocr} that was derived in section 6.2 (equation 6.10) is defined in an apparently different form in section F2 in the 2005 AISC Specification. The following derivation, demonstrating that the two equations are exactly the same, is presented in order to avoid needless head-scratching on the part of the student.

6.7.2 Equation 6.6:

$$M_{ocr} = \frac{\pi}{L}\sqrt{EI_y GJ}\sqrt{1 + \frac{\pi^2 EC_w}{GJL^2}}$$

can be expressed in terms of a critical stress as follows

$$F_{cr} = \frac{M_{ocr}}{S_x} = \frac{\pi^2 E}{L^2}\sqrt{\frac{C_w I_y}{S_x^2}}\sqrt{1 + \frac{G}{E}\frac{JL^2}{\pi^2 C_w}}$$

Introducing a term r_{ts}, that can be thought of as the radius of gyration of a cross-section composed of the compression flange and one-sixth of the compressed part of the web, $r_{ts}^2 = \frac{\sqrt{I_y C_w}}{S_x} = \frac{h_o I_y}{S_x}$, then the critical stress equation can be written as follows (note also that for a doubly-symmetric wide-flange shape $C_w = \frac{h_o^2 I_y}{4}$, where h_o is the distance between the centers of the flanges, $h_o = d - t_f$):

$$F_{cr} = \frac{\pi^2 E}{\left(\frac{L}{r_{ts}}\right)^2} \sqrt{1 + \left(\frac{2}{\pi^2} \times \frac{G}{E}\right) \frac{J}{S_x h_o} \left(\frac{L}{r_{ts}}\right)^2}$$

With $\left(\frac{2}{\pi^2} \times \frac{G}{E}\right) = 0.078$, we obtain the AISC/05 equation F2.4:

$$F_{cr} = \frac{\pi^2 E}{\left(\frac{L}{r_{ts}}\right)^2} \sqrt{1 + 0.078 \frac{J}{S_x h_o} \left(\frac{L}{r_{ts}}\right)^2} \tag{6.67}$$

6.8 SUMMARY

Lateral torsional buckling differs from the column buckling we had previously investigated because it occurs as an out-of-plane buckling mode, unlike the beam column instability we examined in Chapter 4, where the column buckles in the plane of bending. The basic derivation of the critical buckling moment is based on a worst-case loading scenario of uniform moment, loading at the centroid of the cross-section and a simply supported end condition. Numerous researchers have suggested means by which to account for deviations from these initial assumptions may be considered when determining the critical moment.

In the case of singly symmetric beams, the derivation requires consideration of the additional warping stresses that develop due to the lack of symmetry about the bending axis. The solution for critical moment is similar in form to the solution for the basic case of a doubly symmetric section with the addition of an additional cross-sectional property, β_x, which is equal to zero for doubly symmetric sections.

The discussion of inelastic behavior in beams is similar to that of inelastic behavior of columns introduced in Chapter 3. The concepts of tangent and reduced moduli and the effects of imperfections and residual stresses are valid for beams as well as columns. A simple example of a rectangular beam provides insight into the behavior, but the case of wide-flange beams is significantly more complicated. In design the behavior of beams that buckle inelastically is handled using a transition from the plastic moment capacity to the elastic buckling solution.

PROBLEMS

6.1. Calculate the elastic torsional-flexural buckling stress P_{cr}/A (equation 6.40) for two channel section columns of length L. Compare the results with the elastic minor axis buckling stress. Experiment with other channel sections and other lengths. Comment on whether ignoring torsional-flexural buckling for channels is reasonably justified as is done in common design practice.

Given: The ends are torsionally and flexurally simply supported. The two channels are MC6 × 16.3 and MC6 × 12. $E = 29000$ ksi and $G = 0.385E$ and $L = 120$ in.

6.2. Calculate the elastic critical stress of a 200 in. long pinned-end WT12 × 88 column. Investigate the flexural buckling about the x-axis and the flexural-torsional buckling stress.

6.3. Rework Example 6.4, including the calculation of the cross-section properties.

6.4. Determine the maximum length of a simply supported W21 × 44 beam erected so that its vertical axis is the y-axis. Use a factor of safety of 2.5 against elastic lateral-torsional buckling.

CHAPTER SEVEN

BRACING

7.1 INTRODUCTION

In all of the previous chapters, there has been discussion of braced points and braced members; however, the actual details of the bracing requirements have, for the most part, been omitted. In general, the bracing has been considered to be ideal—that is, the brace provides perfect restraint to the member being braced in that it allows no lateral displacement of the braced point. Ideal bracing is neither possible nor necessary, but the importance of bracing in maintaining stability of structures cannot be overemphasized. To aid discussion in the chapter, we use the classification of bracing as one of four types: discrete, continuous, relative and lean-on, as described below:

Discrete bracing resists movement only at the location where it is attached to the member it is bracing, as shown in Figure 7.1. It is also referred to as nodal bracing.

Continuous bracing, such as composite floor decking, is self-explanatory; it provides a continuous restraint to lateral movement (see Figure 7.2).

Relative bracing, rather than preventing lateral movement at one point, controls the relative movement between two braced points, as shown in Figure 7.3. X-bracing and truss bracing are examples of relative bracing.

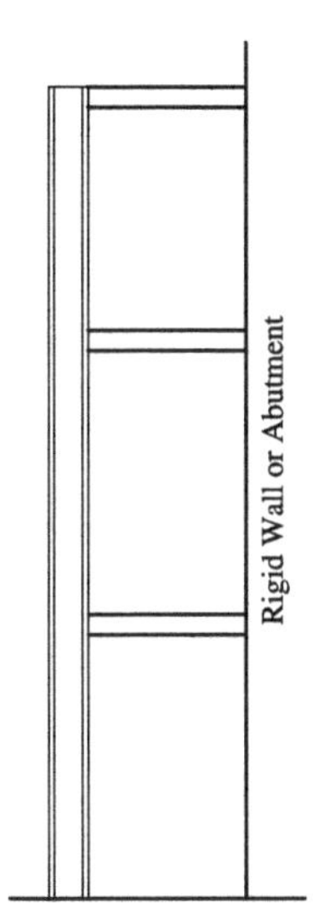

Fig. 7.1 Discrete bracing.

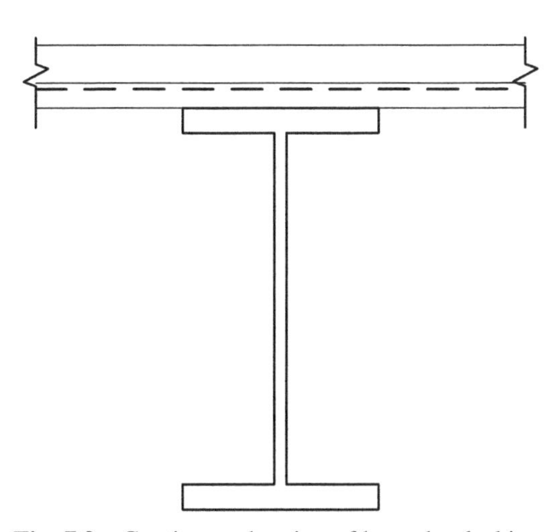

Fig. 7.2 Continuous bracing of beam by decking.

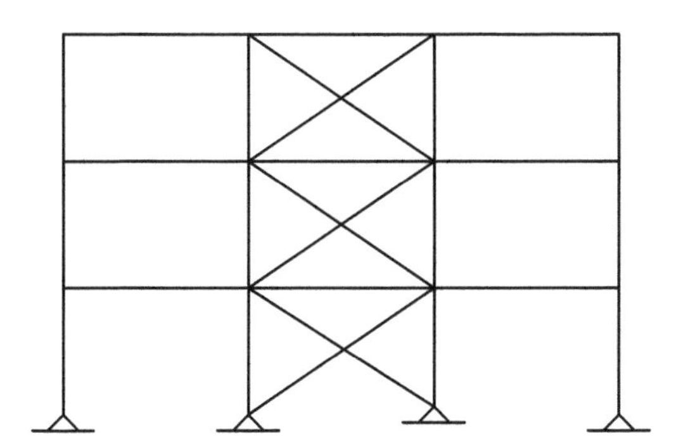

Fig. 7.3 Relative bracing.

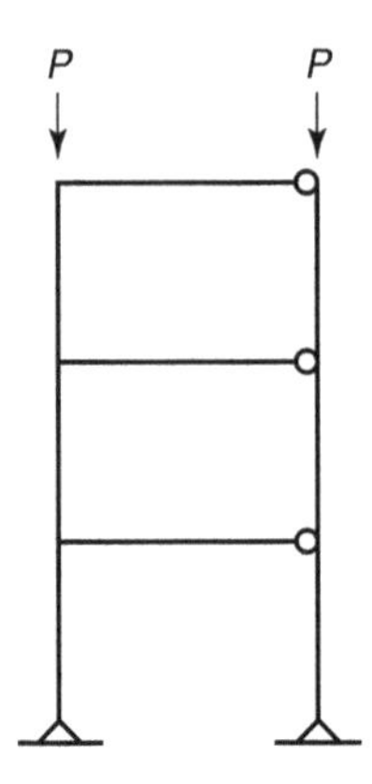

Fig. 7.4 Lean-on bracing.

Lean-on bracing defines the case when one member requires the lateral support of an adjacent member, such as the case of leaning columns (described in Chapter 5), as shown in Figure 7.4.

In order to illustrate basic bracing concepts, the next sections provide examples of columns or frames braced by continuous, lateral, and lean-on bracing. For more information on continuous bracing of a column flange, the reader is referred to Chapter 13 of Galambos (1998). Design requirements for both column and beam bracing, provided in Appendix 6 of the AISC Specification (2005) are discussed with examples provided of the application of the provisions.

In general, this chapter provides an overview of bracing sufficient to provide the student or design professional with the background to understand and correctly apply design requirements for stability bracing. A wealth of literature exists on theoretical development of bracing forces and stiffness requirements, as well as specific bracing applications. For a more thorough treatment of the subject, the reader is referred to the suggested reading at the end of the chapter.

7.2 DISCRETE BRACING

7.2.1 Single Column with Nodal Brace

An example of a column with a discrete brace is shown in Figure 7.5. The brace is assumed to behave elastically and is modeled as a linear spring with stiffness β. The respective boundary conditions at the column top and bottom are also shown in Figure 7.5. The column is assumed to be perfectly vertical and straight.

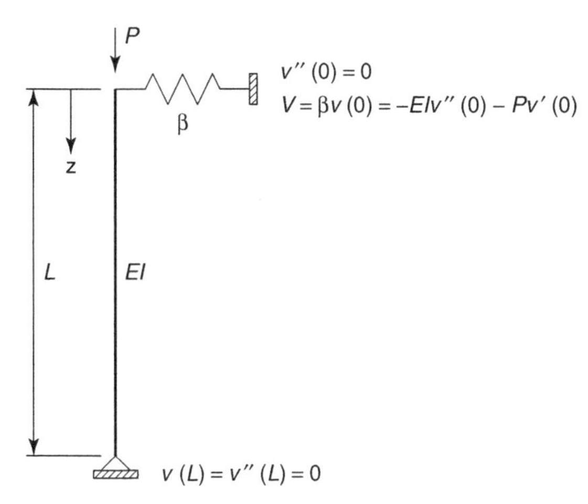

Fig. 7.5 Braced column, loading and boundary conditions.

The column of length L and stiffness EI is pinned at its base, and it is restrained from lateral deflection by a linear spring at the top. At the top, the shear is equal to the deflection times the spring constant β. Upon substitution of the four boundary conditions into the general deflection expression $v = A + Bz + C \sin kz + D \cos kz$ (derived in Chapter 2), we obtain the following buckling determinant:

$$\begin{vmatrix} 0 & 0 & 0 & -k^2 \\ -\beta & -P & 0 & -\beta \\ 1 & L & \sin kL & \cos kL \\ 0 & 0 & -k^2 \sin kL & -k^2 \cos kL \end{vmatrix} = 0 \tag{7.1}$$

As with solutions presented in Chapter 2, $k^2 = \frac{P}{EI}$. Decomposition of the determinant results in the following eigenfunction:

$$k^4 (P - \beta L) \sin kL = 0 \tag{7.2}$$

Since k^4 is not equal to zero, there are two possible buckling conditions, and thus two critical loads:

$$P - \beta L = 0 \rightarrow P_{\text{cr}}^{(1)} = \beta L$$

$$\sin kL = 0 \rightarrow P_{\text{cr}}^{(2)} = \frac{\pi^2 EI}{L^2}$$

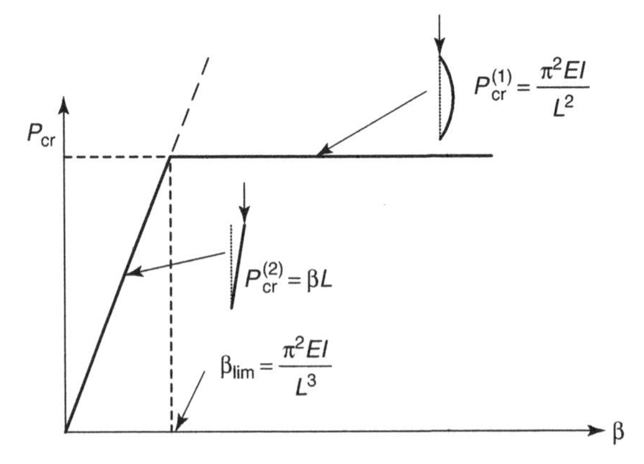

Fig. 7.6 Two buckling modes of the column in Figure 7.5.

The lowest of these critical loads controls. The relationship between these two buckling loads is further explored in the graphs in Figure 7.6, which graphs the relationship between the critical load, P_{cr}, and the spring constant β:

The diagonal straight line corresponds to the rigid-body rotation of the column while the spring constant is relatively small. However, when the second buckling load is reached, it will govern.

The problem illustrates that perfect restraint is not required to provide adequate bracing to achieve a "braced" buckling mode. In order to prevent sway buckling, it is only necessary to have a spring stiffness equal to the limiting value determined when the two critical loads are equal to each other: $\beta_{lim} = \frac{\pi^2 EI}{L^3}$.

7.2.2 Multi-bay Frame with Nodal Brace

Often, multiple columns may be braced by a single discrete brace, so in this problem we investigate of the stability of the two-bay single-story rigid frame when a single linear spring is attached to the top of the structure, as seen in Figure 7.7.

This case is similar to Case II in Chapter 5, a multibay frame with no lateral restraint, where the slope deflection equations are given by

$$M_{BA} = \frac{EI_C}{cL_C}(\theta_B - \rho)$$

$$M_{CD} = \frac{EI_C}{cL_C}(\theta_C - \rho)$$

$$M_{EF} = \frac{EI_C}{cL_C}(\theta_E - \rho)$$

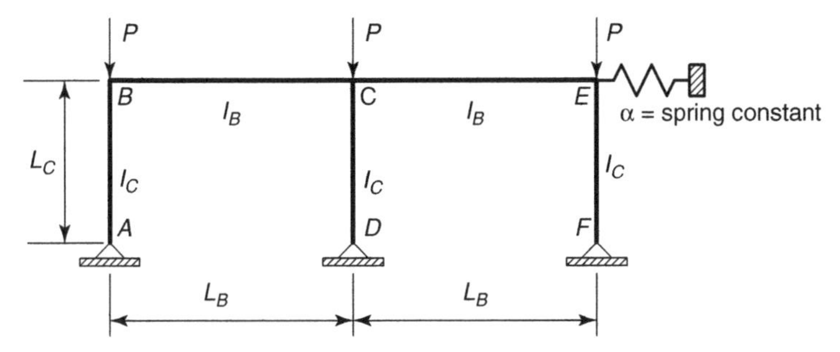

Fig. 7.7 Frame with lateral restraint.

$$M_{BC} = \frac{EI_{\mathrm{B}}}{L_{\mathrm{B}}}(4\theta_B + 2\theta_C)$$

$$M_{CB} = \frac{EI_{\mathrm{B}}}{L_{\mathrm{B}}}(2\theta_B + 4\theta_C)$$

$$M_{CE} = \frac{EI_{\mathrm{B}}}{L_{\mathrm{B}}}(4\theta_C + 2\theta_E)$$

$$M_{EC} = \frac{EI_{\mathrm{B}}}{L_{\mathrm{B}}}(2\theta_C + 4\theta_E)$$

And the equilibrium equations are

$$M_{BA} + M_{BC} = 0$$

$$M_{CB} + M_{CD} + M_{CE} = 0$$

$$M_{EC} + M_{EF} = 0$$

The fourth equilibrium equation here differs from the frame with no lateral restraint in Chapter 5, where the fourth equilibrium equation set the sum of the shears to zero. In this case, the reaction from the spring to the lateral deflection $\Delta = \rho L_{\mathrm{C}}$ is equal to $\alpha\rho L_{\mathrm{C}}$. The fourth equation, the shear equilibrium equation is, therefore:

$$M_{BA} + M_{CD} + M_{EF} + 3PL_{\mathrm{C}}\rho = L_{\mathrm{C}}(\alpha\rho L_{\mathrm{C}}) \tag{7.3}$$

The following characteristic equation is obtained by substituting the slope-deflection equations into the four equilibrium equations and setting the determinant of the coefficients of the unknown rotations equal to zero:

$$\left(\frac{PL_{\mathrm{C}}^2}{EI_{\mathrm{C}}} - \frac{\alpha L_{\mathrm{C}}^3}{3EI_{\mathrm{C}}}\right)\left(\frac{1}{c^2} + \frac{12\gamma}{c} + 24\gamma^2\right) - \frac{8\gamma}{c}\left(\frac{1}{c} + 3\gamma\right) = 0 \tag{7.4}$$

Although we could solve for the critical load for a given spring constant α, we will not. What is of interest is the value of the spring constant α required to force the frame into the non–sway-buckling mode. The governing equation for this buckling mode is $\frac{1}{c} = 2.536\gamma$. Substituting this term into equation 7.4 we can solve for the required spring constant. The resulting equation is:

$$\begin{aligned}\frac{\alpha L_C^3}{EI_C} &= \frac{3PL_C^2}{EI_C} - \frac{(-28.25\gamma^2)}{6.431\gamma^2 - 30.413\gamma^2 + 24\gamma^2} \\ &= \frac{3PL_C^2}{EI_C} + \frac{28.25\gamma^2}{0} \to \infty \end{aligned} \tag{7.5}$$

The required stiffness of the restraining spring is shown to be approaching infinity as the critical load approaches the value that it would attain for the non–sway-buckling mode. It is thus not realistic to aspire to reach this load, although it is possible to come close with a fairly stiff spring. Practically, it is recommended that one should be satisfied with attaining the Euler buckling load, $\frac{P_E L_C^2}{EI_C} = \phi^2 = \pi^2$. This is one of the bases for the bracing provisions of the AISC Specification. Substituting this value into equation 7.4, noting that $\frac{1}{c} = \frac{\phi^2 \tan\phi}{\tan\phi - \phi} = \frac{\pi^2 \tan\pi}{\tan\pi - \pi} = 0$, the following value is obtained for the spring stiness required so that the columns can be designed as pinned-end members:

$$\alpha_{req} = \frac{3P_E}{L_C} \tag{7.6}$$

It is important to notice that the desired brace stiffness is proportional to the amount of axial load on the entire system being braced, rather than simply the adjacent column.

To illustrate the effect of increasing values of the spring stiffness, equation 7.4 is solved for the critical load and the effective length factor for $\gamma = \frac{I_B L_C}{I_C L_B} = 2.0$. The results are shown in Table 7.1. This table again proves that while complete restraint cannot be reached, it is possible to get close enough for practical purposes.

TABLE 7.1 Critical Load and Effective Length for Varying Degrees of Brace Stiffness

$\frac{\alpha L_C^3}{EI_C}$	$\frac{PL_C^2}{EI_C}$	K_{eff}
0	2.13	2.15
π^2	4.88	1.42
$3\pi^2$	9.87	1.00
$10\pi^2$	15.13	0.81
∞	15.32	0.80

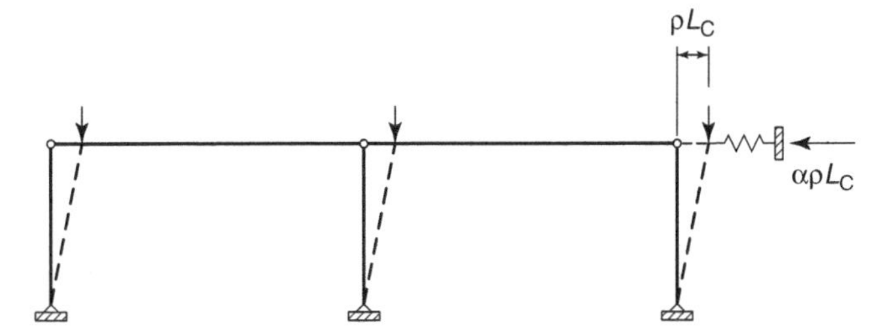

Fig. 7.8 Simplified approach for finding required brace stiffness.

The required spring stiffness can also be found by a much simpler way than by the method above. Knowing that we want to attain $P_E = \frac{\pi^2 EI_C}{L_C^2}$, we can assume that the joints are pinned. The structure to be analyzed is then represented by the free-body-diagram in Figure 7.8. The equilibrium equation is then

$$3P_E \rho L_C - \alpha \rho L_C^2 = 0$$

$$\alpha_{req} = \frac{3P_E}{L_C} = \frac{3\pi^2 EI_C}{L_C^3}$$

If there are m columns in the story, then the required spring stiffness is

$$\alpha_{req} = \frac{mP_E}{L_C} \tag{7.7}$$

In each of the previous examples, a single brace is employed. If multiple intermediate braces are included, the bracing stiffness requirement increases (Winter 1960). The design provisions of AISC 2005 for nodal bracing consider the most conservative case in which multiple intermediate braces are provided.

7.3 RELATIVE BRACING

7.3.1 Frame with X-bracing

When relative bracing is used, a cut at any point along the braced member will also pass through the brace. The most recognizable form of relative bracing of columns is X-bracing, as shown in Figure 7.9.

The frame consists of two pinned-end columns that are connected by a beam and by diagonal braces. These braces are assumed to be acting in tension only (i.e., the brace that is in compression will buckle and will not participate in providing stiffness). The spring constant is determined by subjecting the frame to a force F and calculating the resulting deflection Δ. The spring constant is then

$$\beta = \frac{F}{\Delta}$$

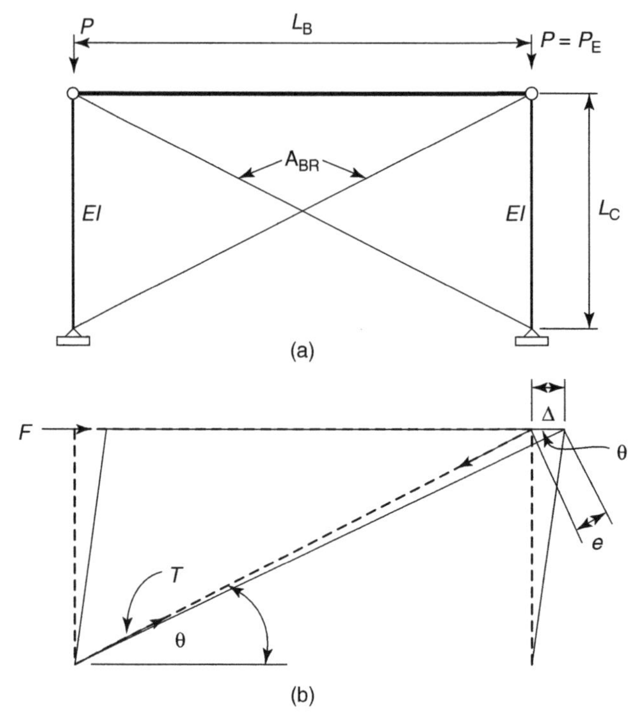

Fig. 7.9 Diagonally braced frame.

The tensile force T in the diagonal is obtained by equilibrium considerations under the assumption of small deflections and rotations.

$$\text{Tensile force in brace} = T = \frac{F}{\cos\theta}$$

$$\text{Length of brace: } L_{BR} = \frac{L_B}{\cos\theta}$$

$$\text{Bar elongation: } e = \frac{TL_{BR}}{EA_{BR}} = \frac{FL_B}{EA_{BR}\cos^2\theta}$$

$$\Delta = \frac{e}{\cos\theta} = \frac{FL_B}{EA_{BR}\cos^3\theta} = \frac{F}{EA_{BR}}\frac{(L_B^2 + L_C^2)^{\frac{3}{2}}}{L_B^2}$$

$$\beta = \frac{EA_{BR}L_B^2}{(L_B^2 + L_C^2)^{\frac{3}{2}}}$$

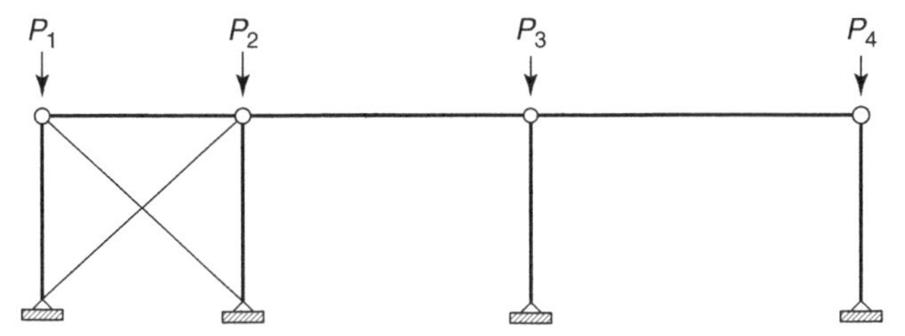

Fig. 7.10 Braced frame with multiple columns.

The required brace area needed to provide minimum stiffness so that the two columns can support the Euler load is obtained by equating the two critical loads:

$$\sum P_E = \frac{2\pi^2 EI}{L_C^2} = \beta L_C = \frac{EA_{BR}L_B^2 L_C}{(L_B^2 + L_C^2)^{\frac{3}{2}}}$$

$$A_{BR} \geq 2\pi^2 I \left[\frac{(L_B^2 + L_C^2)^{\frac{3}{2}}}{L_B^2 L_C^3}\right]$$

For a frame with several columns in a story, as shown in Figure 7.10, the required brace area is

$$A_{BR} \geq \sum P_E \left[\frac{(L_B^2 + L_C^2)^{\frac{3}{2}}}{EL_B^2 L_C}\right]$$

Once again, we see that the required brace force is proportional to the total load on the system being braced.

7.4 LEAN-ON BRACING

The next example illustrates the concept of minimum lateral bracing stiffness required to permit the column to reach the basic simply supported Euler load, P_E. Figure 7.11 shows a pinned-end column that is braced by being laterally connected to a vertical cantilever.

When this restraining member is subjected to a lateral force H it deflects an amount

$$\Delta = \frac{HL^3}{3EI_S} \rightarrow \beta = \frac{H}{\Delta} = \frac{3EI_S}{L^3}.$$

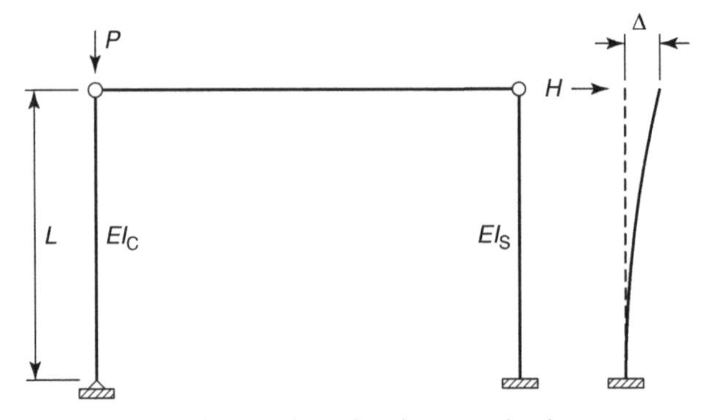

Fig. 7.11 Bracing of a pinned-end column.

(based on the flexural stiffness of the column). The required stiffness is obtained from equating

$$\frac{3EI_S}{L^3} \text{ to } \beta_{\text{lim}} = \frac{\pi^2 EI}{L^3}.$$

This results in the minimum required moment of inertia

$$I_S = \frac{\pi^2 EI}{3} = 3.29EI$$

The column supporting the leaning column does not only require additional stiffness, it also requires additional strength. The means for accounting for leaning columns in determining the required strength of the supporting columns is discussed in greater detail in Chapter 8 in the discussion of story and stiffness based K-factors (Section 8.6.5.).

7.5 EFFECTS OF IMPERFECTIONS

The ideal bracing stiffness requirements for relative and nodal bracing discussed in the previous sections were based on developing the Euler buckling load in perfectly straight columns. However, Winter (1960) showed that bracing requirements include both stiffness and strength, and that the bracing force is a function of the initial imperfection in the system. Consider the relatively braced column of length L in Figure 7.12 with an initial imperfection modeled by a nonverticality (or out-of-plumbness) of Δ_o, displaced by an addition amount Δ.

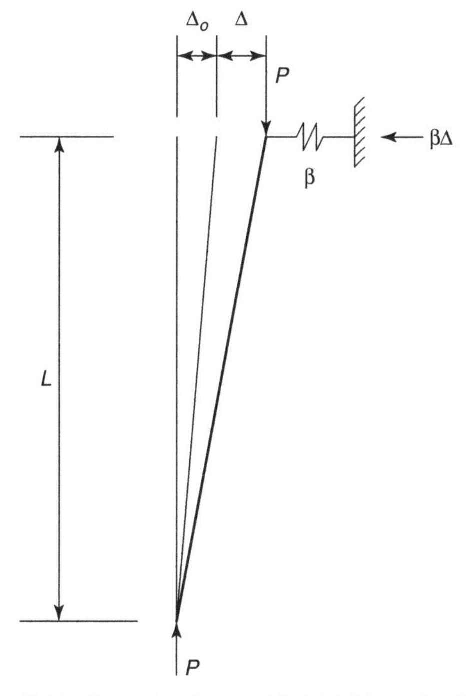

Fig. 7.12 Braced column with initial imperfection.

Summing moments about the base of the column, we obtain

$$\sum M = 0$$

$$P(\Delta_o + \Delta) = \beta\Delta L$$

In the case of no initial imperfection ($\Delta_o = 0$), the critical load in the column (as shown in section 7.2) is given by

$$P_{\rm cr} = \beta L$$

We identify the *ideal* brace stiffness as the brace stiffness required to ensure the column can reach the Euler buckling load, $P_{\rm E}$. The ideal stiffness is given by

$$\beta_i = \frac{P_{\rm E}}{L}$$

If imperfections are included in the system, the ideal bracing stiffness is insufficient to force the column into a non–sway-buckling mode, as shown

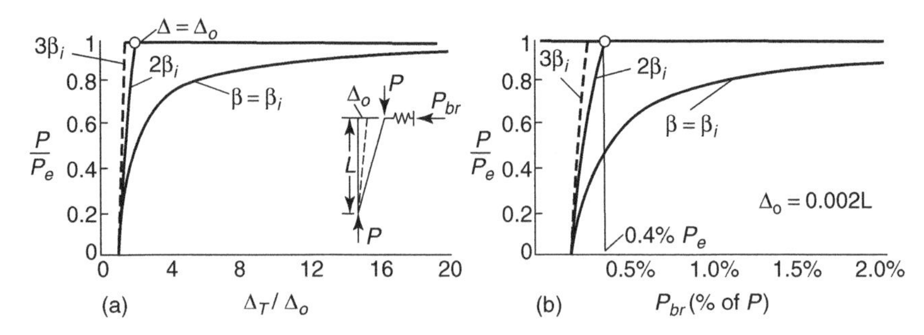

Fig. 7.13 Effects of imperfections on required bracing strength (AISC 2005, Figure C-A-6.3). Copyright © American Institute of Steel Construction, Inc. Reprinted with permission. All rights reserved.

in Figure 7.13. It is also obvious from the graphs that the brace force is a function of the imperfection.

As reported in Yura (1993), early recommendations for bracing provisions did not include the required increase in bracing stiffness due to initial imperfections. The provisions included in the 2005 AISC Specification were developed to include the impact of imperfections, as described in section 7.6.

7.6 COLUMN BRACING PROVISIONS

The 2005 edition of the AISC Specification includes provisions for bracing strength and stiffness in Appendix 6, "Stability Bracing for Columns and Beams." For both beams and columns, bracing requirements depend on whether the bracing is considered to be relative or nodal (i.e., discrete). Requirements for continuous column bracing are not provided, and lean-on bracing of columns is handled in member stability provisions as discussed in Chapter 8.

In all of the provisions for bracing, the brace is assumed to be perpendicular to the member it is bracing. For relative bracing provided by diagonal bracing or the common X-bracing, the forces and stiffness required need to be adjusted to account for the angle of the brace. Based on the effects of imperfections on required bracing force and stiffness (discussed in section 7.5), the AISC specified provisions require twice the ideal brace stiffness and assume an initial out-of-plumbness of $L/500$. This imperfection corresponds to the maximum out-of-plumbness allowed in the AISC *Code of Standard Practice*.

7.6.1 Example 7.4: Relative Column Bracing

In this example, we determine the area required for a tension only X-brace for the frame shown in Figure 7.14 based on stiffness and strength requirements. Minimum stiffness requirement is given as

$$\beta_{br} = \frac{1}{\phi}\left(\frac{2P_r}{L_b}\right) \qquad \text{(AISC A-6-2)}$$

where L_b = Distance between braces = 15 ft.

For this frame, four columns are restrained by a single X-brace, and P_r = the total axial load on the frame:

$$P_r = 10P = 1000\,\text{kips}$$
$$\phi = 0.75$$
$$\beta_{br} = \frac{1}{\phi}\left(\frac{2P_r}{L_b}\right) = \frac{1}{0.75}\left[\frac{2(1000)}{15'(12''/\text{ft.})}\right] = 14.8\,\text{kips/in.}$$

In order to design the brace, it is necessary to determine the stiffness that is provided by the brace. Since the brace is an axially loaded member, the stiffness is given by

$$\beta_{br} = \frac{P}{\Delta} = \frac{A_b E}{L_b}\cos^2\theta = \frac{A_b(29{,}000)}{15'(12)}\cos^2\left[\arctan\left(\frac{15}{35}\right)\right] = 136.1A_b$$

based on the fact that the force and displacement assumed in the design requirements are assumed to be perpendicular to the column. The cosine term comes from the orientation of the brace (and hence, the brace force and displacement) relative to the assumed orientation perpendicular to the column.

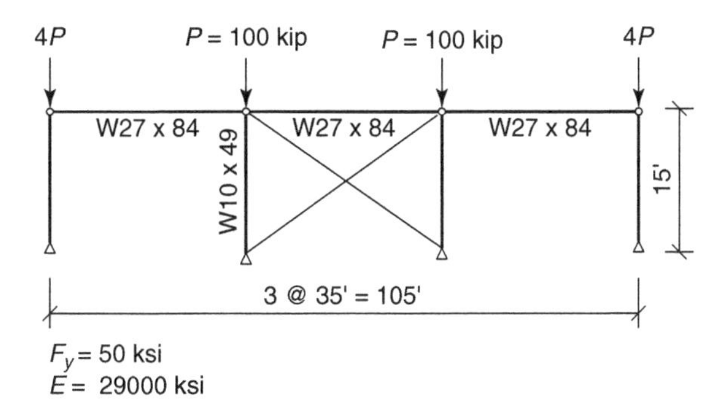

Fig. 7.14 X-bracing design example.

Once the provided brace force is calculated, the required area is given by

$$136.1A_b = 14.8\,\text{kips/in.}$$
$$A_b \geq 0.108\,\text{in}^2.$$

Minimum brace strength is given in Appendix 6 as

$$P_{br} = 0.004\,P_r = 0.004\,(1000) = 4\,\text{kips} \qquad \text{(AISC A-6-1)}$$

The strength of the brace is governed by the design rules for tension members. Since the connection detail is not provided (and outside the scope of the example), we determine the area based on the limit state of yield on gross area:

$$\phi P_n = \phi F_y A_g = 0.9(50)A_g \geq 4\,\text{kips}$$
$$A_g \geq 0.089\,\text{in}^2.$$

The area required for stiffness controls the design for the stability bracing of the frame.

7.6.2 Nodal Column Bracing

In order to consider design requirements for nodal bracing, we reconsider the frame in Example 7.4 with a nodal brace rather than an X-brace as shown in Figure 7.15. For nodal bracing, the required brace strength is:

$$P_{br} = 0.01P_r$$
$$P_{br} = 0.01(1000\,\text{kips}) = 10\,\text{kips} \qquad \text{(AISC A-6-3)}$$

And the required brace stiffness is:

$$\beta_{br} = \frac{1}{\phi}\left(\frac{8P_r}{L_b}\right) \qquad \text{(AISC A-6-4)}$$

At this point, it is useful to remember that the stiffness requirement for nodal bracing of columns is based on the instances of many nodal braces. The most conservative instance of nodal bracing was adopted in the specification rather than providing a more complicated provision that is a function of the number of braces. The commentary to Appendix 6 provides

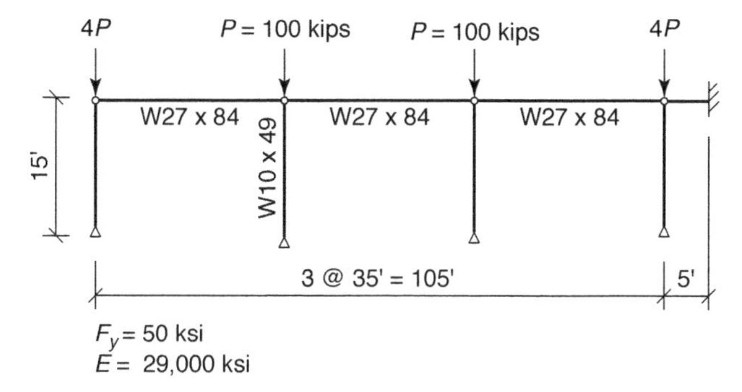

Fig. 7.15 Column nodal bracing design example.

a reduction in the required bracing stiffness to account for the exact number of braces. The bracing stiffness may be reduced by

$$\frac{N_i}{4}$$

where

$$N_i = 4 - \frac{2}{n} \quad \text{and} \quad n = \text{number of braces.}$$

In this instance, we are only using one brace, so

$$N_i = 4 - \frac{2}{1} = 2$$
$$\frac{N_i}{4} = 1/2$$

So

$$\beta_{br} = \frac{N_i}{4}\left[\frac{1}{\phi}\left(\frac{8P_r}{L_b}\right)\right] = \frac{1}{2}\left[\frac{1}{0.75}\left(\frac{8(1000)}{15'(12''/\text{ft})}\right)\right] = 29.6\,\text{kips/in.}$$

The stiffness provision requires an area of

$$\beta_{br} = \frac{P}{\Delta} = \frac{A_b E}{L_{\text{brace}}} = \frac{A_b(29{,}000)}{5'(12''/\text{ft})} = 483A_b$$
$$A_b \geq 0.06\,\text{in}^2.$$

In applying the design provisions, it is useful to remember that L_b = distance between braces. This should not be confused with the length of the brace itself.

In this instance, detailing of the brace will likely by dictated by architectural or functional requirements. A small W section (W6 × 12) or the smallest of Square HSS sections (HSS 2 × 2 × 1/8) has more than the required axial compressive strength. For longer braces, or those with larger stiffness and strength requirements, the axial strength of the brace will need to be checked using the provisions of Chapter E of the AISC Specification.

7.7 BEAM BRACING

In the previous sections, we were able to develop straightforward closed-form solutions for the required bracing strength of lateral bracing for columns. From these basic cases, we were able to adjust them (as shown in Figure 7.13) to include the effects of imperfections and from there examine the basis for the AISC specification equations.

The topic of beam bracing is substantially more complicated than column bracing due to the nature of beam instability. Just as lateral torsional buckling of beams is more computationally difficult than the in-plane buckling of columns, beam bracing is a more difficult topic as prevention of both lateral and torsion displacements are required by the brace to prevent instability. Much of the work done on beam bracing is based on experimental investigations and finite element modeling rather than on developing closed form analytical solutions. In considering beam bracing, we discuss the basic requirement of such bracing—that it must prevent both lateral displacement and twisting of the beams. In doing so we discuss the basic attributes that significantly effect the required strength and stiffness of the bracing. Since both lateral displacement and twisting must be prevented, we consider beam bracing in two separate categories: lateral bracing (section 7.7.1) and torsional bracing (section 7.7.2). It is useful to note that some bracing details prevent both lateral displacement and twist, such as a composite slab on the compression flange of a beam in positive bending.

7.7.1 Lateral Beam Bracing

As previously mentioned, a well-detailed and positioned lateral brace can perform double duty; that is, it can resist both lateral displacement and twisting of the beam. Consequently, the best location for a lateral brace is to place it in the location that best prevents twisting of the cross-section. Braces located close to the center of twist are not effective. Consequently, braces placed at the centroid of a cross-section are significantly less effective than those placed at the top flange of a simply supported beam. Yura (1993) demonstrates that for a W16 × 26 beam with a concentrated load at the midspan, the ideal stiffness required by a centroidally located brace is 44 times greater than that of a top flange brace.

In general, the lateral bracing should be attached to the compression flange. The exception is a cantilever beam, where the more effective location of the brace is on the top, or tension flange. This is because the center of twist is located below the compression flange, and a top flange brace provides better resistance against both lateral displacement and torsion of the beam.

Bracing of beams in double curvature bending presents a more complicated issue. Bracing placed on a single flange does not prevent buckling of the beam. Also, a common misconception is that the inflection point can be considered a brace point. This is completely false, as there is no prevention of lateral movement or twist at this location. Consequently, if bracing is required it must be provided on both flanges near the point of inflection, and on the compression flange at relevant locations along the length of the beam.

Similar to column bracing, an increase in the number of braces increases the ideal bracing requirements. The derived specification equations are developed for the conservative case of multiple intermediate braces.

7.7.2 Torsional Beam Bracing

Torsional systems are designed and detailed to prevent the twist of the cross-section rather than the lateral displacement. While lateral braces are idealized as linear springs, torsional braces are modeled as rotational springs with a rotational stiffness. The stiffness depends on the mode of deformation of the bracing system. For example, the stiffnesses of torsional bracing bent in single and double curvature are shown in Figure 7.16 and Figure 7.17, respectively. Single curvature bending may be associated with through girders, while double curvature bending is

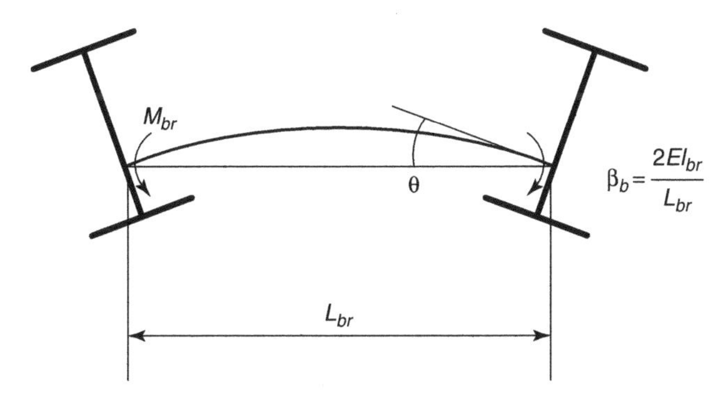

Fig. 7.16 Torsional diaphragm stiffness in single curvature bending.

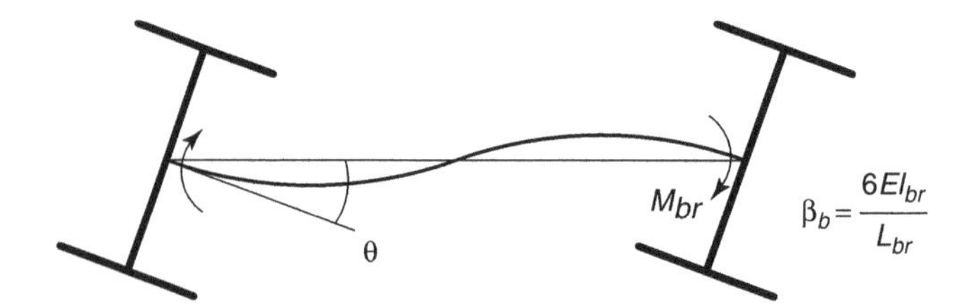

Fig. 7.17 Torsional diaphragm stiffness in double curvature bending.

the likely mode of floor decks or diaphragm braces. Torsional bracing may also be provided by cross-frames.

Work by Taylor and Ojalvo (1966) and Yura (1993) indicate that, unlike lateral beam bracing, torsional bracing stiffness requirements are not significantly affected by load location, brace placement, or the overall number of braces. However, the stiffness is highly dependent on the amount of cross-sectional distortion that occurs at the point of bracing. The provisions developed for the AISC specification therefore include the effects of web distortion of the beam being braced.

For a doubly symmetric beam with continuous bracing, Taylor and Ojalvo derived the following equation:

$$M_{\mathrm{cr}} = \sqrt{M_o^2 + \overline{\beta}_b E I_y}$$

where M_o = moment capacity of the unbraced length and $\overline{\beta}_b$ = torsional brace stiffness.

Yura (1993) modified the formula to account for the effects of cross-sectional distortion:

$$M_{\mathrm{cr}} = \sqrt{(C_b M_o)^2 + \frac{C_b^2 E I_y \overline{\beta}_T}{2C_{tt}}}$$

The first term under the radical represents the strength of the beam in the absence of a torsional brace. The factor C_{tt} accounts for the load location, and β_T is the stiffness of the torsional bracing per unit length of the beam. This equation was used in developing the torsional bracing provisions of the AISC specification.

7.8 AISC DESIGN PROVISIONS FOR BEAM BRACING

The AISC 2005 design provisions for beam are strongly based on the recommendations of Yura (1993). As is the case with column bracing, separate provisions are provided for nodal and relative lateral braces. Also as in the case of column bracing, the braces are assumed to be horizontal to the member being braced. Provisions are also provided for torsional bracing, based on the previously referenced works of Taylor and Ojalvo and Yura.

7.8.1 Lateral Beam Bracing Provisions

As reported in the commentary to Appendix 6 of the 2005 AISC Specification, the lateral bracing provisions are based on the approach originally developed by Winter, in which the bracing stiffness is given by:

$$\beta_{br} = \frac{2N_i(C_b P_f)C_t C_d}{\phi L_b} \qquad \text{(AISC C-A-6-3)}$$

where

N_i = 1.0 for relative bracing, $4 - \frac{2}{n}$ for discrete (nodal) bracing
N = number of intermediate braces
P_f = beam compression flange force $= \frac{\pi^2 E I_{yc}}{L_b^2}$
I_{yc} = out-of-plane moment of inertia of the compression flange
C_b = modification factor for nonuniform moment from Chapter F
C_t = 1.0 for centroidal loading $= 1 + (1.2/n)$ for top flange loading
C_d = double curvature moment factor $= 1 + \left(\frac{M_S}{M_L}\right)^2$
M_S = smallest moment causing compression in each flange
M_L = largest moment causing compression in each flange

C_d is only applied to the brace closest to the inflection point

Equation C-A-6-3 was simplified to equations A-6-6 and A-6-8 in Appendix F by conservatively assuming multiple braces and using the approximation

$$C_b P_f = \frac{M_u}{h}$$

As previously discussed, the stiffness requirements for lateral bracing are highly sensitive to the location of the brace. For this reason, the provisions require that bracing be attached as follows:

- Near the compression flange in all beams except cantilevered beams
- On the tension flange in cantilevered beams
- To both flanges in beams subjected to double curvature bending for the brace located closest to the inflection point

For relative bracing, the required strength and stiffness are given as follows:

Required brace strength:

$$P_{br} = 0.008 M_r \frac{C_d}{h_o} \qquad \text{(AISC A-6-5)}$$

Required brace stiffness:

$$\beta_{br} = \frac{1}{\phi}\left(\frac{4 M_r C_d}{L_b h_o}\right) \qquad \text{(AISC A-6-6)}$$

And for nodal bracing, the requirements are given as Required brace strength:

$$P_{br} = 0.02 M_r \frac{C_d}{h_o} \qquad \text{(AISC A-6-7)}$$

Required brace stiffness:

$$\beta_{br} = \frac{1}{\phi}\left(\frac{10 M_r C_d}{L_b h_o}\right) \qquad \text{(AISC A-6-8)}$$

where

M_r = Required flexural strength of the braced member(s)
h_o = distance between flange centroids (tabulated in the Design Manual)
C_d = 1.0 for single curvature bending
= 2.0 for double curvature bending applied to the brace
L_b = Laterally unbraced length of the beam (in.)

7.8.2 Nodal Bracing Example

In this example, we calculate the required bracing stiffness such that a simply supported W12 × 79 girder can reach its plastic moment capacity. The beam is braced by floor beams that frame into the girder in line with the top flange at third points as shown in Figure 7.18. The floor beams transfer the load to the girder as shown. We will assume that the floor beams terminate to a fixed location, such as a shear wall, and that they only brace the beam shown. The plastic moment capacity of the girder is 5,590 in-kips, and L_p for the beam is 10.6 ft.

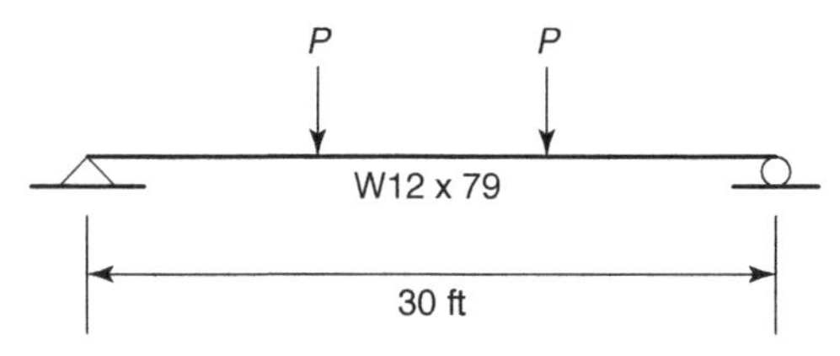

Fig. 7.18 Nodal bracing example.

The required strength of the bracing member is given by

$$P_{br} = 0.02M_r \frac{C_d}{h_o} = 0.02(5590)\frac{1.0}{11.6} = 9.6 \text{ kips}$$

where $C_d = 1.0$ for single curvature bending, and h_o is a tabulated section property of the beam.

The required bracing stiffness is given by

$$\beta_{br} = \frac{1}{\phi}\left(\frac{10M_rC_d}{L_bh_o}\right) = \frac{1}{0.75}\left(\frac{10(5590)(1.0)}{(10')(12''/\text{ft})(11.6)}\right) = 53.5 \text{ kip/in.}$$

Since a nodal brace is being employed, the actual stiffness of the brace is given by P/Δ, so the bracing stiffness is equal to

$$\beta_{br} = \frac{AE}{L}$$

Assuming a brace length of 15 ft. (180 in.), the required brace area would be

$$A \geq \frac{53.5(180'')}{29,000} = 0.33 \text{ in.}^2$$

The bracing member therefore needs to be greater than 0.33 in^2 in area with a compressive strength of 9.6 kips. These criteria can be met by any of a large number of beam shapes.

The calculated brace stiffness assumes that the framing member, in this case a beam, is perpendicular to the girder being braced. In the next example, we calculate the required strength and stiffness of a relative brace oriented at an angle to the girder being braced.

7.8.3 Relative Beam Bracing Example

Consider the deck support system shown in Figure 7.19. In this instance we have three W12 × 72 girders braced by the top flange truss shown. The

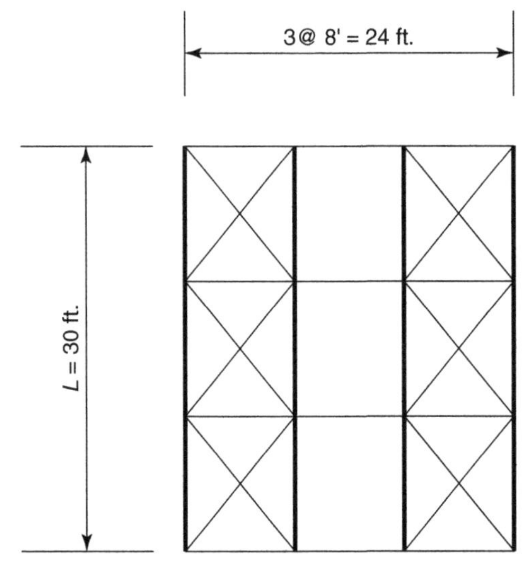

Fig. 7.19 Relative bracing design example.

braces are considered to be tension-only. Each truss braces two girders. Again, we assume that we wish to develop the full plastic moment capacity of the girders.

For a relative bracing system, the required brace strength is given by

$$P_{br} = 0.008 M_r \frac{C_d}{h_o} \qquad \text{(AISC A-6-5)}$$

Since each brace is stabilizing two girders, the required strength is given by

$$P_{br} = 0.008 M_r \frac{C_d}{h_o} = \frac{0.008(2)(5590)(1)}{11.6} = 7.71\,\text{kips}$$

The required brace stiffness is

$$\beta_{br} = \frac{1}{\phi}\left(\frac{4M_r C_d}{L_b h_o}\right) = \frac{1}{0.75}\left[\frac{4(2)(5590)(1.0)}{8'(12''/\text{ft})(11.6)}\right] = 53.5\,\text{kips/in.}$$

The brace is not perpendicular to the braced element, and that is taken into account in the calculation of the required area to meet the design criteria. In this case, the brace meets the girder at an angle equal to

$$\theta = \arctan\frac{8}{10} = 38.7^\circ$$

The bracing stiffness is then given by

$$\beta = \frac{P}{\Delta} = \frac{AE}{L_b}\cos^2\theta$$

And the required area is calculated as

$$A \geq \frac{53.5(8)(12)}{29000(\cos^2 38.7)} = 0.29\,\text{in.}^2$$

Based on yielding of the tension brace, the required area to meet the strength criterion would be

$$A \geq \frac{1}{0.9}\left(\frac{7.71}{50}\right) = 0.17\,\text{in.}^2$$

7.8.4 Torsional Beam Bracing Provisions

The strength and stiffness requirements of torsional braces were developed to provide a minimum of twice the ideal stiffness required based on assumed initial out-of-twist equal to 2 percent of the length to depth ratio of the beam and a torsional brace stiffness (excluding web distortion) equal to

$$\beta_T = \frac{2.4LM_u^2}{nEI_yC_b^2} \qquad \text{(AISC C-A-6-6)}$$

Algebraic manipulation leads to the required brace moment capacity equal to

$$M_{br} = \frac{0.024M_rL}{nC_bL_b} \qquad \text{(AISC A-6-9)}$$

And the required cross-frame or diaphragm bracing stiffness

$$\beta_{Tb} = \frac{\beta_T}{\left(1 - \frac{\beta_T}{\beta_{\text{sec}}}\right)} \qquad \text{(AISC A-6-10)}$$

where

$$\beta_T = \frac{1}{\phi}\left(\frac{2.4LM_r^2}{nEI_yC_b^2}\right) \qquad \text{(AISC A-6-11)}$$

$$\beta_{\text{sec}} = \frac{3.3E}{h_o}\left(\frac{1.5h_ot_w^3}{12} + \frac{t_sb_s^3}{12}\right) \qquad \text{(AISC A-6-12)}$$

$$\phi = 0.75$$

L = span length

N = $f$number of nodal braced point

β_T = brace stiffness excluding web distortion (kip-in/rad)

β_{sec} = web distortional stiffness including the effect of web transverse stiffeners (kip-rad/in)

Because torsional bracing is sensitive to distortion of the cross-section, particular attention must be paid to web stiffness. If the web distortional stiffness, given by β_{sec} is less than the brace stiffness, the stiffener will be ineffective. This is evident in that the equation for required stiffness will be negative if this occurs. Web stiffeners may be required at torsional bracing locations.

7.9 SUMMARY

The bracing provisions of the 2005 AISC specification are substantially based on the early work of Winter, who demonstrated that braces require both minimum strength and stiffness to be effective, and that the brace force is dependent on the amount of initial imperfection in the system. Bracing requirements are based on the type of brace being designed, whether it is nodal or relative, and for beams, whether it is a lateral brace or a torsional brace.

It is useful to remember that the bracing provisions for columns and lateral beam bracing are conservatively based on the assumption of multiple braces. Modifications may be made for single brace locations as described in the Commentary to Appendix 6. In addition, the accurate computation of brace stiffness is necessary to correctly apply the design equations. In many instances, the designer may need to refer to a basic textbook in structural analysis to recall the correct formulation of a brace stiffness, whether it be lateral, where stiffness $= F/\Delta$, or torsional, where stiffness $= M/\theta$. Because the provided stiffness is dependent on the detail of the brace, the specification cannot provide the designer with standard formulae for actual brace stiffness. Some degree of knowledge of structural response must be brought to the table in applying the provisions.

Detailing can be essential in the success of bracing in providing stability. Poorly located or detailed braces can be as ineffective as if no bracing were provided at all. The provisions provide significant guidance in placement of bracing. A quick review of bracing literature can also help the designer avoid poor bracing details. In general, remembering the basic behavior of bracing and the motion is it intended to prevent can go a long way in helping the designer determine if a brace will be effective.

SUGGESTED READING

Galambos, T. V. 1998. *Guide to Stability Design Criteria For Metal Structures*, (5th ed.). John Wiley and Sons, New York, Chapter 12, "Bracing."

"Is Your Structure Suitably Braced?" Proceedings of the 1993 Conference. Structural Stability Research Council, Milwaukee, WI, April 1993.

PROBLEMS

7.1. Determine the ideal bracing stiffness for the perfectly straight column shown as a function of P_E and L_b, the elastic buckling load of the column. Plot P_{cr} versus $\frac{\beta}{\beta_i}$.

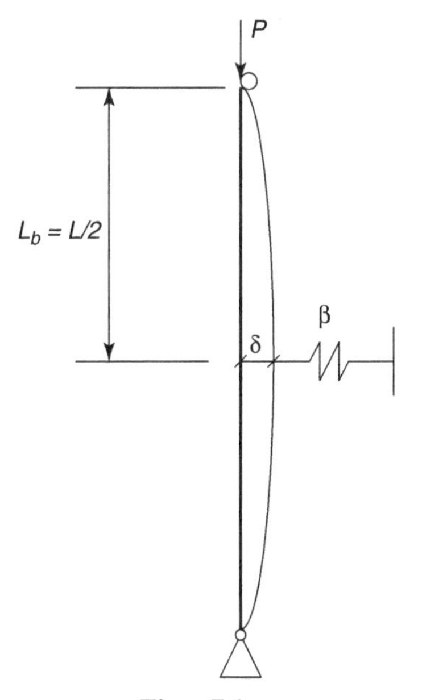

Fig. p7.1

7.2. Redo problem 7.1 with an initial out-of-straightness of $\delta_o = 0.002L$.

7.3. Design the relative brace for the frame shown. Consider stability of the brace in the design.

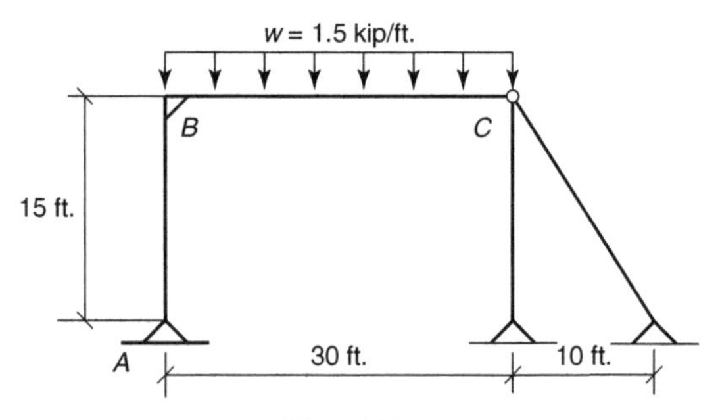

Fig. p7.3

7.4. Design the nodal brace for the frame shown. Compare the results of the design to the design in problem 7.3.

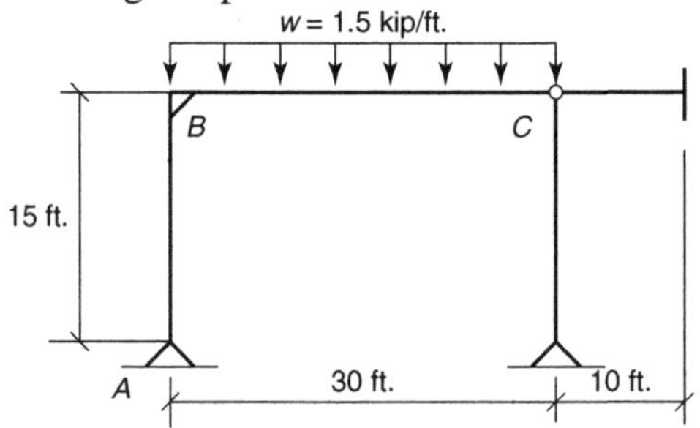

Fig. p7.4

7.5. Determine the required area of the tension only X-bracing.
Given: $P = 35\,\text{kips}$
$L = 18\,\text{ft.}$

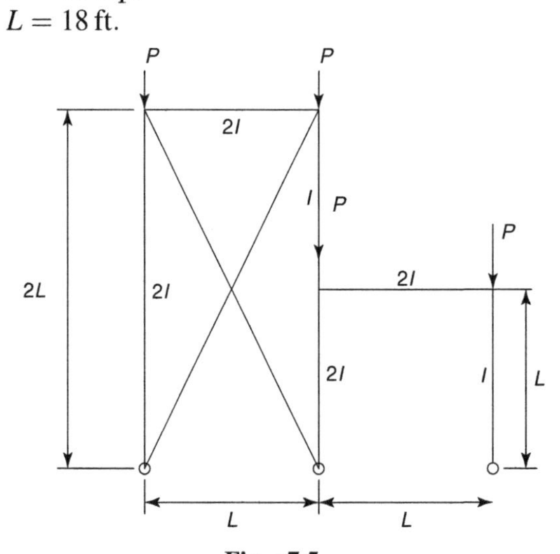

Fig. p7.5

7.6. Determine the required stiffness and strength of the diaphragm brace for the beam system below. Consider the requirements if it is designed

a. As a lateral brace (attached to a rigid abutment)

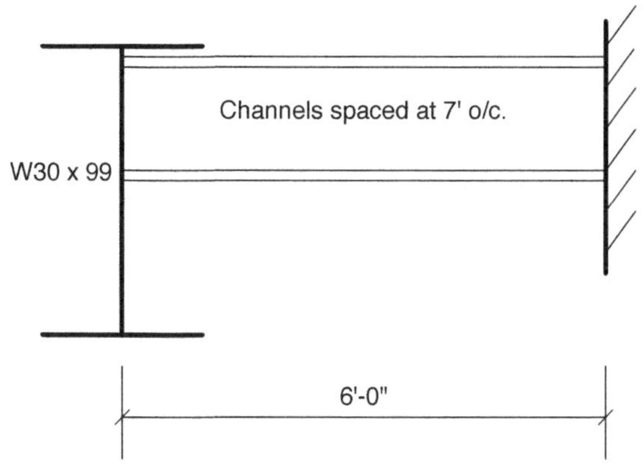

Fig. p7.6a

b. As a torsional brace between two girders

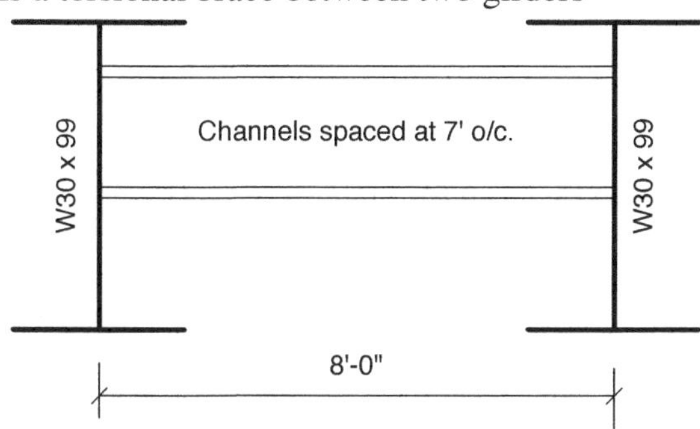

Fig. p7.6b

7.7. Determine the required area of the cross-bracing for the girders shown.

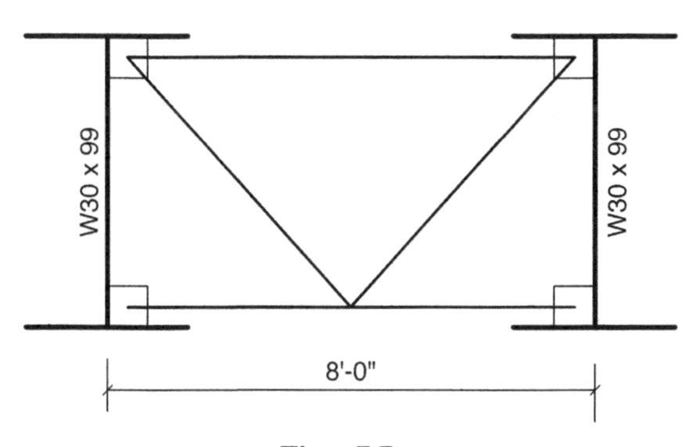

Fig. p7.7

CHAPTER EIGHT

SPECIFICATION-BASED APPLICATIONS OF STABILITY IN STEEL DESIGN

8.1 INTRODUCTION

The previous chapters have discussed the theoretical models that describe stability behavior of members and framing systems. Obviously, analytically or numerically deriving the behavior of a system for every design is entirely impractical, so stability theory has been developed into practical methods of assessing the strength of real structures. Specification-based approaches all require some level of generalization or simplification to be applicable to the wide variety of practical cases that arise, and it is important to understand the assumptions built into the design equations in order to apply them in a correct and economical manner. This chapter focuses on the most recent AISC specification (AISC 2005) and provides discussion of the basis of the design equations included in the specification as well as examples of their applications. A brief discussion of international design codes is also included. We presume that the reader is familiar with the *Load and Resistance Factor Design* (LRFD) philosophy. While allowable stress design (ASD) is allowed in the current specification, the focus in this chapter is on design applications using LRFD.

System stability is difficult to assess directly without the assistance of analysis capabilities that are not often available for design office use.

Consequently, in the AISC Specification, system stability is checked on a member by member basis. If the most critical member in a framing system is able to meet strength criteria, then the structure is considered to have sufficient strength and stability to withstand the given loading. *Framing members* in this chapter include beams, columns, beam-columns, and bracing members. The chapter first discusses the three primary equations used to assess member strength as related to stability: the beam-column interaction equation (AISC Chapter H), the column strength curve (AISC Chapter E), and the flexural strength equations (AISC Chapter F). Derivation of the column and moment equations is discussed in Chapters 2 and 5, respectively, and are only discussed briefly here. A background on the development of the beam-column interaction equations is presented in section 8.2; this discussion gives some insight into the guidelines set forth for developing specification based equations and why second-order analysis forces and moments are required when using the equations. Sections 8.3 and 8.4 present the equations for axial strength and flexural strength. Section 8.5 presents the three approaches for addressing frame stability provided in the specification, as well as an overview of the level of analysis required for each method. In section 8.6, effective length factors are reviewed, and the corrections that must be made to the alignment chart–based approach are presented, along with examples. Section 8.7 provides examples of frame stability assessment by the three approaches outlined in section 8.5. Finally, section 8.8 provides an overview of Canadian and European design requirements.

8.2 DEVELOPMENT OF THE BEAM-COLUMN INTERACTION EQUATIONS

For structures in which moment connections or moment frames are used, stability checks are typically governed by the beam-column interaction check. Current assessment of beam-column strength by the AISC Specification involves the determination of the axial strength of the member in the absence of moment (P_c) and flexural strength of the member in the absence of axial force (M_c). Once established, these values are used as anchor points for an equation that models the interaction between axial forces and moments in the beam-column.

The AISC interaction equations were developed based on the following series of guidelines (ASCE 1997):

1. The equations should be general and applicable to a wide range of problems . . .

2. The equations should be based on the load effects obtained from second-order elastic analysis . . .
3. The equations should distinguish clearly between the second-order elastic load effects and the resistances such that the calculation of second-order forces . . . can be clearly separated from the interpretation and design application of the equations.
4. The equations should predict identical ultimate strengths for problems in which the strengths are the same . . .
5. The equations should not necessarily be required to consider strength and stability separately, since in general, all columns of finite length fail by some combination of inelastic bending and stability effects.
6. The equations should be capable of capturing the limit state of pure stability under axial load, including the effects of restraint provided by *elastic* beam members to the *elastic* or *inelastic* columns.
7. The equations should not be more than five percent unconservative when compared to strengths obtained from "exact" second-order inelastic solutions.

The interaction equations were established in large part based on a fit to "exact" second-order inelastic analyses of sensitive benchmark frames developed by Kanchanalai (1977). In Kanchanalai's studies, inelastic response, including the effects of residual stresses, was explicitly captured, but geometric imperfections were not considered. The developers of the AISC LRFD beam-column equations accounted for geometric imperfections in the manner illustrated in Figure 8.1. Figure 8.1 shows the results for a perfectly straight W8 × 31 column bent about the weak-axis in which the first-order solution, (P/P_y vs. M_1/M_p, representing the applied loads) is converted to the second-order curve (P/P_y vs. M_2/M_p, representing the maximum second order elastic design forces) using an elastic amplification factor.

The normalized LRFD column axial capacity for this member is $P_n/P_y = 0.591$; the "exact" solutions provided by a rigorous second-order, inelastic (or plastic zone) analysis gives a value of $P/P_y = 0.71$. The reduction in strength is due to the second-order moment caused by initial imperfections. The first-order imperfection moment is assumed to vary linearly with the axial load, as shown in Figure 8.1, and the second-order imperfection moment is established by elastic amplification of the first-order moment. The moment capacity is established by subtracting these imperfection moments from the original curves, and the resulting capacities are given by the net M_1/M_p and M_2/M_p values. The final curves, including the imperfection effects, are presented in Figure 8.2. The AISC beam-column equations are based on

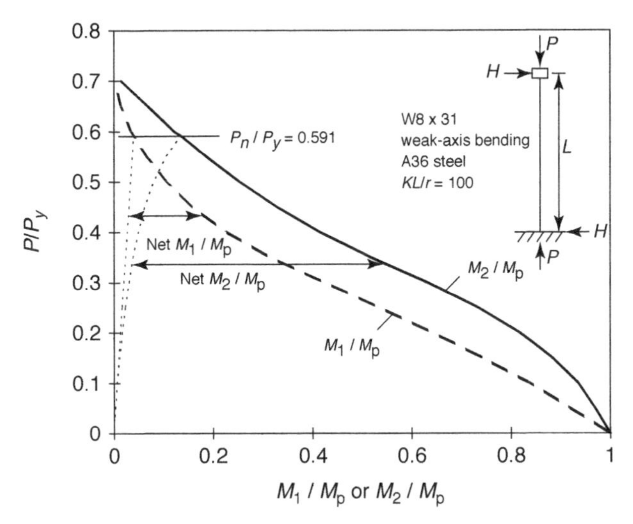

Fig. 8.1 Procedure for determining net M_1/M_p and M_2/M_p curves.

curve fitting to the lower bound of the net M_2/M_p curves based on a large parametric study of beam columns. Despite response differences in strong and weak axis bending, a single interaction equation is used to simplify the design process.

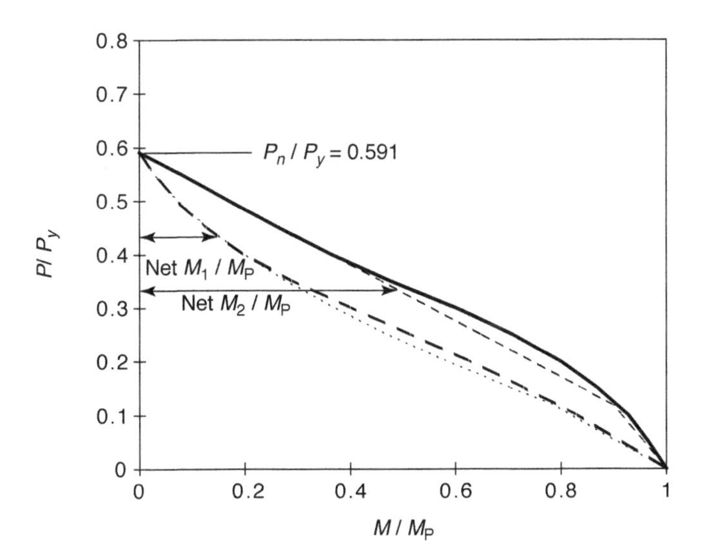

Fig. 8.2 Curve fit of AISC-LRFD beam-column equations to "exact" strength curves.

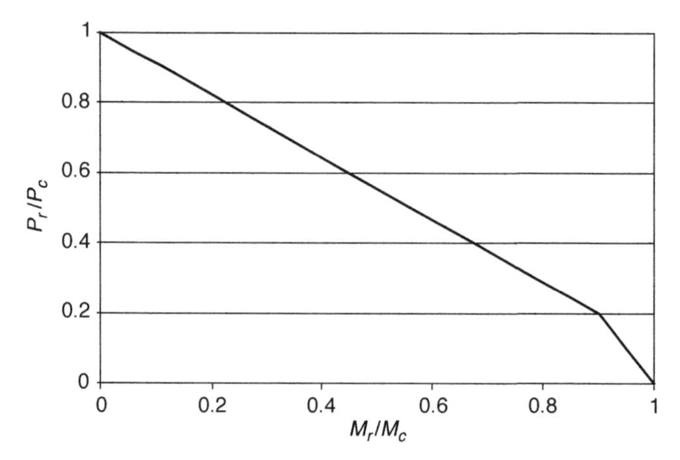

Fig. 8.3 AISC interaction curve.

The current AISC beam-column interaction equations have been simplified to a bilinear approximation based on the best fit of the curves developed from the benchmark frames. These equations are provided in Chapter H of the AISC Specification. The interaction equations produce the bilinear curve shown in Figure 8.3, and are given by

For $\frac{P_r}{P_c} \geq 0.2$:

$$\frac{P_r}{P_c} + \frac{8}{9}\left(\frac{M_{rx}}{M_{cx}} + \frac{M_{ry}}{M_{cy}}\right) \leq 1.0 \qquad \text{(AISC H1-1a)}$$

H1-1a

For $\frac{P_r}{P_c} < 0.2$:

$$\frac{P_r}{2P_c} + \left(\frac{M_{rx}}{M_{cx}} + \frac{M_{ry}}{M_{cy}}\right) \leq 1.0 \qquad \text{(AISC H1-1b)}$$

H1-1b

where

P_r = Required second-order axial strength
P_c = Available axial compressive strength
M_r= Required second-order flexural strength
M_c= Available flexural strength

As previously discussed, application of the interaction equation requires that the axial capacity (P_c) and moment capacity (M_c) be calculated for the individual member. These calculations per the AISC Specification are discussed in sections 8.3 and 8.4, respectively.

8.3 ASSESSMENT OF COLUMN STRENGTH

Chapter 4 discussed that members rarely are loaded only along their centroidal axis. Most members have some moment, torsion, and/or shear associated with the axial load. However, the behavior of columns that are laterally supported by other columns, bracing, or shear walls and are pinned at the ends are dictated by the axial component. The characterization of their behavior as axially loaded columns is perfectly appropriate if the strength equation accounts for those factors that can affect strength, including residual stresses, imperfections, and boundary conditions. Determination of the axial capacity in the absence of moment is also essential to establish one of the anchor points for the beam-columns interaction equation, as described in section 8.2.

The development of the curve, based on studies performed at Lehigh University in the 1960s, is discussed in Chapter 3, section 3.7. The curve itself includes the effects of inelasticity and initial out-of-straightness of the column. The effects of boundary conditions have traditionally been included through the calculation of effective length factors, or K-factors (although the current 2005 AISC Specification provides two additional methods by which the actual member length may be used in determining column strength regardless of end condition).

The AISC column curve is divided into two sections. For columns with high slenderness ratios, the failure mode is based on the elastic bucking strength. The 0.877 multiplier of the Euler buckling load accounts for the initial imperfection and is used to establish the nominal column strength. For shorter columns, the curve is based on both experimental and analytical results and it includes the effects of both incidental moments caused by eccentric loads or initial out-of-straightness and inelasticity due to residual stresses. The point of transition between the two curves is a function of slenderness ratio and yield stress. The curve is described by the following AISC equations:

For inelastic buckling:

For $\dfrac{KL}{r} \leq 4.71\sqrt{\dfrac{E}{F_y}}$

$$F_{cr} = \left[0.658^{\frac{F_y}{F_e}}\right] F_y \qquad \text{(AISC E3-2)}$$

For elastic buckling:

For $\frac{KL}{r} > 4.71\sqrt{\frac{E}{F_y}}$

$$F_{cr} = 0.877F_e \quad \text{(E3-3)}$$

where

$$F_e = \frac{\pi^2 E}{\left(\frac{KL}{r}\right)^2} \quad \text{(E3-4)}$$

The effective length factor accounts for any variation in end restraint from the pinned–pinned condition for isolated columns. The K-factor is discussed in great detail in section 8.6.

For beam-columns, once the column strength is established, it is necessary to determine the moment capacity of the member in order to use the interaction equations of Chapter H in the Specification. The moment capacity, including the potential for lateral torsional buckling, is described in the next section.

8.4 ASSESSMENT OF BEAM STRENGTH

The theoretical development of the elastic critical moment due to lateral torsional buckling is provided in Chapter 6 (equation 6.7). The factors that most affect the resistance of a beam to LTB include the unsupported length between brace points (L_b), the cross-sectional properties of the beam, and the moment gradient. For compact sections, the beam may fail in three ways:

1. The beam experiences elastic lateral torsional buckling.
2. The beam experiences inelastic lateral torsional buckling.
3. The plastic moment capacity of the cross-section is reached.

The three failure modes are described by three portions of the buckling curve shown in Figure 8.4.

8.4.1 Plastic Moment Capacity

The plastic moment capacity is given by

$$M_p = F_y Z_x$$

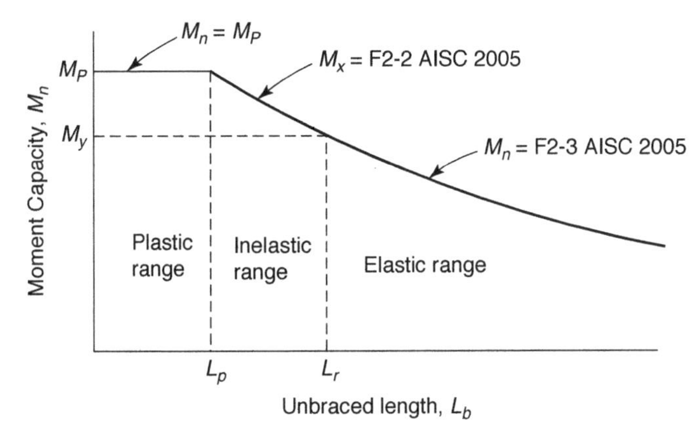

Fig. 8.4 The AISC flexural strength curve.

And applies to beams where the unbraced length, $L_b < L_p$ given by

$$L_p = 1.76 r_y \sqrt{\frac{E}{F_y}} \quad \text{(AISC F2-5)}$$

8.4.2 Elastic LTB

Equation F2-3 describes elastic lateral torsional buckling, and is given by

$$M_n = F_{\mathrm{cr}} S_x \le M_p \quad \text{(AISC F2-3)}$$

where the critical stress is given as F_{cr}:

$$F_{\mathrm{cr}} = \frac{C_b \pi^2 E}{\left(\frac{L_b}{r_{ts}}\right)^2} \sqrt{1 + 0.078 \frac{Jc}{S_x h_o} \left(\frac{L_b}{r_{ts}}\right)^2} \quad \text{(AISC F2-4)}$$

This capacity applies to beams where the unbraced length, $L_b > L_r$, given by

$$L_r = 1.95 r_{ts} \frac{E}{0.7 F_y} \sqrt{\frac{Jc}{S_x h_o}} \sqrt{1 + \sqrt{1 + 6.76 \left(\frac{0.7 F_y}{E} \frac{S_x h_o}{Jc}\right)^2}} \quad \text{(AISC F2-6)}$$

The value c is dependent on the type of cross-section and is equal to 1 for doubly symmetric sections.

8.4.3 Inelastic LTB

For unbraced lengths that fall between these two limits, the inelastic buckling strength is given by the linear interpolation of the elastic and plastic regions:

$$M_n = C_b\left[M_P - (M_P - 0.7F_yS_x)\left(\frac{L_b - L_p}{L_r - L_p}\right)\right] \leq M_p \qquad \text{(AISC F3-1)}$$

The 2005 specification includes slight modifications from the 1999 specifications (AISC 2001) in the elastic buckling equations, by replacing M_{cr} with the product of F_{cr} and S_x, and by eliminating the X_1 and X_2 from the L_r equations. Terms were further rearranged and simplified by adding and defining the r_{ts} term

$$r_{ts}^2 = \frac{\sqrt{I_yC_w}}{S_x} \qquad \text{(AISC F2-7)}$$

and adding the factor c to account for differences between I-sections and channels. The 2005 and 1999 LTB equations are identical in value when $c = 1.0$, which is true for all doubly symmetric sections.

8.4.4 Variations from Initial Assumptions

The previous equations are based on the assumptions built into the derivation of the elastic moment capacity, namely:

- Simply supported beam (warping is unrestrained)
- Uniform moment gradient
- Compact, doubly symmetric cross-section
- Small in-plane deflections and angles of twist
- No local buckling or distortion of the cross-section
- Load applied at the shear center

As discussed in Chapter 6, a uniform moment is the most conservative loading for lateral torsional buckling. Although numerous researchers have proposed factors to account for moment gradient, the equation developed by Kirby and Nethercot (1979) is used in the specification since it applies to any general moment diagram within the unbraced length. The nominal capacity in the elastic and inelastic buckling ranges is multiplied by C_b, which is given by

$$C_b = \frac{12.5M_{\max}}{2.5M_{\max} + 3M_A + 4M_B + 3M_C}R_m \leq 3.0 \qquad \text{(AISC F1-1)}$$

The specification provides no specific modifications for end restraint (in-plane or torsional) or location of the load on the cross-section. Again, researchers have proposed different options to account for these deviations from the original model; some of these have been discussed in Chapter 6. Additional provisions are provided for noncompact sections. Provisions for beams with noncompact flanges are given in section F3, and those for noncompact or slender webs are given in sections F4 and F5, respectively. Coverage of these provisions is outside the scope of this chapter.

The following example illustrates the application of the three primary equations that determine stability of members in framing systems, the column strength, flexural strength, and interaction equations.

8.4.5 Example 8.1: Beam-column Interaction Check for an Isolated Beam-column

In the following example, we assess the stability of an isolated beam-column. The beam column shown in Figure 8.5 is subjected to the factored second-order loads shown. It is not braced along the length in either direction. Because the section is compact, we do not need to check for failure modes due to local buckling.

The first step is to determine the axial load capacity using equations E3-2 through E3-4 of the specification. Since the end conditions are the same in

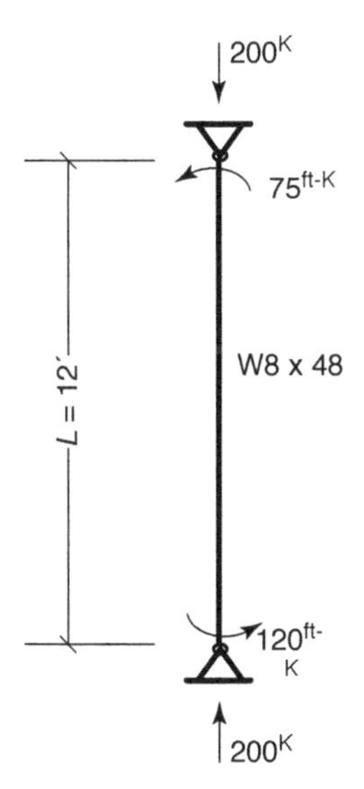

Fig. 8.5 Beam-column strength check example.

either direction, we know that weak-axis buckling governs. First, we must check whether the column would buckle elastically or inelastically:

$$\frac{KL}{r} = \frac{1.0(12')(12\,\text{in./ft.})}{2.08''} = 69.2 \leq 4.71\sqrt{\frac{E}{F_y}} = 4.71\sqrt{\frac{29{,}000}{50}} = 113.4$$

This dictates which equation is used for column strength. The critical stress is then given by

$$F_{\text{cr}} = \left[0.658^{\frac{F_y}{F_e}}\right]F_y$$

Where $F_e = \dfrac{\pi^2 E}{\left(\frac{KL}{r}\right)^2} = \dfrac{\pi^2(29,000)}{(69.2)^2} = 59.7\,\text{ksi}$

$$F_{\text{cr}} = \left[0.658^{\left(\frac{50}{59.7}\right)}\right]50\,\text{ksi} = 35.2\,\text{ksi}$$

The axial capacity is then given by

$$P_n = F_{\text{cr}}A_g = 35.2\,\text{ksi}(14.1\,\text{in.}^2) = 496\,\text{kips}$$

Next, the moment capacity is determined. The AISC Design Manual provides tabulated values for many of the variables required in the equations for flexural capacity; however, for the sake of completeness, we calculate them here. First, we calculate the values of L_p and L_r:

$$L_p = 1.76r_y\sqrt{\frac{E}{F_y}} = 1.76(2.08'')\sqrt{\frac{29{,}000}{50}} = 88.16'' = 7.35'$$

$$L_r = 1.95r_{ts}\frac{E}{0.7F_y}\sqrt{\frac{Jc}{S_x h_o}}\sqrt{1+\sqrt{1+6.76\left(\frac{0.7F_y}{E}\frac{S_x h_o}{Jc}\right)^2}}$$

$$= 1.95(2.35)\frac{29{,}000}{0.7(50)}\sqrt{\frac{1.69}{(43.2)(7.82)}}$$

$$\times\sqrt{1+\sqrt{1+6.76\left(\frac{0.7(50)}{29000}\frac{(43.2)(7.82)}{1.96}\right)^2}} = 422.7'' = 35.2'$$

since $L_p < L < L_r$

The flexural capacity is based on the portion of the curve in which inelastic LTB occurs, and is given by

$$M_n = C_b\left[M_p - (M_p - 0.7F_yS_x)\left(\frac{L_b - L_p}{L_r - L_p}\right)\right] \leq M_P$$

where

$$M_p = Z_xF_y = (49\,\text{in.}^3)(50\,\text{ksi}) = 2450\,\text{in.} - \text{kip} = 204\,\text{ft.} - \text{k}$$

Remember that the C_b factor takes into account moment gradient. Since the basis of the formulation is a beam with a uniform moment (and hence uniform compressive force in the compression flange), this represents the most conservative solution, Any other load condition proves to be less conservative. For this loading, the moment diagram is linear, and the absolute values of the quarter point moments are given by

$$\begin{aligned} M_{\max} &= 120 \\ M_A &= 71.25 \\ M_B &= 22.5 \\ M_C &= 26.5 \end{aligned}$$

And $R_m = 1$ for a doubly symmetric member

therefore

$$C_b = \frac{12.5M_{\max}}{2.5M_{\max} + 3M_A + 4M_B + 3M_C}$$

$$R_m = \frac{12.5(120)}{2.5(120) + 3(71.25) + 4(22.5) + 3(26.5)}(1.0) = 2.20$$

The flexural capacity is then given by

$$M_n = 2.20\left[2450 - ((2450 - 0.7(50)(43.2))\left(\frac{12' - 7.35'}{35.2' - 7.35'}\right)\right]$$

$$= 5045\,\text{in-kips} \geq M_p$$

Since the calculated $M_n > M_p$,

$$M_n = M_p = 2450\ \text{in-k}(204\ \text{ft.-kip})$$

Once the moment and axial capacities are independently calculated, the interaction equation can be checked as follows:

$$\frac{P_r}{P_c} = \frac{180}{0.9(496)} = 0.403 > 0.2 \quad \text{use AISC Eqation H1} - 1\text{a}$$

$$\frac{P_r}{P_c} + \frac{8}{9}\left(\frac{M_{rx}}{M_{cx}} + \frac{M_{ry}}{M_{cy}}\right) = 0.403 + \frac{8}{9}\left(\frac{120}{0.9(204)} + 0\right) = 0.984 < 1.0$$

The column is sufficiently strong to handle the applied loads.

8.5 SPECIFICATION-BASED APPROACHES FOR STABILITY ASSESSMENT

A significant number of changes were made from previous AISC *Specifications* in the way that member and system stability are checked in the 2005 AISC *Specification.* The most notable of these is the explicit discussion of analysis requirements for design assessment included in Chapter C, "Stability Analysis and Design." The 2005 AISC specification language states that any analysis/design procedure is acceptable as long as it takes into account the following effects:

- Flexural, shear, and axial deformations in members
- Member stiffness reduction from residual stresses
- Geometric imperfections due to initial out-of-plumbness (Δ_o) and out-of-straightness (δ_o)
- Second-order effects $P - \Delta$ and $P - \delta$, which are the effects of the axial load (P) acting through the displacement at the member ends (Δ) and displacement relative to the member (δ).

The specification presents three approaches that may be used to determine required strengths of members, connections, and other elements:

1. The second-order analysis approach (AISC C2.2a) described in section 8.5.2, also called the critical load or K-factor approach
2. The first-order analysis approach (AISC C2.2b) described in section 8.5.3
3. The direct analysis approach (AISC Appendix 7) described in section 8.5.4

The methods differ by the means and the extent to which they directly model those items that affect the beam-column strength. In addition, the rigor of analysis required also varies from method to method. For the first-order and second-order methods, both have minimum frame stiffness limits they must meet. These requirements are measured as a ratio of first order to second-order drift and are used as a measure of the amplification of the second-order moments and axial forces in the system. The direct analysis approach is the only method outlined in the specification that has no limit on frame stiffness. The other principal differences in the approaches include the method of accounting for the effects of member inelasticity and the means by which the effects of geometric imperfections due to fabrication and erection tolerances are included in the strength check.

In section 8.5.1, we discuss the attributes of the three approaches with some discussion of their development. Section 8.7 includes examples in which each method is applied.

8.5.1 Analysis Approaches

The solutions for column and frame stability presented in this text are based on equilibrium of the column or frame on the *deformed configuration;* these represent a *second-order* solutions. Current specification equations require second-order forces and moments. A brief review of analysis approaches is presented here.

- In *first-order elastic analysis*, the material is modeled as linear-elastic, and equilibrium is only satisfied on the undeformed configuration of the structure. The relationship between loads and displacements is linear.
- In *second-order elastic analysis*, the material is still considered linear-elastic, but the equilibrium is calculated on the deformed geometry of the structure. Equilibrium of the deflected shape of a member results in larger internal moments and forces due to second-order ($P-\Delta$ and $P-\delta$) effects.

The second order results may be obtained from one of two approaches:

1. A direct second-order analysis (e.g, using a commercial or academic software algorithm that directly calculates second-order forces and moments)
2. The NT–LT approach in which linear elastic analysis results are used and amplification factors are applied.

For the second method, the moments and axial loads should be found under two conditions: no translation and lateral translation (NT and LT, respectively). M_{nt} is the maximum moment and P_{nt} is the maximum axial load assuming that story sidesway is prevented. M_{lt} is the maximum moment and P_{lt} is the maximum axial load caused by sidesway from either lateral loads or unbalanced gravity loads. The maximum design moment and axial load are then be found using equations C2-1a and C2-1b:

$$M_r = B_1 M_{\text{nt}} + B_2 M_{\text{lt}} \qquad \text{(AISC C2-1a)}$$

$$P_r = P_{\text{nt}} + B_2 P_{\text{lt}} \qquad \text{(AISC C2-1b)}$$

The B_1 and B_2 factors are the second-order amplification factors, which account for the displaced geometry of the frame. The B_1 amplification factor accounts for amplification of moments due to $P-\delta$ effects, or the amplification due to displacements between brace points, and is given by

$$B_1 = \frac{C_m}{1 - \dfrac{\alpha P_r}{P_{e1}}} \geq 1.0 \qquad \text{(AISC C2-2)}$$

where

$$C_m = 0.6 - 0.4\left(\frac{M_1}{M_2}\right) \qquad \text{(AISC C2-4)}$$

M_1 and M_2 are the smaller and larger end moments, respectively, for beam-columns not subjected to transverse loading between supports. For beam-columns subjected to transverse loading, C_m may conservatively be taken equal to 1.0. The moments are calculated from a first-order analysis. The ratio $\frac{M_1}{M_2}$ is positive for double curvature bending and negative for single curvature bending. C_m is similar to the C_b factor used in calculating resistance to lateral torsional buckling in that it accounts for the moment gradient in the beam-columns; the case of a uniform moment, in which $C_b = 1.0$, is the most conservative. Other gradients produce values of $C_m < 1$ and lower amplification factors.

P_{e1} is the elastic critical buckling load of the member and is given by

$$P_{e1} = \frac{\pi^2 EI}{(K_1 L)^2} \qquad \text{(AISC C2-5)}$$

There are two important points in calculating P_{e1}

- The critical load is calculated *in the plane of bending*, which is not necessarily the critical axis in compression.
- K_1 is calculated based on the assumption of no sidesway and is less than or equal to 1.

It is recommended that, unless a smaller value is warranted based on analysis, K_1 should be set equal to 1.

P_r is the required axial strength of the beam-column based on factored loads. For this reason, $\alpha = 1.0$ for LRFD analysis and 1.6 if ASD is used. The reasoning for the 1.6 factor is described in greater detail in section 8.5.3. Although P_r is strictly defined as the required second-order axial capacity, and under strict interpretation should be equal to $P_{\text{nt}} + B_2 P_{\text{lt}}$, the first-order approximation, $P_r = P_{\text{nt}} + P_{\text{lt}}$, can be used in the B_1 equation.

The $B_1 - B_2$ approach uncouples the effect of the P–Δ and P–δ effects. In most frames, this is a perfectly reasonable approximation. However, in instances where B_2 is greater than 1.2, a rigorous second-order analysis is recommended in the specification. Many methods in commercial software do not include P–δ effects, but are instead based on P–Δ approximations. In fact, if a direct second-order algorithm method is used, it is important for the designer to understand how the commercial software package implements the second-order analysis (e.g., whether an approximate method or a rigorous approach is used, and if the former, the limits of the approximation). Commercial software packages often use approximate second-order analysis methods than can be sensitive to the type of framing being analyzed. In general, it is important to understand the methodology for second-order results that a software package utilizes and verify that the results are valid prior to assuming that a first-order analysis with amplifiers does not need to be used.

The B_2 factor accounts for amplification of moments due to P–Δ effects, and is given by

$$B_2 = \frac{1}{1 - \dfrac{\alpha \sum P_{\text{nt}}}{\sum P_{e2}}} \geq 1.0 \qquad \text{(AISC C2-3)}$$

$\sum P_{\text{NT}}$ = Total vertical load supported by the story

$$\sum P_{e2} = \sum \frac{\pi^2 EI}{(K_2 L)^2} = R_M \frac{\sum HL}{\Delta_H} \qquad \text{(AISC C2-6)}$$

$\sum P_{e2}$ = Elastic critical buckling resistance of the story

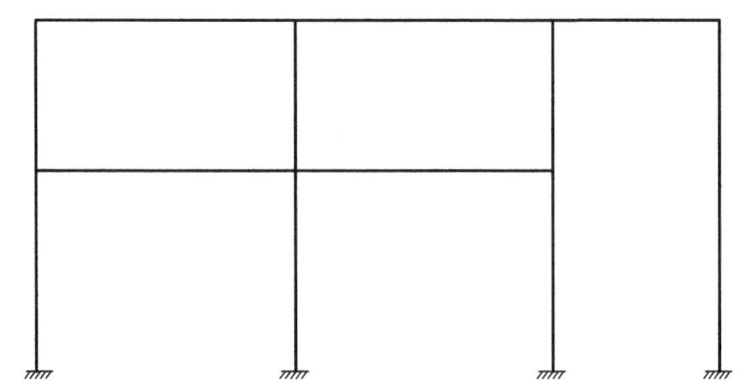

Fig. 8.6 Framing system with nonuniform stories.

It is useful to note that while B_1 is based on member properties, B_2 is a story-based stiffness. In moment frames, the sidesway instability is a modeled as a story phenomenon rather than a member phenomenon. The basis of the stability provisions for sway frames is that no single column can buckle in a sidesway mode; instead, all columns in a story buckle simultaneously. Thus B_1 must be calculated for each member, while B_2 is calculated at each story level. Unfortunately, no provisions are given in the specification for buildings in which the stories do not line up—for example, the frame shown in Figure 8.6. In this type of frame, the B_2 factor for the right side column cannot be easily calculated from the specification approach. In frames such as this, a direct second-order analysis is typically the more prudent choice.

Since many frames are drift controlled, it is possible to calculate P_{e2} based on a target drift limit rather than a first order analysis. For example, if the target drift is $\frac{L}{\Delta} = 500$, then $P_{e2} = 0.85(500) \sum H$, which can be used to approximate B_2. Since B_2 is a story stiffness, any lateral load may be used to determine its value in a first-order analysis. It is often convenient to simply use a predetermined percentage of the total gravity load.

It is important to note that the second-order amplification affects not only the beam column, but also the moments in any adjoining members and connections as required by equilibrium. Per the commentary to Chapter C:

> The associated second-order internal moments in connected members can be calculated satisfactorily in most cases by amplifying the moments in the members of the lateral load resisting system, in other words, the columns and the beams, by their corresponding B_1 and B_2 values. For beam members, the larger of the B_2 values from the story above or below is used.

While more time-consuming, a better estimate of the moments in adjoining members can be achieved by considering free body diagrams of the joints, and appropriate distribution of the amplified moments. In general, the approximate approach is preferable for design purposes, unless a more

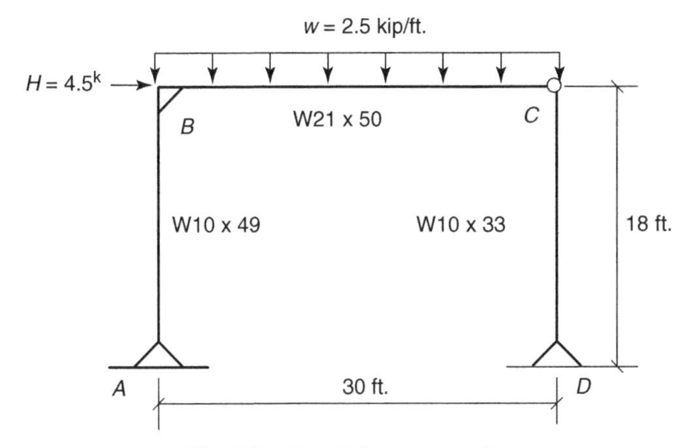

Fig. 8.7 Portal frame example.

exacting analysis is required. Of course, in those instances, it may be more appropriate to simply use a direct second-order approach.

8.5.2 Example 8.2: Second-order Amplified Moments

The simple portal frame shown in Figure 8.7 is used in a number of examples throughout the chapter to illustrate the application of the stability provisions in the specification. The loads given are the factored loads. We start by determining the second-order axial forces and moments in the columns from the NT–LT approach using amplification factors B_1 and B_2. We then compare these to the second order forces and moments in the system using a direct second-order analysis.

To begin the NT–LT process, we first analyze the nonsway frame by fixing joint C against translation and applying the factored loads as shown in Figure 8.8. After a first-order analysis is run, we record the moments and axial loads in the columns, as well as the reaction at joint C. The forces and moments are shown in Table 8.1, and the reaction $R = 9.42$ kips.

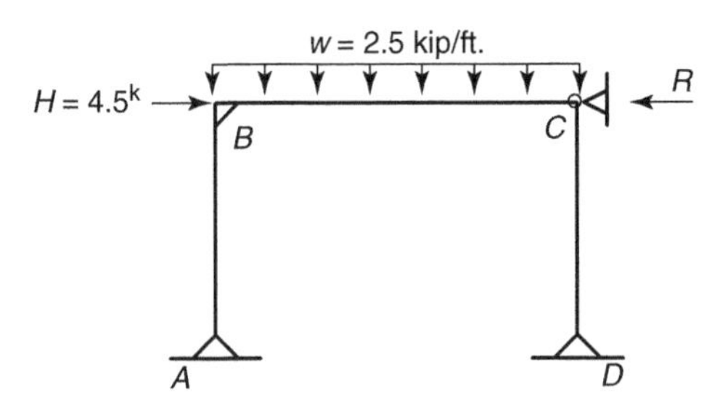

Fig. 8.8 NT analysis model.

TABLE 8.1 NT–LT Analysis Results

	NT Analysis		LT Analysis	
	P_{NT} (kip)	M_{NT} (in-kip)	P_{LT} (kip)	M_{LT} (in-kip)
Col *AB*	–40.5	$-1,063$	5.7	2035
Col *CD*	–34.5	0	–5.7	0

For the LT analysis, an equal and opposite reaction at C is applied to the sway permitted model of the frame, as shown in Figure 8.9.

One the analyses are complete, we can calculate the B_1 and B_2 factors.

B_1 is given by

$$B_1 = \frac{C_m}{1 - \dfrac{\alpha P_r}{P_{e1}}} \geq 1.0$$

where

$$C_m = 0.6 - 0.4\left(\frac{M_1}{M_2}\right) = 0.6 - 0.4 \cdot \left(\frac{0}{303.8}\right) = 0.6$$

and

$$P_{e1} = \frac{\pi^2 EI}{(K_1 L)^2} = \frac{\pi^2 \cdot (29,000)(272)}{[(1.0)(18)(12)]^2} = 1{,}670\,\text{kips (using}\,K_1 = 1.0)$$

For column *AB*

$$P_r = P_{\text{nt}} + P_{\text{lt}} = -40.5 + 5.7 = 34.8\,\text{kips}$$

$$B_1 = \frac{0.6}{1 - \dfrac{(1.0)34.8}{1670}} = 0.61 \rightarrow B_1 = 1.0$$

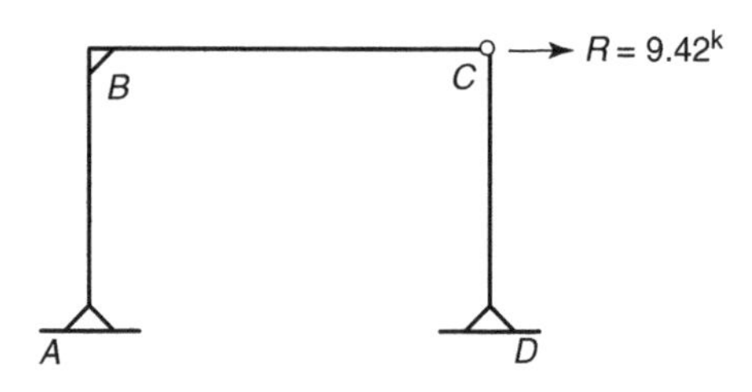

Fig. 8.9 LT analysis model.

A value of $B_1 < 1$ indicates that no amplification of the M_{NT} moment is required.

Next, we determine the second-order amplification of the sway moments by calculating

$$B_2 = \frac{1}{1 - \frac{\alpha \sum P_{NT}}{\sum P_{e2}}} \geq 1.0, \text{ where } \alpha = 1.0 \text{ (factored loading used)}$$

$$\sum P_{NT} = (40.5 + 35.5) = 75 \text{ kips}$$

$$\sum P_{e2} = \sum \frac{\pi^2 EI}{(K_2 L)^2} = R_M \frac{\sum HL}{\Delta_H}$$

where $R_m = 0.85$ for moment frames

Using the second form of the $\sum P_{e2}$ equation based on the LT analysis results, where $\Delta_H = 5.87$ in.

$$\sum P_{e2} = \frac{\sum HL}{\Delta_H} = \frac{0.85(9.42)(18')(12''/\text{ft.})}{5.87''} = 295 \text{ kips}$$

$$B_2 = \frac{1}{1 - \frac{75}{295}} = 1.34$$

The final forces are given by

$$M_r = B_1 M_{nt} + B_2 M_{lt}$$
$$P_r = P_{nt} + B_2 P_{lt}$$

and are tabulated in Table 8.2, where it can be seen that, in this case, they compare conservatively to the direct second-order elastic analysis results.

TABLE 8.2 Second-order Moment and Axial Load Values

	B_1-B_2 Analysis		Second-order Analysis	
	P_r (kip)	M_r (in-kip)	P_r (kip)	M_r (in-kip)
Col *AB*	−33.4	1664	−33	1534
Col *CD*	−42.1	0	−42	0

8.5.3 Design by Second-order Analysis, the Critical-load Method (C2.2a)

The critical-load approach should be familiar to designers as the traditional effective length (K-factor) approach from previous specifications. There are two differences in the 2005 specification from previous specifications:

1. The ratio of second-order to first-order drift must be less than or equal to 1.5; that is:

$$\Delta_2/\Delta_1 \leq 1.5$$

2. All gravity-only load combinations must include a minimum lateral load at each level equal to

$$N_i = 0.002Y_i$$

where

$$Y_i = \text{design gravity load at each level}$$

Physically, the lateral load, often referred to as a *notional load*, accounts for the possible out-of-plumbness that may be present in the erected structure based on the limits set in the AISC *Manual of Standard Practice*. Philosophically, it is often considered to be a minimum lateral load to be placed on the system. Regardless, it provides a small amount of perturbation to a system, which allows for a more robust solution in the case that a direct second-order (or nonlinear) analysis algorithm is employed. It also ensures that there is some amplification of moments in symmetric systems. This amplification exists in realistic (imperfect) structures, but an analysis would not capture any of this amplification under gravity load. For example, the beam in the frame in Figure 8.10 would not have an internal moment if the perfect system were analyzed, but the imperfection creates a moment of $P - \Delta_0$ as well as beam shears. The minimum horizontal load ensures that these forces and moments are accounted for in the design.

Second-order analysis results must be used in this approach. This is regardless of whether the ASD or LRFD philosophy is applied to the design. Any of the previously discussed means of determining second-order moments and axial loads is acceptable.

The focus on the examples in this chapter is application of the LRFD philosophy, although the specification allows design using either ASD or LRFD. There is one important point to make if the ASD checks are used. The methods outlined in the 2005 specification have, for the most part, been

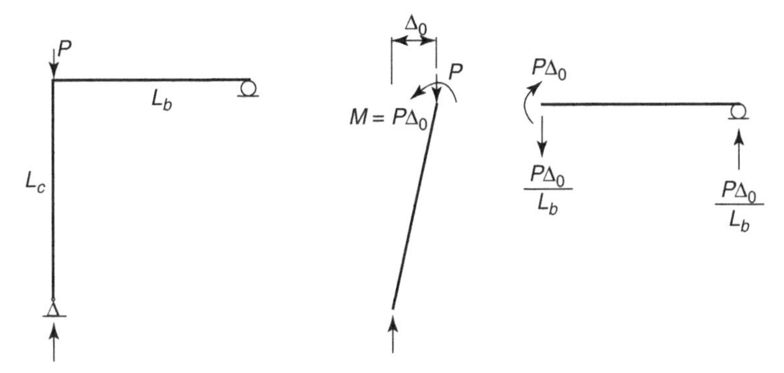

Fig. 8.10 Second-order moments and shears developed from imperfections.

validated using factored load levels. Because second-order analysis methods are nonlinear in nature, the system response (or more particularly, the second-order amplification) is sensitive to whether factored loads cases or service load cases are used. Therefore, for ASD load cases, it is necessary to follow the requirements of section C2.2a.2:

> For design by ASD, analyses shall be carried out using 1.6 times the ASD load combinations, and the results shall be divided by 1.6 to obtain the required strengths.

For the second-order method, if $\Delta_2/\Delta_1 \leq 1.1$, the *K*-factor may be set equal to 1.0. Although not explicitly stated in the specification, it is presumed here that the second-order drift limits must be met at *all* levels of the building. That is, a uniform analysis and design check approach should be used over the entire building or structure and not varied from level to level based on individual story drift ratios. If the drift ratio is greater than 1.1, then the *K*-factor must be calculated to find the axial capacity using equations E3-2 through E3-4 (section 8.3).

The interaction formulas are then used to check the capacity of the beam-column in order to ensure that they can withstand the combined load. When biaxial bending and compression occur and are dependent on the magnitude of the axial load, equations H1-1a and H1-1b can be used to check the interaction. When bending occurs along only one axis, equation H1-2 should be used.

The second-order *critical load approach* can be summarized as follows:

Step 1: Verify that the second- to first-order drift ratio for all stories is less than 1.5.

Step 2: For the gravity load case(s), calculate the minimum horizontal loads to be applied at each level, equal to 0.2 percent of the factored gravity load at each level (for ASD, the factored gravity load = 1.6x the ASD load combination)

Step 3a (Direct second-order approach): Run a second-order analysis for each load case, and determine the required axial loads (P_r) and moments (M_r) for each member.

Step 3b (Approximate second-order approach): Begin by running NT and LT analyses for the structure and calculate the B_1 and B_2 factors. Run all required load cases. Use equations C2-1a and C2-1b to find the required second-order moments (M_r) and axial loads (P_r).

Step 4: Calculate K-factors for each member (see section 8.6).

Step 5: Determine the member axial and moment capacities of each member. Check the capacity of each member versus the required loads.

8.5.4 First-order Elastic Analysis with Amplification Factors (C2.2b)

The first-order approach is similar to notional load approaches currently allowed in many international specifications (e.g., CSA, Eurocode). The use of this approach in the AISC specification is limited to stiff framing systems in which the amplification is less than 10 percent, as established by the first- to second-order drift ratio:

$$\frac{\Delta_2}{\Delta_1} \leq 1.1$$

This ratio is similar in magnitude to the B_2 amplifier discussed in section 8.5. In addition, for members in the lateral resisting system, moment amplification along the length of the member (not captured by a B_2 amplification) is limited by placing a maximum value on the axial load of the column to less than 50 percent of the squash load. This approach also allows for the simplification that the actual member length may be used in the column strength curve, and the calculation of K-factors is not required. Specifically, this approach may be used if the following three criteria are satisfied as described in C2.2b of the specification:

1. The required axial compressive strength, P_r, must be less than or equal to one-half of the yield strength, P_y, of the member:

$$\alpha P_r \leq 0.5 P_y \qquad \text{(AISC C2-7)}$$

This relationship needs only to be satisfied by members that contribute to the lateral stability of the structure.

2. All load combinations must include an additional load, N_i, at each level.

$$N_i = 2.1\left(\frac{\Delta}{L}\right)Y_i \geq 0.0042Y_i \qquad \text{(AISC C2-8)}$$

3. The B_1 moment amplification factor must be used with the total moments in order to account for nonsway members. B_1 can be calculated according to equation C2.2.

As previously mentioned, it is not explicitly stated, but is presumed that the three criteria listed must apply to *all* members and/or all stories in the structure. The application of multiple approaches to the same structure is not allowed.

The first-order approach can be summarized as follows:

Step 1: Verify that the second- to first-order drift ratio for all stories is less than 1.1.

Step 2: Calculate the minimum horizontal loads to be applied at each level:

$$N_i = 2.1\left(\frac{\Delta}{L}\right)Y_i \geq 0.0042Y_i$$

Step 3: Run the first-order analysis, calculate B_1 factors, and find the required axial loads (P_r) and moments $(M_r = B_1 M_1)$.

Step 4: Determine the member axial and moment capacities of each member. Check the capacity of each member versus the required loads. Axial capacities are calculated using the actual member length; that is, $K = 1$.

As with any method requiring the calculation of story based parameters (in this case, $\frac{\Delta_2}{\Delta_1}$), it can be difficult to prove satisfaction of the first criterion for structures in which there is variation in the story heights, such as multiple-story columns or mismatched floor heights in the same story. However, for relatively stiff structures with regular geometry, this provides a simple approach to assess the frame strength.

8.5.5 Direct Analysis (Appendix 7)

The most comprehensive of the three approaches is discussed in Appendix 7 of the AISC *Specification*. It is applicable to any type of framing system and for any stiffness of framing (as measured by the second-order to first-order drift ratio), and eliminates the need for the calculation of story-based factors. The *direct analysis* approach is based on a simple principle: If the parameters that affect member and system strength can be directly modeled, to the extent possible, within an elastic analysis, the overall simplicity and rationality of the elastic analysis-design calculations is improved. The parameters include in the approach are residual stresses, initial geometric imperfections, and boundary conditions. These are based on the *Structural Stability Research Council Technical Memorandum 5* (Galambos 1998), which states the stability design of metal structures must account for all the attributes that "influence significantly the load-resisting capacity of a frame, member or element." These include (but are not limited to the following):

1. "Experimentally determined physical characteristics, such as residual stresses, material nonlinearities, . . . , rationalized as may be appropriate.
2. A statistically appropriate combination of acceptable characteristics that are specified in supply, fabrication, and erection standards, such as out-of-straightness . . .
3. Effect of boundary conditions, such as restraint applied to the end of members."

Effects that are difficult to model or are not easily captured in the analysis, such as member out-of-straightness, are accounted for in the member strength equations. Two modifications are made that allow the analysis to more closely predict the internal forces that would be obtained from a rigorous nonlinear, inelastic analysis. These include a reduction of the member stiffness and direct modeling of the initial out-of-plumbness. Boundary conditions are captured in the second-order analysis.

The imperfection can either be directly modeled as an $L/500$ out-of-plumbness at each story level, or by means of an equivalent notional load, N_i, at each level given by

$$N_i = 0.002Y_i$$

This is the same notional load required for gravity load cases in the critical-load approach.

The reduction in flexural stiffness accounts for both the general yielding that may occur at design load levels, as well as the effects of residual stresses on column inelasticity. In particular, the reduction due to the residual stress is handled on a member-by-member basis through the use of the inelastic stiffness reduction factor, or tau factor. For all members, the stiffness is reduced by 0.8. This can be most easily handled by reducing the elastic modulus to a value of $0.8E$. In general, the flexural stiffness is given by

$$EI^* = 0.8\tau_b EI \qquad \text{(AISC A-7-2)}$$

The inelastic stiffness reduction factor used in this approach is given as follows:

$$\text{for} \quad \frac{\alpha P_r}{P_y} \leq 0.5, \ \tau_b = 1.0$$

$$\text{for} \quad \frac{\alpha P_r}{P_y} > 0.5, \ \tau_b = 4\left[\frac{\alpha P_r}{P_y}\left(1 - \frac{\alpha P_r}{P_y}\right)\right]$$

where $\alpha = 1.0$ for LRFD and $\alpha = 1.6$ for ASD.

For most members in lateral load resisting systems, the axial load will be less than $0.5P_y$; only in rare instances will a tau factor be required for a specific member in the lateral resisting system. In reality, only members involved in the lateral resisting system need to have their flexural stiffness reduced, but it is usually easier to simply set $E = 0.8(29{,}000)$ than to apply the reduction on a member-by-member basis. In the notional load approaches used in many international specifications, the inelastic stiffness reduction is included through an increased lateral load rather than applied as-needed on a member-to-member basis. Although checking the axial load on each beam-column is more labor intensive, it provides the benefit of only applying the stiffness reduction to the locations where it is necessary. This provides better accuracy with respect to the distribution of forces and moments in the system and, on average, less error than a general notional load approach (Surovek-Maleck and White 2003).

The direct analysis approach can be summarized as follows:

Step 1: For each load case, calculate the minimum horizontal loads to be applied at each level, equal to 0.2 percent of the factored gravity load at each level:

$$N_i = 0.002\Sigma Y_i$$

Alternatively, model the frame with an $H/500$ out-of-plumbness.

Step 2: Run a second-order analysis for each load case, and determine the required axial loads (P_r) and moments (M_r) for each member. Run the analysis using a reduced stiffness of $0.8EI$.

Step 3: Check the axial load levels in the columns. If necessary, calculate τ_b and apply an additional reduction of the flexural stiffness of columns in the lateral resisting system that have an axial load greater than 50 percent of the squash load ($EI^* = 0.8\tau_b EI$). This step should not be needed often, and no further iterations of the analysis are required past this point.

Step 4: Determine the member axial and moment capacities of each member. Check the capacity of each member versus the required loads using $K = 1$.

It is allowed in Appendix 7 of the 2005 AISC *Specification* to treat the notional load as a minimum lateral load that can be neglected in the presence of a larger lateral load, provided that $\frac{\Delta_2}{\Delta_1} < 1.5$.

For frames that have lower second-order amplification and $\frac{\sum H_i}{\sum Y_i} \geq 0.002$, the impact of the notional load is negligible to within tolerable error levels. It is worth noting, though, that if a direct second-order analysis is being used, it is easier to simply include the notional loads rather than to perform separate first- and second-order analyses to see if it can be neglected, and inclusion of the load had limited impact on the economy of the structure.

From a purist standpoint, the imperfections being modeled would be in place regardless of whether a lateral load were present, and should be modeled. If the designer wishes to avoid calculation of the notional loads for multiple load cases, the geometry of the frame can be altered to directly model an *H*/500 out-of-plumbness. This is of particular benefit in frames where the story load is not easily calculated, as previously discussed.

There is one caveat that must be placed on use of the direct analysis approach. The method can be sensitive to $P - \delta$ effects when direct second-order analysis is used. If an algorithms that neglects $P - \delta$ effects is to be used, the beam-column axial forces must meet the following limit:

$$\alpha P_r < 0.15 P_e$$

This limit should be met by most beam-columns in practical lateral resisting systems. Only those columns whose stiffnesses contribute to the lateral stability of the system must meet this limit.

8.6 EFFECTIVE LENGTH FACTORS, K-FACTORS

In order to compute the correct *K*-factors for a structure, it is necessary to run either an elastic or inelastic buckling analysis of the structure. Few (if any) commercial structural design programs have this capability, so approximate methods for determining *K*-factors are provided in the commentary to

Chapter C of the Specification. The theoretical development of the alignment charts was discussed in Chapter 2, section 2.8. There are two charts, one for braced frames (sidesway inhibited) and one for sway frames (sidesway uninhibited). These are shown in Figures 8.11 and 8.12, respectively. The charts are a graphical representation of the equations developed for K that include the G factors at the top and bottom of the columns. The G factors are a measure of the relative stiffness of the beams and columns at the joint, and thus the restraint provided by the adjacent members to the column being considered.

For braced frames:

$$\frac{\left(\frac{\pi}{K}\right)^2 G_A G_B}{4} - 1 + \frac{G_A + G_B}{2}\left(1 - \frac{\frac{\pi}{K}}{\tan\frac{\pi}{K}}\right) + \frac{2\tan\frac{\pi}{2K}}{\frac{\pi}{K}} = 0$$

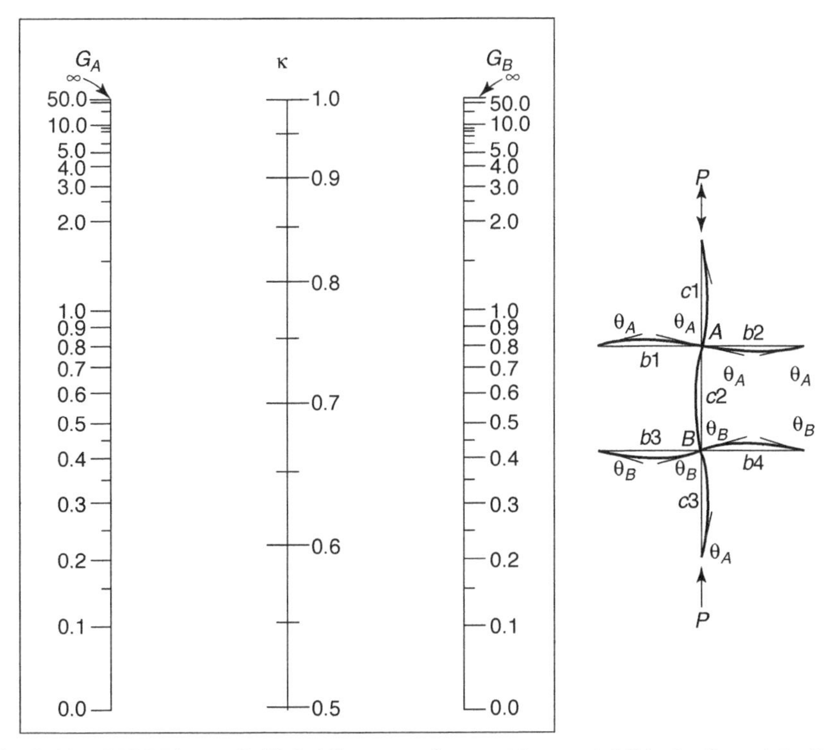

Fig. 8.11 AISC Figure C-C2.3 Alignment chart—sidesway inhibited.

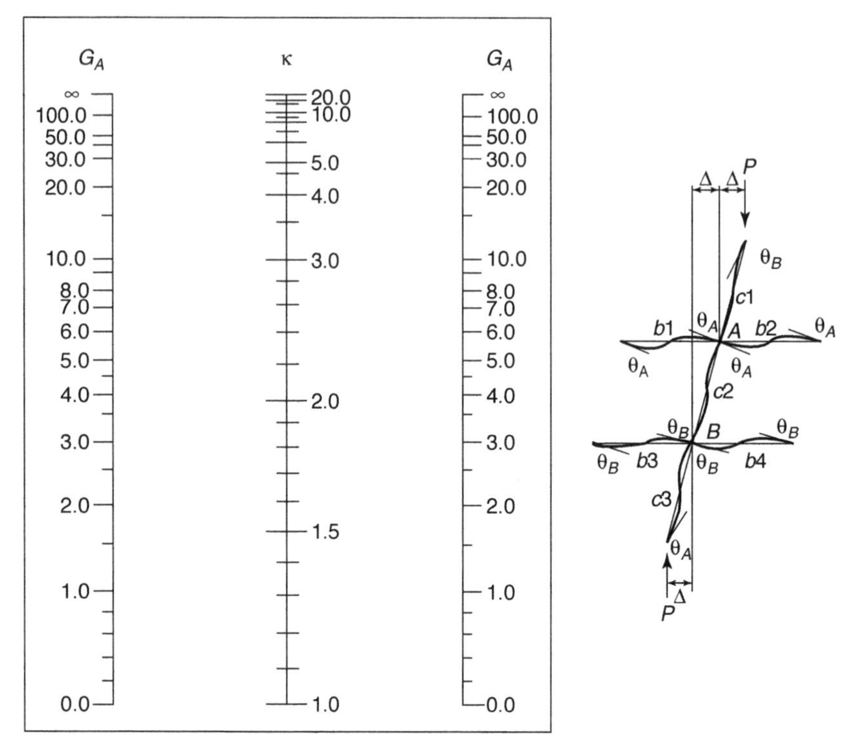

Fig. 8.12 AISC Figure C-C2.4 Alignment chart—sidesway uninhibited. Copyright © American Institute of Steel Construction, Inc. Reprinted with permission. All rights reserved.

For sway frames:

$$\frac{\dfrac{\pi}{K}}{\tan\left(\dfrac{\pi}{K}\right)} - \frac{\left(\dfrac{\pi}{K}\right)^2 G_T G_B - 36}{6(G_T + G_B)} = 0$$

The G values represent a measure of the restraint at the top and bottom of the column and are given by:

$$G = \frac{\Sigma\left(\dfrac{I}{L}\right)_c}{\Sigma\left(\dfrac{I}{L}\right)_g}$$

Once the values of G are calculated at the top and bottom joints, K can be calculated directly from the transcendental equations, or they can be determined graphically from the alignment charts.

In applying the alignment charts to determine the effective length factor K, it is important to recognize the assumptions that are inherent in these equations and the accompanying charts. These are provide in the Commentary to Chapter C and include the following:

1. The behavior is purely elastic.
2. All members have constant cross-section.
3. All joints are rigid.
4. For columns in sidesway inhibited frames, rotations at opposite ends of the restraining beams are equal in magnitude and opposite in direction, producing single curvature bending (see Figure 8.11).
5. For columns in sidesway uninhibited frames, rotations at opposite ends of the restraining beams are equal in magnitude and direction, producing reverse curvature bending (see Figure 8.12).

These assumptions are almost never strictly met by realistic frames. Accurately using the alignment chart requires modifications to account for variations in the frame behavior from those in the assumptions.

8.6.1 Behavior Is Purely Elastic

The first assumption is that all behavior is elastic. If the column is yielding at the onset of buckling, then the column stiffness is reduced. In this instance, the beams provide more relative restraint to the columns than in the elastic case, causing a lower G-factor and, consequently, a lower K-factor. Thus the following modification to G considers inelasticity in the column due primarily to residual stresses:

$$G = \frac{\Sigma(\tau_a EI/L)_c}{\Sigma(EI/L)_g}$$

The variable τ_a is the inelastic stiffness reduction factor. It is interesting to note that the specification actually contains two different formulas for inelastic stiffness reduction, the one previously discussed in section 8.5 that is applied in the direct-analysis approach, and the one provided here for column stiffness reduction. This inelastic stiffness reduction factor is also based on the ratio of applied axial load to the squash load of the column, as follows:

$$\text{For} \quad P_n/P_y \leq 0.39 \text{ (elastic)} \qquad \text{(AISC C-C2-12)}$$

$$\tau_a = 1.0$$

$$\text{For} \quad P_n/P_y > 0.39 \text{ (inelastic)}$$

$$\tau_a = -2.724(P_n/P_y)\ln(P_n/P_y)$$

Because the stiffness reduction factor is a function of P_n, the process of finding the inelastic K-factor is iterative. A conservative approach is to use $\frac{\alpha Pu}{\phi}$ in place of P_n when calculating τ_a. Inelastic K-factors are always lower than the elastic K-factor for cases where $\tau_a < 1$. For this reason, it is conservative to calculate the elastic K-factor in columns that experience inelasticity prior to buckling.

8.6.2 Rotations at Opposite Ends of the Restraining Beams Are Equal in Magnitude and Direction

Only in perfectly symmetric sway frames are the rotations at the ends of the beams equal in magnitude and direction. Because the restraint provided by the beam to the column is affected by the end rotation of the beams, it is prudent to account for variations from the original assumption of perfect double curvature bending. Again, the effect is on the beam restraint, and it is accounted for in the value of G as follows:

$$G = \frac{\sum\left(\frac{I}{L}\right)_c}{\sum\left(\frac{I}{L'}\right)_g}$$

where

$$L'_g = L_g\left(2 - \frac{M_F}{M_N}\right) \qquad \text{(AISC Commentary p. 243)}$$

The values M_F and M_N are the moments from the far and near end of the beam, respectively, based on a first-order *lateral* analysis. The ratio M_F/M_N is positive in double-curvature bending. Certain special cases are easily calculated:

$$\text{If the far end is pinned } L'_g = 2L_g$$
$$\text{If the far end is fixed, } L'_g = 1.5L_g$$

For instances in which $M_F > 2M_N$, L'_g becomes negative and the equation (equation 2.56) for K should be used.

For sway inhibited frames, two adjustments to the girder length are provided:

$$\text{If the far end is fixed, } L'_g = 0.5L$$
$$\text{If the far end is pinned, } L'_g = 2/3L.$$

It is also important to note that the original derivation assumes that all connections are fixed. If the girder to column connection is a shear tab or otherwise classified as a pinned connection, the girder stiffness does not contribute to the restraint of the column, and $(\frac{EI}{L})_g = 0$.

8.6.3 Example 8.3 Inelastic Effective Length Factor Calculation

Figure 8.13 shows a W12 × 58 column with heavy axial load connected to a W14 × 22 beam. The inelastic *K*-factor for the beam can be determined by using the alignment chart approach. First, we adjust the girder length to account for the pinned end:

$$L_g' = L_g\left(2 - \frac{M_F}{M_N}\right) = 20(2) = 40'$$

The adjustment in girder length accounts for the stiffness the girder provides in restraining the column due to the end condition. We can then find the inelastic stiffness reduction factor for the column based on the ratio of the axial load to the squash load of the column:

$$\frac{\left(\dfrac{\alpha P}{\phi}\right)}{P_y} = \frac{\left(\dfrac{1.0\,(450\,\text{kips})}{0.9}\right)}{(17\,\text{in.}^2)(50\,\text{ksi})} = 0.588 > 0.39$$

$$\tau_a = -2.724\left(\frac{P}{P_y}\right)\ln\left(\frac{P}{P_y}\right) = -2.724(0.588)\ln(0.588) = 0.85$$

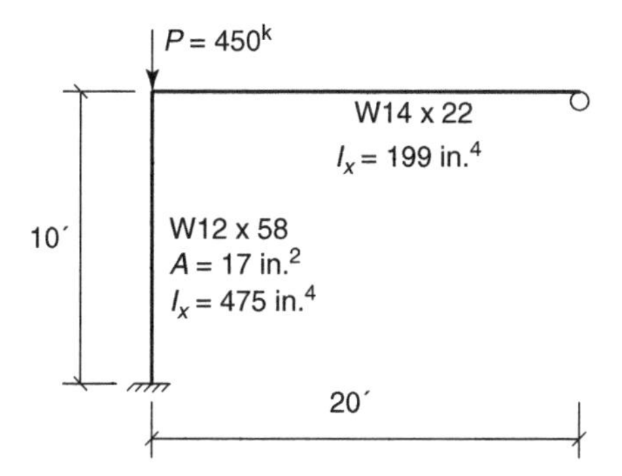

Fig. 8.13 *K*-factor example using inelastic stiffness.

The G values are then calculated as

$$G_T = \frac{\sum\left(\frac{\tau_a I}{L}\right)_c}{\sum\left(\frac{I}{L'}\right)_g} = \frac{\frac{(0.85)475}{10}}{\frac{199}{40}} = 8.12 \qquad G_B = 1.0$$

From the alignment chart, the approximate value of K can be found to be $K_x = 1.85$ If the inelastic stiffness reduction factor is neglected, then

$$G_T = \frac{\sum\left(\frac{I}{L}\right)_c}{\sum\left(\frac{I}{L'}\right)_g} = \frac{\frac{475}{10}}{\frac{199}{40}} = 9.5 \qquad G_B = 1.0$$

And $K_x = 1.9$

8.6.4 Girders Containing Significant Axial Load

The girder stiffness is reduced for high levels of axial load. In these instances, the value $(\frac{EI}{L})_g$ should be multiplied by the factor

$$\left(1 - \frac{Q}{Q_{\text{cr}}}\right)$$

where

$$Q = \text{axial load in the girder}$$

$$Q_{\text{cr}} = \text{in-plane buckling load of the girder based on } K = 1.0$$

8.6.5 Effect of Nonadjacent Framing on the Column Restraint—the Story Buckling Approach

The value G takes into account only the contribution of the adjacent members to the column, and this approach is based on the stability of the individual column. As previously discussed in section 8.5, the sidesway instability of sway frame is a story phenomenon. It is not possible for one column to buckle in a sidesway instability mode without the entire story buckling simultaneously. If each column acted independently to support its own axial load and second-order moments, then modifications due to nonadjacent framing would not be required. However, many framing systems redistribute P–Δ effects to members in the story based on individual member stiffness, and it is necessary to design lateral-resisting columns in a story to compensate for leaner columns (see Case V in Chapter 5). As stated in the Commentary to Chapter C:

> This approach focuses the engineer's attention on the most fundamental stability requirement in building frames, providing adequate overall story stiffness in relation to the total vertical load, $\alpha\Sigma P_r$, supported by the story.

Two approaches are presented for modifying K-factors of beam-columns in sway frames to account for the redistribution of P–Δ moments based on member stiffness, the story-stiffness approach (LeMessurier 1976) and the story-buckling approach (Yura 1971). The story-stiffness approach calculates K-factors based on the contributed stiffness of the column to the overall stiffness of the story, as follows:

$$K_2 = \sqrt{\frac{\sum P_r}{(0.85 + 0.15R_L)P_r}\left(\frac{\pi^2 EI}{L^2}\right)\left(\frac{\Delta_H}{\sum HL}\right)} \geq \sqrt{\frac{\pi^2 EI}{L^2}\left(\frac{\Delta_H}{1.7HL}\right)}$$

where

R_L = vertical load in all leaning columns/vertical load in all columns

For systems with a large percentage of leaning columns, this ratio approaches 1.0.

It is possible that the value $K_2 < 1.0$, depending on the relative stiffness of the column in the framing system. The minimum value on the right-hand side of the equation ensures that a nonsidesway buckling mode is prevented. This method requires the calculation of the first-order drift, but only requires that the K-factor be calculated once.

In the story-buckling approach, the sideway uninhibited K-factor, K_{n2}, is first calculated from the alignment chart approach, and then adjusted as follows:

$$K_2 = \sqrt{\frac{\left(\frac{\pi^2 EI}{L^2}\right)}{P_r} \cdot \frac{\sum P_r}{\sum\left[\frac{\pi^2 EI}{(K_{n2}L)^2}\right]}} \geq \sqrt{\frac{5}{8}}K_{n2}$$

Again, K_2 may be less than 1 for some beam-columns, but is limited by the right-hand side of the equation to prevent nonsidesway buckling. In calculating the story-buckling K, it is important to only include the stiffness of columns in the lateral-resisting system in the calculation of K_2. It is important to note that *neither* of these values for K_2 should be used in calculation P_{e2} to determine B_2.

8.6.6 Example 8.4 Effective Length Calculation with Leaner Column

Reconsidering the frame in Figure 8.7, we determine the K-factor for column AB. For this frame, we consider three possible variations from the assumptions built into the alignment chart:

1. The behavior is purely elastic.
2. All joints are rigid.
3. For columns in sidesway inhibited frames, rotations at opposite ends of the restraining beams are equal in magnitude and opposite in direction, producing single curvature bending.

From Example 8.3, we know that $P_r = 33^k$, so we can check whether an inelastic stiffness reduction factor is applicable

$$\frac{\left(\dfrac{\alpha P_r}{\phi_c}\right)}{(F_y A_g)} = \frac{\left(\dfrac{1.0(33)}{0.9}\right)}{50(14.4)} = 0.051 \leq 0.39$$

therefore, $\tau = 1.0$, and no inelastic stiffness reduction is applied.

The pinned joint at C requires the length of the beam to be modified before using the alignment chart approach:

$$L_g' = L_g\left(2 - \frac{M_F}{M_N}\right) = 30\left(2 - \frac{0}{77.76}\right) = 60'$$

Next we calculate the sway K-factor:

$$G_{\text{top}} = \frac{\sum\left(\dfrac{\tau_a I}{L}\right)_c}{\sum\left(\dfrac{I}{L}\right)_g} = \frac{\left(\dfrac{1.0(272)}{18}\right)}{\left(\dfrac{984}{60}\right)} = 0.92$$

$$G_{\text{bottom}} = 10.0(\text{pinned-end})$$

From alignment charts with $G_{\text{top}} = 0.92$ and $G_{\text{bottom}} = 10.0$, we obtain

$$K_x \cong 1.9$$

The K-value can be calculated more accurately using the transcendental equation that defines the sway-permitted alignment chart

$$\frac{\dfrac{\pi}{K}}{\tan\left(\dfrac{\pi}{K}\right)} - \frac{\left(\dfrac{\pi}{K}\right)^2(10)(0.95) - 36}{6(10.95)} = 0$$

which gives a value of $K_x = 1.89$.

This initial value of K_x must be adjusted to account for the leaning column, column CD. This can be done using either the story-stiffness or the story-buckling approach in the commentary. Using story stiffness, where K_2 is given by

$$K_2 = \sqrt{\frac{\sum P_r}{(0.85 + 0.15R_L)P_r}\left(\frac{\pi^2 EI}{L^2}\right)\left(\frac{\Delta_H}{\sum HL}\right)}$$

The ratio $\frac{\Delta_H}{\sum HL}$ was previously calculated to find B_2 in Example 8.2.

$$\frac{\Delta_H}{\sum HL} = \frac{1}{P_{e2}} = \frac{1}{295} = 0.0034$$

The term $\sum P_r$ is the summation of the total axial load on all columns in the story.

$$\sum P_r = 75\,\text{kips}$$

$$R_L = \frac{\sum P_r\text{\_leaning\_columns}}{\sum P_r\text{\_all\_column}} = \frac{42\,\text{kips}}{75\,\text{kips}} = 0.56$$

$$K_2 = \sqrt{\frac{\sum P_r}{(0.85 + 0.15R_L)P_r}\left(\frac{\pi^2 EI}{L^2}\right)\left(\frac{\Delta_H}{\sum HL}\right)}$$

$$= \sqrt{\frac{75}{(0.85 + 0.15(0.56))33}\left(\frac{\pi^2(29{,}000)(272)}{(18 \times 12)^2}\right)(0.0034)} = 3.7$$

The K-factor calculated by the story-buckling approach is given by

$$K_2 = \sqrt{\frac{\left(\frac{\pi^2 EI}{L^2}\right)}{P_r}\left(\frac{\sum P_r}{\sum\left(\frac{\pi^2 EI}{(K_{n2}L)^2}\right)}\right)} = \sqrt{\frac{\left(\frac{\pi^2(29000)(272)}{(216^2)}\right)}{33}\frac{75}{\left(\frac{\pi^2 29000(272)}{(1.95(216))^2}\right)}}$$

$$= \sqrt{\frac{(1.95)^2(75)}{33}} = 2.94$$

Based on an elastic buckling analysis of the frame, the elastic critical buckling load, $P_{cr} = 165$ kips, and the actual K-factor for column, is given by

$$K = \sqrt{\frac{P_e}{P_{cr}}} = \sqrt{\frac{\left(\frac{\pi^2 EI}{L^2}\right)}{P_{cr}}} = \sqrt{\frac{1669}{165.3}} = 3.18$$

The actual value lies between the calculated story-stiffness and story-buckling values.

8.7 DESIGN ASSESSMENT BY TWO APPROACHES

In this section, we compare the direct-analysis approach to the critical load or K-factor approach using the portal frame introduced in Example 8.2. In this example, we cannot use the first-order approach because the B_2 factor (calculated in Example 8.2) is greater than 1.1. While Δ_2/Δ_1 and B_2 are not the exact same value (drift amplification versus moment amplification), B_2 is allowed as an approximation of the drift ratio.

8.7.1 Critical Load Approach

In order to perform the strength check, we use the results of Example 8.2 with the B_1 and B_2 amplifiers and the K-factor values determined in Example 8.4. For this example, we assume the frame is braced in the out-of-plane direction. We check the sidesway stability of the frame by checking the lateral resisting member, Column AB. As with the previous example, we focus on the factored lateral load case. The steps are provided in section 8.5.3:

Step 1: Verify that the second- to first-order drift ratio for all stories is less than 1.5.

From Example 8.2, $B_2 = 1.28$, so this approach is permitted.

Step 2: Calculate the minimum horizontal loads to be applied at each level for the gravity load case, equal to 0.2 percent of the factored gravity load at each level.

For this example, we are checking the lateral load case only, and this calculation is not required. We perform this calculation when using the direct-analysis method.

Step 3a: Determine the axial loads (P_r) and moments (M_r) for each member. From Example 8.2, we know that internal axial load and moment for member *AB* are

$$P_r = 33\,\text{kips} \quad M_r = 1534\,\text{in-kips}$$

Step 4: Calculate *K*-factors for each member
we use the *K*-factor value determined from the elastic buckling analysis in Example 8.3, although either K_2 value could be used.

$$K_y = 1.0\ (\text{assuming out-of-plane-bracing})$$
$$K_x = 3.18$$

Step 5: Determine the member axial and moment capacities of each member. Check the capacity of each member versus the required loads.

Axial Load Capacity

$$\frac{K_y L}{r_y} = \frac{1.0(18')(12''/\text{ft.})}{2.54} = 85$$
$$\frac{K_x L}{r_x} = \frac{3.18(18')(12'/\text{ft.})}{4.35} = 158$$

Sidesway buckling in-plane controls:

$$4.71\sqrt{\frac{E}{F_y}} = 113.4 \quad \text{therefore } F_{cr} = 0.877 F_e$$

$$F_{cr} = 0.877\frac{\pi^2 E}{\left(\frac{KL}{r}\right)^2} = 0.877\frac{\pi^2(29000)}{(157.9)^2} = 10.1\,\text{ksi}$$

$$\phi P_n = 0.9(10.1\,\text{ksi})(14.4\,\text{in.}^2) = 130\,\text{kips}$$

Moment Capacity

Using the AISC Beam Tables (AISC Section 3), for a W10 × 49

$$L_p = 8.97' \quad L_r = 31.6' \quad \phi M_p = 227\,\text{ft-kip} \quad \phi M_r = 143\,\text{ft-kip}$$

We calculate C_b based on a linear moment diagram (from $M = 0$ to $M = 1542$ in-kips)

$$C_b = 1.67$$

We calculate the moment capacity to be

$$\phi M_n = 1.67\left[227 - (227 - 143)\left(\frac{18 - 8.97}{31.6 - 18}\right)\right] = 285\text{ ft-kip} > M_p$$

therefore:

$$\phi M_n = 227\,\text{ft-kip}$$
$$\phi M_n = 2724\,\text{in-kip}$$

Interaction Check

$$\frac{P_r}{P_c} = \frac{33}{130} = 0.25 > 0.2$$

We use equation H1-1a

$$\frac{P_r}{P_c} + \frac{8}{9}\left(\frac{M_{rx}}{M_{cx}} + \frac{M_{ry}}{M_{cy}}\right) \le 1.0$$

$$0.25 + \frac{8}{9}\left(\frac{1534}{2724}\right) = 0.75 < 1.0$$

8.7.2 Direct Analysis Approach

The primary difference between the critical load approach and the direct analysis approach is the need to calculate K-factors in the first approach. We check the same column following the steps laid out in section 8.5.5.

Step 1: Calculate the minimum horizontal loads to be applied at each level, equal to 0.2 percent of the factored gravity load at each level.

The option is provided in Appendix 7 of the AISC Specification that if the drift ratio is less than 1.5, then the lateral load may be neglected in the presence of a higher lateral load. We include the notional load

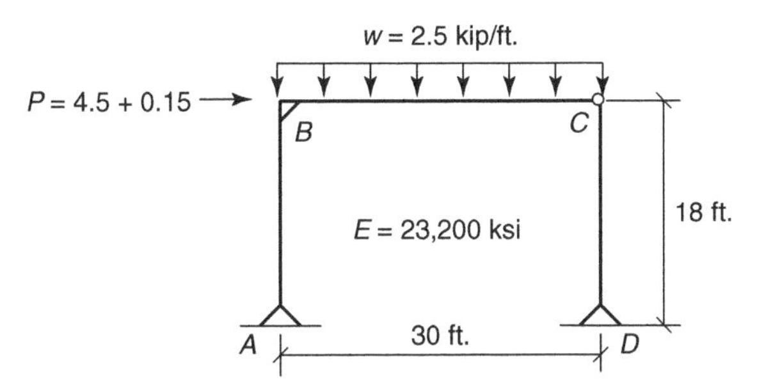

Fig. 8.14 Direct analysis model with notional load.

in this example in order to have an example demonstrating the method. In addition, we use a direct second-order analysis to calculate the moments and axial forces, rather than the NT–LT approach. In this instance, we would not calculate the B_2 factor as part of the analysis. The effort involved in calculating the notional load is substantially less than that involved in calculating a B_2 factor, and since the effect in the presence of a higher lateral load is nominal, it is does not have a significant effect on the economy of the system.

The notional load is given as 0.2 percent of the total gravity load on the story:

$$N_i = 0.002(2.5\,\text{k/ft.})(30') = 0.15\,\text{kip}$$

Step 2: Run a second-order analysis for each load case, and determine the required axial loads (P_r) and moments (M_r) for each member. Run the analysis using a reduced stiffness of $0.8EI$.

Our model for the frame is given in Figure 8.14. It is also appropriate to model the imperfection effect directly, rather than applying a notional load. This alleviates the need to calculate different notional loads for different load cases, although it requires a somewhat greater effort when modeling the system. It is also beneficial when considering frames, such as the example in Figure 8.6 that has nonuniform story levels. An example of the model with the imperfection directly modeled is shown in Figure 8.15.

Results from a second-order elastic analysis are

$$M_r = 1{,}770\,\text{in-kips}$$
$$P_r = 32.3\,\text{kips}$$

Step 3: Check the axial load levels in the columns to see if an inelastic stiffness reduction is required.

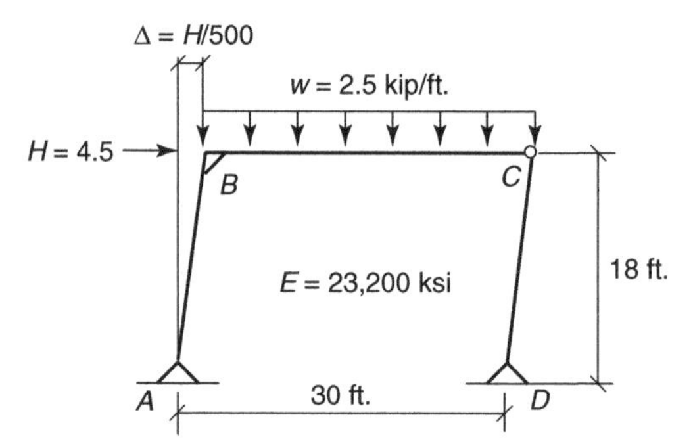

Fig. 8.15 Direct-analysis model with imperfection directly modeled.

In order to verify that a greater reduction in stiffness is not required, we check the ratio P/P_y for the columns in the lateral-resisting system. This can be estimated prior to running the analysis, or the check can be made after the analysis is run, and if necessary, a second analysis is run with the additional stiffness reduction. In most practical moment frames, this is necessary.

$$\frac{P}{P_y} = \frac{32.3}{(50)(14.4)} = 0.045 \ll 0.5$$

Since this ratio is less than 0.5, no additional stiffness reduction needs to be applied to the lateral-resisting columns.

Step 4: Determine the member axial and moment capacities of each member. Check the capacity of each member versus the required loads using $K = 1$.

Since the actual member length is used in the axial strength calculation (i.e., $K_x = 1.0$), we find that $\frac{K_y L}{r_y}$ controls the axial capacity check, and we can use the tabulated values from the AISC design manual. The moment capacity is the same as the previous calculation for the critical load approach, giving us

$$\phi P_n = 383 \text{ kips}$$
$$\phi M_n = 2724 \text{ in.} = \text{kips}$$

Performing the interaction check:

$$\frac{P_r}{P_c} = \frac{33.4}{383} = 0.08 < 0.2$$

So we use equation H1-1b:

$$\frac{P_r}{2P_c} + \left(\frac{M_{rx}}{M_{cx}} + \frac{M_{ry}}{M_{cy}}\right) \leq 1.0$$

$$\frac{1}{2}(0.087) + \left(\frac{1770}{2724}\right) = 0.69$$

8.8 FRAME DESIGN REQUIREMENTS IN CANADA AND EUROPE

In the previous sections of this chapter, methods of modern frame design according to the requirements of the AISC (2005) Specification are presented. The proportioning of framed structures in other modern design standards is very similar, with minor variations. In the following, we briefly examine the similarities and differences in two modern comparable codes: The Canadian S16.1 Standard (2004) and Eurocode 3 (2004).

The basis for the determination of the design forces in each member is an elastic second-order analysis, identical to the type of analysis already presented. Equilibrium is formulated on the deformed configuration of the frame in order to capture the destabilizing effects of the changed geometry of the structure and the unavoidable initial out-of-plumb. Each of these two codes prefers that the second-order analysis is performed directly with a computer program that incorporates these effects. However, it is permitted to calculate the design forces by a first-order analysis if they are then amplified to account approximately for the destabilizing second-order effects, much as the B_1–B_2 method presented. The method of obtaining the design forces is thus identical in the three specifications.

Each of the standards account for the unavoidable initial out-of-plumb by prescribing a notional lateral force that is applied at each story level. In the Canadian S16.1 code and in Eurocode 3, the magnitude of the required notional lateral load is equal to $0.005\Sigma P_{\text{gravity}}$, or 0.5 percent of the gravity load acting above the story under consideration. This is larger than the 0.2 percent in the AISC specification. However, the structural analysis in the AISC is performed on frames where 80 percent of the flexural and axial stiffness is used, while the other two codes analyze the frame with its actual stiffness values. This fact is one of the major differences between the three codes. The other difference is in the form of the interaction equations used to make the check of the beam-columns. The AISC interaction equations are presented in Chapter 4 and in this chapter. Following are the interaction equations of the three codes; these equations apply to in-plane bending of compact section beam-columns that are subject to flexure about the major axis of the member. The required design axial forces P_u and bending moments M_u are determined by a second-order analysis for the factored loads, including the effects of the initial out-of-plumb. The corresponding strength terms are P_n and M_n that are calculated with the appropriate strength equations in the respective standards. These expressions have been presented earlier in Chapters 3 and 6 for columns and beams, respectively. The resistance factor in the interaction equations is represented by the symbol ϕ. The plot of the interaction equations in Figure 8.16 shows the comparisons between the interaction equations. The AISC plot lies between the Canadian and the Eurocode plots. However, the differences are not dramatic. Comparative frame designs using the three different codes do not show significant differences in the required sections.

AISC interaction equations:

$$\frac{P_u}{\phi P_n} + \frac{8}{9}\frac{M_u}{\phi M_n} \leq 1.0 \quad \text{if} \quad \frac{P_u}{\phi P_n} \geq 0.2$$

$$\frac{P_u}{2\phi P_n} + \frac{M_u}{\phi M_n} \leq 1.0 \quad \text{if} \quad \frac{P_u}{\phi P_n} < 0.2$$

Canadian interaction equations:

$$\frac{P_u}{\phi P_n} + \frac{0.85 M_u}{\phi M_n} \leq 1.0$$

$$\frac{M_u}{\phi M_n} \leq 1.0$$

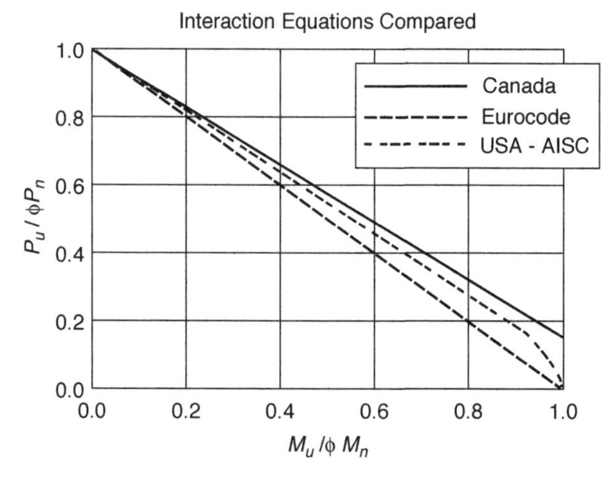

Fig. 8.16 Comparison of international interaction equations.

Eurocode 3 interaction equation:

$$\frac{P_u}{\phi P_n} + \frac{M_u}{\phi M_n} \leq 1.0$$

8.9 SUMMARY

The stability provisions in the AISC specification have been developed to provide a conservative or safe design based on theoretical behavior as well as experimental and analytical studies. The general approach is to assess frame stability on a member-by-member basis using the axial, flexural, and beam-column provisions.

Three approaches have been provided to assess frame or structural stability: the first-order method, the critical load or effective length approach and the direct analysis method. The three differ in the rigor of analysis required and the means by which imperfections and inelasticity are considered in the analysis and member strength assessment. Differing from past specifications, the effective length approach is limited to frames that meet a second- to first-order drift limit, and a notional load representing an initial imperfection is required in addition to the gravity loads in load cases absent of lateral loads.

PROBLEMS

8.1. Using the provisions of AISC Chapter F, determine the allowable uniform load that may be placed on a simply supported beam with a length of 20*d*, a depth of 16 in., a width of 1 in.

8.2. Determine the anchor points (P_n and M_n) and the transition point for the interaction diagrams for the following beam-columns:

- **a.** W 8 × 48, $L = 18$ ft.
- **b.** W14 × 120, $L = 16$ ft.
- **c.** HSS6 × 6 × 1/4, $L = 12$ ft.

8.3. The frame shown represents an 11-bay frame in which all but the two center columns are pinned. The applied force 4P on the exterior columns represents four bays of gravity load. Assess the adequacy of the W10 × 49 column in the frame shown using

- **a.** the second-order method using the NT–LT approach
- **b.** the second-order method using a direct second-order analysis
- **c.** the direct analysis approach

Compare results of the internal forces and moments from the three approaches and discuss how these affect the strength check using the interaction equation.

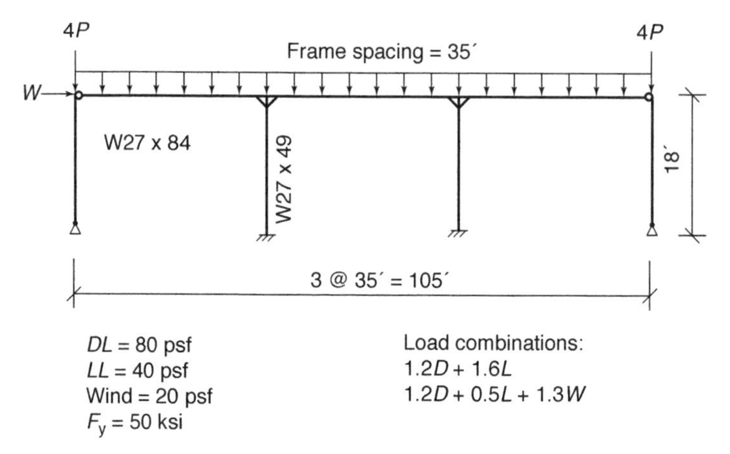

Fig. p8.3

8.4. What is the allowable factored load P that can be applied to the column? Compare results for the K-factor and direct-analysis approaches.

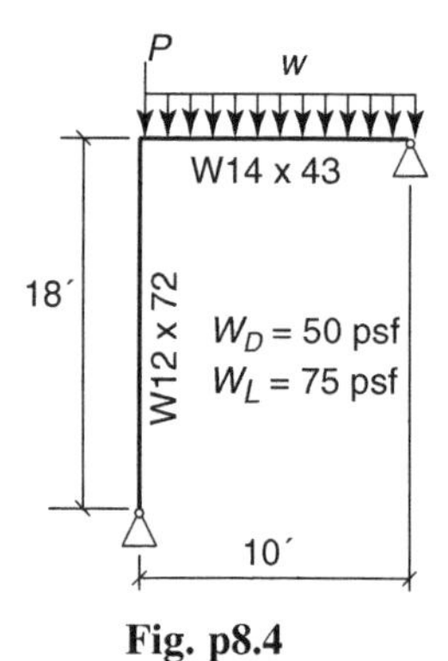

Fig. p8.4

8.5. Perform an interaction check for column C2 using the critical load and direct analysis approaches. Note the orientation of the columns, and use the story buckling modification for the column C2. Assume the frame is braced out-of-plane and that the x-bracing and beams are pinned.

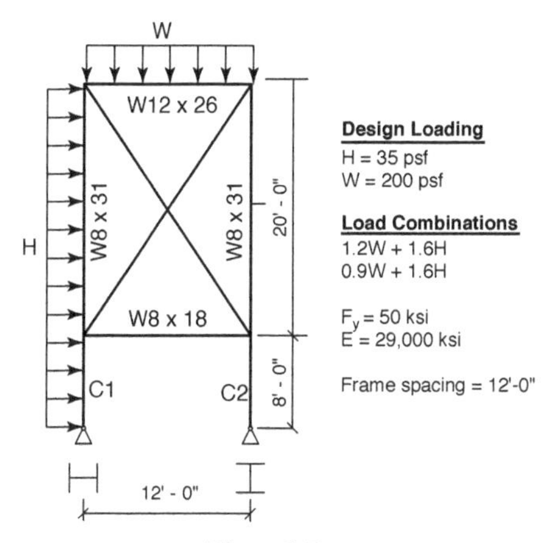

Fig. p8.5

8.6. Design column C1 and B1 for the frame shown. Verify that design using the direct analysis and critical load approaches. Compare the results.

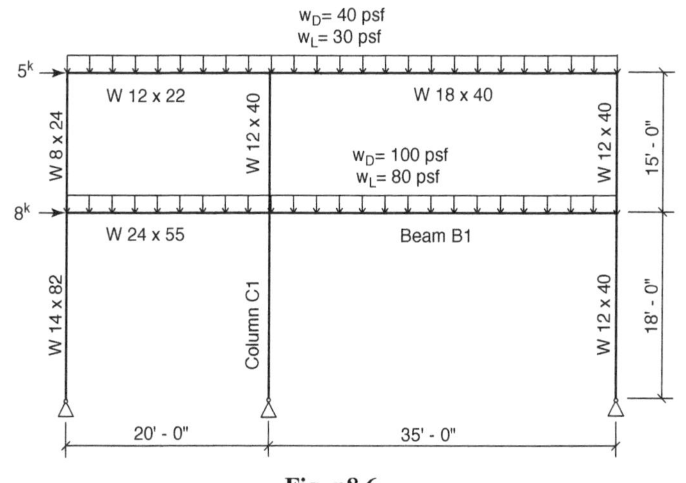

Fig. p8.6

8.7. Compare and contrast the methods by which initial imperfections and inelasticity are accounted for in the three frame stability assessment methods outlined in the AISC Specification.

REFERENCES

CHAPTER 1

Galambos, T. V., ed. 1998. *Guide to Stability Design Criteria for Metal Structures.* New York: John Wiley and Sons.

Royal Commission. 1971. *Report of Royal Commission into the Failure of West Gate Bridge.* Melbourne: C. H. Rixon, Government Printer.

CHAPTER 2

Allen, H. G., and P. S. Bulson. 1980. *Background to Buckling.* New York: McGraw-Hill.

AISC. 2005. *Specification for Structural Steel Buildings.* Chicago: American Institute of Steel Construction.

Bazant, Z., and L. Cedolin. 1991. *Stability of Structures.* New York: Oxford University Press.

Bleich, F. 1952. *Buckling Strength of Metal Structures.* New York: McGraw-Hill.

Brush, D. O., and B. O. Almroth. 1975. *Buckling of Bars, Plates and Shells.* New York: McGraw-Hill.

Chajes, A.1974. *Principles of Structural Stability Theory.* Englewood Cliffs, NJ: Prentice-Hall.

Chen, W. F., and E. M. Lui. 1987. *Structural Stability, Theory and Implementation.* New York: Elsevier.

Churchill, R. V. 1948. *Introduction to Complex Variables and Applications.* New York: McGraw-Hill.

Hetenyi, M. 1946. *Beams on Elastic Foundation.* Ann Arbor: University of Michigan Press.

Simitses, G. J. 1976. *An Introduction to the Elastic Stability of Structures.* Englewood Cliffs, NJ: Prentice-Hall.

Timoshenko, S. P., and J. M. Gere. 1961. *Theory of Elastic Stability.* New York: McGraw-Hill.

Winter, G. 1960. "Lateral Bracing of Columns and Beams." *Trans. ASCE* 125 (Part 1): 809–825.

CHAPTER 3

AISC. 2005. *Specification for Structural Steel Buildings*. Chicago: American Institute of Steel Construction.

Beedle, L. S., and L. Tall. 1960."Basic Column Strength." *ASCE J. Struct. Div.* 86 (ST7) (July): 139–173.

Bjorhovde, R.1978."The Safety of Steel Columns." *ASCE J. Struct. Div.* 104 (ST3): 463–477.

Galambos, T. V. (ed.). 1998. *Guide to Stability Design Criteria for Metal Structures* (5th ed.). New York: John Wiley and Sons.

Johnston, B. G. 1961."Buckling Behavior above the Tangent Modulus Load." *Journal of the Engineering Mechanics Div.* 87 (EM6), (December): 192.

Johnston, B. G. 1964."Inelastic Buckling Gradient." *Journal of the Engineering Mechanics Div.* ASCE 90 (EM6), (December): 31–48.

Osgood, W. R. 1951."The Effect of Residual Stress on Column Strength." Proceedings of the First Nat. Congress of Applied Mechanics, June: 415.

Ramberg, W., and W. R. Osgood. 1943. "Description of Stress-Strain Curves by Three Parameters," Technical Note No. 902, National Advisory Committee for Aeronautics, Washington, D.C.

Shanley, F. R. 1947."Inelastic Column Theory." *Journal of Aeronautical Science* 14(5) (May): 261.

Timoshenko, S. P. 1953. *History of the Strength of Materials.* New York: McGraw-Hill. (Reprinted by Dover Books on Engineering, 1983.)

CHAPTER 4

AISC. 2005. *Specification for Structural Steel Buildings*. Chicago: American Institute of Steel Construction.

ASCE. 1971, *Plastic Design in Steel, A Guide and Commentary.* ASCE Manuals and Reports on Engineering Practice, No. 41.

Austin, W. F. 1961."Strength and Design of Metal Beam-Columns," *ASCE J. Struct. Div.* 87 (ST4):1–34.

Chen, W. F., and Atsuta, T. 1976. *Theory of Beam-Columns*, Vols. 1 and 2. New York: McGraw-Hill.

Galambos, T. V. (ed.). 1998. *Guide to Stability Design Criteria for Metal Structures*, Chapter 8. New York: John Wiley and Sons.

Gaylord, E. H., Jr., and C. N. Gaylord. 1968. *Structural Engineering Handbook,* Chapter 7. New York: McGraw-Hill.

SSRC. 1991. *Stability of Metal Structures: A World View.* Bethlehem, PA: Structural Stability Research Council.

Timoshenko, S. P., and J. M. Gere. 1961. *Theory of Elastic Stability* (2nd ed). New York: McGraw-Hill.

Van Kuren, R. C., and T. V. Galambos. 1964."Beam-Column Experiments." *ASCE J. Struct. Div.* 90 (ST2): 223–256.

Winter, G. 1958."Lateral Bracing of Columns and Beams." *ASCE J. Struct. Div.* 84 (ST2): 1561. (Also published in the 1961 *ASCE Transactions* 125 (Part 1): 809.

CHAPTER 5

AISC.(2005). *Specification for Structural Steel Buildings.* Chicago: American Institute of Steel Construction.

Chen, W. F., Y. Goto, and R. Liew. 1996. *Stability Design of Semi-Rigid Frames.* New York: John Wiley and Sons.

Galambos, T. V. 1960. "Influence of Partial Base-Fixity on Frame Stability." *Journal of the Structural Division*, ASCE 86 (ST5) (May).

Galambos, T. V. (ed.). 1998. *Guide to Stability Design Criteria for Metal Structures* (5th ed.). New York: John Wiley and Sons.

Yura, J. A. 1971."The Effective Length of Columns in Unbraced Frames." *Engineering Journal*, AISC 8 (2) (April): 37–40.

CHAPTER 6

AISC. 1963. *Specification for the Design, Fabrication and Erection of Structural Steel Buildings.* New York: American Institute of Steel Construction.

AISC. 1993. *Load and Resistance Factor Design for Structural Steel Buildings.* Chicago: American Institute of Steel Construction.

AISC. 2005. *Specification for Structural Steel Buildings.* Chicago: American Institute of Steel Construction.

AISC. 2005. *Steel Construction Manual* (13th ed.) Chicago, American Institute of Steel Construction.

Australian Standard 2004. *Steel Structures.* AS4100. Standards Australia, North Sydney,- New South Wales, Australia.

Canadian Standards Association. 2004. *S16-01 Limit States Design of Steel Structures.* Toronto, Ontario, Canada: Canadian Standards Association.

European Standard prEN 1993-1-1. 2004 *Eurocode 3: Design of Steel Structures.* European Committee for Standardisation, Brussels, Belgium.

Galambos, T. V.1968. *Structural Members and Frames.* Upper Saddle River, NJ.: Prentice-Hall.

Galambos, T. V. (ed.). 1998. *Guide to Stability Design Criteria for Metal Structures*. (Chapter 5, "Beams," pp. 192–217). Structural Stability Research Council. New York: John Wiley & Sons.

Kirby, P. A., and D. A. Nethercot, 1979. *Design for Structural Stability*. New York: John Wiley & Sons.

Kitipornchai, S., and N. S. Trahair. 1980."Buckling Properties of Mono-Symmetric I-Beams." *ASCE J. Struct. Div.* 106 (ST 5): 941–958.

Nethercot, D. A. 1983."Elastic Lateral Buckling of Beams." In *Beams and Beam-Columns: Stability and Strength*, ed. R. Narayanan. Barking, Essex, England: Applied Science Publishers.

Nethercot, D. A., and K. C. Rocky. 1972."A Unified Approach to the Elastic Lateral Buckling of Beams." *AISC Eng. J.* 9(3): 96–107.

Ojalvo, M., and R. S. Chambers. 1977. "Effects of Warping Restraints on I-Beam Buckling." *ASCE J. Struct. Div.* 103 (ST 12): 2351–2360.

Salmon, C. G., and J. E. Johnson. 1996. *Steel Structures, Design and Behavior* (4th ed.) (Chapter 8, "Torsion," p. 424–478). New York: HarperCollins College Publishers.

Salvadori, M. G. 1955."Lateral Buckling of I-Beams." *Trans. ASCE* 120,p. 1165.

Timoshenko, S. P., and J. M. Gere. 1961. *Theory of Elastic Stability*. New York: McGraw-Hill.

Trahair, N. S. 1969."Elastic Stability of Continuous Beams." *ASCE J. Struct. Div.* 95 (ST 6): 1295–1312.

Trahair, N. S. 1977."Lateral Buckling of Beams and Beam-Columns." In *Theory of Beam-Columns*, Vol. 2, ed. W. F. Chen and T. Atsuta. New York: McGraw-Hill, Chapter 3.

Trahair, N. S. 1983."Inelastic Lateral Buckling of Beams." In *Beams and Beam-Columns: Stability and Strength*, ed. R. Narayanan. Barking, Essex, England: Applied Science Publishers.

Trahair, N. S. 1993. *Flexural-Torsional Buckling of Structures*. Boca Raton, FL: CRC Press.

Vlasov, V. Z. 1961. *Thin-Walled Elastic Beams*. Jerusalem: Israel Program for Scientific Translation.

CHAPTER 7

AISC 2005. *Steel Construction Manual* (13th ed.). Chicago: American Institute of Steel Construction.

Taylor, A. C., and M. Ojalvo. 1966."Torsional Restraint of Lateral Buckling." *Journal of the Structural Division,* ASCE 92 (ST2): 115–129.

Winter, G. 1960."Lateral Bracing of Columns and Beams." *ASCE Transactions* 125: 809–825.

Yura, J. A. 1993. "Fundamentals of Beam Bracing." In *Is Your Structure Suitably Braced?* Proceedings of the 1993 Annual Technical Session and Meeting, Structural Stability Research Council, Milwaukee, WI, April 1993.

CHAPTER 8

AISC. 2005. *Steel Construction Manual* (13th ed.). Chicago: American Institute of Steel Construction.

ASCE.1997. *Effective Length and Notional Load Approaches for Assessing Frame Stability: Implications for American Steel Design.* American Society of Civil Engineers Structural Engineering Institute's Task Committee on Effective Length under the Technical Committee on Load and Resistance Factor Design, 442 pp.

Canadian Standards Association. 2004. *Limit States Design of Steel Structures, S16.1.* Missisauga, Ontario: CSA.

European Standard, Eurocode 3. 2004. *Design of Steel Structures, Part 1-1, General Rules and Rules for Buildings,* prEN 1993-1-1:2004. Brussels, Belgium.

Kanchanalai, T. 1977."The Design and Behavior of Beam-Columns in Unbraced Steel Frames."AISI Project No. 189, Report No. 2, Civil Engineering/Structures Research Lab., University of Texas, Austin, TX, 300 pp.

Kirby, P. A., and D. A. Nethercot. 1979. *Design for Structural Stability.* New York: John Wiley & Sons.

Galambos, T. V. (ed.). 1998. *Guide to Stability Design Criteria for Metal Structures* (Chapter 5, "Beams," pp. 192–217), Structural Stability Research Council. New York: John Wiley & Sons.

LeMessurier, W. J. 1977."A Practical Method of Second Order Analysis. Part 2: Rigid Frames." *Engineering Journal,* AISC, 13(4): 89–96.

Yura, J. A. 1971."The Effective Length of Columns in Unbraced Frames." *Engineering Journal,* AISC 8(2) (April): 37–42.

Surovek-Maleck, Andrea E., and Donald W. White. 2003. "Direct Analysis Approach for the Assessment of Frame Stability: Verification Studies." Proceedings of the 2003 SSRC Annual Technical Sessions and Meeting, Baltimore., pp. 423–441.

INDEX

Printed in the USA/Agawam, MA
August 8, 2022

796851.008

Printed in the USA/Agawam, MA
August 8, 2022

796851.008